Th. Rabilloud (Ed.)

Proteome Research: Two-Dimensional Gel Electrophoresis and Identification Methods

With 45 Figures

Dr. THIERRY RABILLOUD
CEA-Laboratoire de Bioénergétique Cellulaire et Pathologique EA 2019,
DBMS/BECP
CEA-Grenoble, 17 rue des Martyrs
38054 Grenoble Cedex 9, France

ISBN 3-540-65792-4 Springer Verlag Berlin Heidelberg New York

Library of Congress Cataloging-in-Publication Data
Proteome research: two-dimensional gel electrophoresis and Identification methods / Th. Rabilloud (ed.). p.cm. – (Principles and practice)
Includes bibliographical references and index.
ISBN 3-540-65689-8. – ISBN 3-540-65792-4 (pbk.)
1. Proteins-Analysis. 2. Gel electrophoresis. I. Rabilloud, Th. (Thierry), 1961- . II, Series.
QP551.P7564 2000 572'.633--dc21

Springer-Verlag Berlin Heidelberg New York
a member of BertelsmannSpringer Science+Business Media GmbH

Printed in Germany

Cover design: d&p, D-69121 Heidelberg
Production: ProEdit GmbH, D-69126 Heidelberg
SPIN 10879744 39/3111 5 4 3 2 1 – Printed on acid-free paper

Foreword

This book appears at a fascinating juncture in the history of the biomedical sciences. For the first time we can contemplate the possibility of preparing complete molecular descriptions of living cells, including complete genomic sequence data, lists and structures of the various RNAs, catalogues of all protein gene products and their derivatives, maps showing the intracellular locations of each macromolecule, and an index of all metabolites.

It is from this perspective that it is useful to examine the role of global high resolution protein analysis, which is the topic of this book. A detailed analysis of the proteins of living cells is an important activity, but can most of the information to be gained not be inferred from genomics? If the complete plans for an organism, and all of the basic data required to express those plans are encoded in DNA, can we not deduce all we need to know about cells from that information?

A long-term goal of biology must certainly be to attempt to characterize man from genomic sequence data, as was suggested in the first proposal for the human genome project (Anderson and Anderson 1985). However, this is not yet possible, nor will it be in the foreseeable future. The complexity of living cells defeats such efforts at present, and we must ask instead, what information is required regarding the emerging field of proteomics which cannot now be inferred from nucleic acid sequences or mRNA abundance data, and is unlikely to be obtained in the foreseeable future?

It is now generally agreed that global protein analyses are necessary for the following reasons: there is poor correlation between mRNA abundance and proteins coded for (Tew et al. 1996; Anderson and Seilhamer 1997; Gygi et al. 1999); almost all proteins are post-translationally modified; the passage of a gene product from site of synthesis to site of activity cannot always deduced from sequence data; and function cannot be reliably inferred for all proteins (or fragments thereof) from sequence information.

Hence, from a purely anatomical point of view, it is essential to do as complete an analysis of the protein complement of individual cell types as possible, and many of the chapters of this book address this objective.

However, if complete molecular inventories of cells become possible, including detailed structural data and subcellular localizations, will that end our quest regarding molecular anatomy? Will we understand the key differences between living and non-living systems?

An essential aspect of biochemistry has been the reconstruction of processes and pathways in vitro. These processes include glycolysis, respiration, protein synthesis, synthesis of DNA and RNA, and transport through isolated membrane vesicles, among others.

We have postulated that no protein has a fixed rate of synthesis, but that all are dynamically controlled and in a constant state of flux (Anderson and Anderson 1996), and that a cell may be considered to be a kanban or "just-in-time" system with minimum excess inventory. At one extreme, one may postulate that most components are grouped into a relatively small number of co-regulated sets, and at the other, that the number of factors affecting the expression of each gene is so large that each gene is effectively under separate control. Only by devising means for up- and down-regulating individual genes experimentally can these alternatives be explored in detail.

Good quantitation of large sets of proteins is essential for one to begin to answer these questions. When a given drug interacts with a target, is the target up- or down-regulated, or unchanged? It has been assumed, largely on the basis of in vitro studies where excess drug is present, that antisense compounds can and do down-regulate protein production by hybridization arrest. However, studies of intact animals have not clearly demonstrated down-regulation of a normal cellular gene by an antisense pharmaceutical followed by a useful pharmacological effect. Systematic studies on global changes in protein abundance in antisense-treated tissues remain to be done. Further, a great deal of work remains to be done on the effects of known enzyme inhibitors on the abundance of the inhibited enzyme in intact animals.

The most interesting question, however, concerns co-regulation and the degree of interconnectedness between metabolic processes and pathways. Is it possible to delete or inhibit any protein without affecting the abundance of others? The fact that many knockout mice survive deletions that whould be expected to be fatal suggests a very large array of compensatory changes in gene expression, which remain to be explored in detail. In such response arrays may lie many of the answers to the question of what is unique about living systems. Only with high-resolution, sensitive quantitative, and repeatable analyses which can be done on very large numbers of samples (i.e., with complete automation) can these questions be explored.

Unlike genomics, which involves the analysis of simple repeating polymers, proteomics involves the study of very complex heteropolymers having a wide range of sizes, physical properties, secondary and tertiary structures, solubilities, and functions. Hence, proteomics is and will continue to be a much more difficult field than genomics, will be more costly, and will require continued innovation and extensive support.

Genomics, for a range of organisms, now approaches completion. Proteomics, and epigenomics (i.e., the science of how genomes and proteomes interact), are, in contrast, only in their infancy. This book reviews the classes of interdisciplinary research required for this new science.

Norman G. Anderson
Large Scale Biology Corporation
Rockville, Maryland

References

Anderson NG, Anderson NL (1985) A policy and program for biotechnology. American Biotechnology Laboratory, Sept/Oct 1985

Anderson NG, Anderson NL (1996) Twenty years of two-dimensional electrophoresis: past, present and future. Electrophoresis 17:443–453

Anderson NL, Seilhammer J (1997) A comparison of selected mRNA and protein abundances in human liver. Electrophoresis 18:533–537

Gygi SP, Rochon Y, Franza BR, Aebersold R (1999) Correlation between protein and mRNA abundance in yeast. Mol Cell Biol 1999 19:1720–30

Tew KD, Monks A, Barone L, Rosser D, Akerman G, Montali JA, Wheatly JB, Schmidt DE Jr (1996) Glutathione-associated enzymes in the human cell lines of the National Cancer Institute Drug Screening Program. Mol Pharmacol 50:149–159

References

[illegible]

Contents

List of Contributors

RUEDI AEBERSOLD
Department of Molecular Biotechnology
University of Washington
Seattle, WA 98195, USA
e-mail: ruedi@u.washington.edu

LUCA BINI
Dipartimento Biologia Molecolare
Università degli Studi di Siena
Via Fiorentina 1, 53100 Siena, Italy
e-mail: bini@unisi.it

HELIAN BOUCHERIE
Institut de Biochimie et Génétique Cellulaires
UPR CNRS 9026
1, rue Camille Saint-Saéns
33700 Bordeaux Cedex, France
e-mail: H. Boucherie@ibgc.u-bordeaux2.fr

STÉPHANE CHARMONT
CEA-Laboratoire de Bioénergétique Cellulaire et Pathologique
EA 2019, DBMS/BECP
CEA-Grenoble, 17 rue des Martyrs
38054 Grenoble Cedex 9, France
e-mail: Stephane@sanrafael.ceng.cea.fr

MIREILLE CHEVALLET
CEA-Laboratoire de Bioénergétique Cellulaire et Pathologique
EA 2019, DBMS/BECP
CEA-Grenoble, 17 rue de Martyrs
38054 Grenoble Cedex 9, France
e-mail: Mireille@sanrafael.ceng.cea.fr

Garry L. Corthals
The Garvan Institute of Medical Research
St Vincent's Hospital
Sydney, Australia
e-mail: corthals@u.washington.edu

Christoph Eckerskorn
TopLab, Company for Applied Biotechnology
Innovation Center of Biotechnology
82152 Martinsried, Germany
e-mail: eckerskorn@toplab.de

Angelika Görg
Technical University of Munich
Department of Food Technology
85350 Freising-Weihenstephan, Germany
e-mail: Angelika.Georg@Irz.tu-muenchen.de

Steven P. Gygi
Department of Molecular Biotechnology
University of Washington
Seattle, WA 98195, USA
e-mail: ruedi@u.washington.edu

Ian Humphery-Smith
The University of Sydney
Centre for Proteome Research and Gene-Product Mapping
National Innovation Centre
Australian Technology Park
Eveleigh, Australia
e-mail: ian@proteome.usyd.edu.au

Jean Labarre
Service de Biochimie et Génétique Moléculaire
CEA-Saclay
91191 Gif-sur-Yvette-Cedex, France

Sabrina Liberatori
Dipartimento Biologia Molecolare
Università degli Studi di Siena
Via Fiorentina 1, 53100 Siena, Italy
e-mail: bini@unisi.it

Friedrich Lottspeich
Max-Planck-Institute for Biochemistry
82152 Martinsried, Germany
e-mail: lottspei@biochem.mpg.de

Barbara Magi
Dipartimento Biologia Molecolare
Università degli Studi di Siena
Via Fiorentina 1, 53100 Siena, Italy
e-mail: bini@unisi.it

Barbara Marzocchi
Dipartimento Biologia Molecolare
Università degli Studi di Siena
Via Fiorentina 1, 53100 Siena, Italy
e-mail: bini@unisi.it

Christelle Monribot
Institut de Biochimie et Génétique Cellulaires
UPR CNRS 9026
1, rue Camille Saint-Saéns
33700 Bordeaux Cedex, France

Vitaliano Pallini
Dipartimento Biologia Molecolare
Università degli Studi di Siena
Via Fiorentina 1, 53100 Siena, Italy
e-mail: bini@unisi.it

Scott Patterson
Amgen Inc.
Protein Structure, Amgen Centre
Thousand Oaks, CA 91320, USA
e-mail: spatters@amgen.com

Michel Perrot
Institut de Biochimie et Génétique Cellulaires
UPR CNRS 9026
1, rue Camille Saint-Saéns
33700 Bordeaux Cedex, France

Thierry Rabilloud
CEA-Laboratoire de Bioénergétique Cellulaire
et Pathologique EA 2019, DBMS/BECP
CEA-Grenoble, 17, rue de Martyrs
38054 Grenoble Cedex 9, France
e-mail: Thierry@sanrafael.ceng.cea.fr

Roberto Raggiaschi
Dipartimento Biologia Molecolare
Università degli Studi di Siena
Via Fiorentina 1, 53100 Siena, Italy
e-mail: bini@unisi.it

Kerstin Strupat
Institut for Medical Physics and Biophysics
University of Münster
48149 Münster, Germany

Margaret Tyler
Macquarie University Centre
for Analytical Biotechnology (MUCAB)
School of Biological Sciences
Macquarie University
Sydney, New South Wales 2109, Australia
e-mail: mwilkins@proteome.org.au

Walter Weiss
Technical University of Munich
Department of Food Technology
85350 Freising-Weihenstephan, Germany
e-mail: Angelika.Georg@Irz.tu-muenchen.de

Marc Wilkins
Macquarie University Centre
for Analytical Biotechnology (MUCAB)
School of Biological Sciences
Macquarie University
Sydney, New South Wales 2109, Australia
e-mail: mwilkins@proteome.org.au

CHAPTER 1

Introduction: The Virtue of Proteomics

T. Rabilloud[1] and I. Humphery-Smith[2]

The term 'proteome', first introduced in 1995 (Wasinger et al. 1995), means the protein complement of a genome. In the cascade of regulatory events leading from the gene to the active protein (Scherrer 1974), the proteome can be seen as the end product of the genome. While the genome is comparatively static, the proteome is a highly dynamic entity, as the protein content of a given cell will vary with respect to changes in the surrounding environment, physiological state of the cell (e.g. position in the cell cycle), stress, drug administration, health and disease. Moreover, different cell types within a multicellular organism will of course have different proteomes, while the genome is held relatively constant. Although the proteome of a given cell at any moment represents only the expression of part of the genome, proteomes are very complex. This is probably an inherent feature of living organisms, as the simplest self-replicating, non-parasitic living organisms show more than 1500 genes. In addition, there exists much potential for molecular variation to occur between a gene and its corresponding active product. This includes suppression, promotion, splicing and both co- and post-translational modifications. This complexity explains why proteomics, i.e. the study of proteomes, must be considered as a similarly large-scale science as genomics, if not even more so. It is not the purpose of this chapter to give an extensive review of proteomes and proteomics. This has been done in many recent review articles (e.g. Humphery-Smith et al. 1997; Humphery-Smith and Blackstock 1997; James 1997; Anderson and Anderson 1998) and in a recent book in this series (Wilkins et al. 1997). As this book is more technology-oriented, we would like to focus on the technological differences existing between proteomics and other holistic analyses of biological macromolecules, i.e. trying to understand the origins and the consequences of these technological differences on the biological results which can be extracted.

Concomitantly with the first appearance of the word 'proteome', the first complete DNA sequence of a self-replicating, non-parasitic organism was published (Fleischmann et al. 1995). Genomics was demonstrated to be feasible after a significant lag phase since the initial sequencing of an entire bacteriophage some 13 years earlier (Sanger et al. 1982). This formidable achievement was followed by

[1] CEA- Laboratoire de Bioénergétique Cellulaire et Pathologique, EA 2019, DBMS/BECP CEA-Grenoble, 17 rue des martyrs, F-38054 GRENOBLE CEDEX 9, France.
[2] The University of Sydney, Centre for Proteome Research and Gene-Product Mapping, National Innovation Centre, Australian Technology Park, Eveleigh, Australia.

Table 1.1. Fully sequenced genomes of non viral organisms (Fall 1998)

Organism	Genome siz	No of ORFs	Reference
Aquifex aeolicus	1.5 Mbp	1570	Deckert et al. 1998
Archaeoglobus fulgidus	2.2 Mbp	2436	Klenk et al. 1997
Bacillus subtilis	4.2 Mbp	4100	Kunst et al. 1997
Borrelia burgdorferi	0.9 + 0.5*Mbp	853 + 430*	Fraser et al. 1997
Caenorhabditis elegans	97 Mbp	19100	The C.elegans sequencing consortium 1998
Escherichia coli	4.6 Mbp	4288	Blattner et al. 1997
Haemophilus influenzae	1.8 Mbp	1743	Fleischmann et al. 1995
Helicobacter Pylori	1.7 Mbp	1590	Tomb et al. 1997
Methanobacterium thermoautotrophicum	1.75 Mbp	1855	Smith et al. 1997
Methanococcus jannaschii	1.66 Mbp	1738	Bult et al. 1996
Mycobacterium tuberculosis	4.4 Mbp	4000	Cole et al. 1998
Mycoplasma genitalium	0.58 Mbp	470	Fraser et al. 1995
Myoplasma pneumoniae	0.81 Mbp	677	Himmelreich et al. 1996
Pyrococcus horikoshii	1.7 Mbp	2061	Kawarabyasi et al. 1998
Saccharomyces cerevisiae	12 Mbp	5885	Goffeau et al. 1996
Synechocystis sp. strain PCC6803	3.6 Mbp	3168	Kaneko et al. 1996

* first number corresponds to chromosome, second number to the total of extrachromosomal elements.

the release of complete DNA sequences for several organisms, prokaryotes or lower eukaryotes (see Table 1.1) and the complete sequence of human DNA is expected within a few years. However, by the virtue of the irony so commonly encountered in science, the release of these complete DNA sequences immediately pointed out the limitations of genomics (Nowak 1995). These limitations were not technological deficiencies associated with the construction of contiguous DNA sequence maps, but rather they were linked more to the limitations of this information with respect to our understanding of cellular molecular biology. In a cell, DNA is a relatively static memory molecule, remaining as such at least over the lifespan of a living organism. Consequently, the knowledge of the presence of a gene within a DNA molecule does not tell us if or when this gene is transcribed and translated, at what rate, what the products of the gene are and what the degree or nature of its functional competence is. Other additional problems arise at the DNA level, for example, delimiting the boundaries of small genes (Rudd et al. 1998; Wasinger and Humphery-Smith 1999) or active genes within "junk DNA" in higher eukaryotes. Such questions, which are at the core of what is now termed "functional genomics", must be answered downstream to the level of DNA sequence, i.e. at the level of gene products. Here, too, the situation is complicated by the fact that most genes have a primary product, the mRNA template, and an end-product, the polypeptide. Often, this end-product can be modified to give rise to several active proteins for a single gene. Proteins are the "molecular workhorses" of a cell and as such are likely to ensure that proteomics becomes the mainstay of functional genomics. However, as we will see later, working with proteins is the most difficult of the molecular triade DNA, RNA, proteins.

This is further accentuated by the absence of an amplification tool comparable to PCR in the protein world, so that the study of less abundant proteins becomes more difficult. Already the analysis of complete mRNA complements of genomes (i.e. transcriptomes) can be quantitatively studied (DeRisi et al. 1997; Wodicka et al. 1997; Blattner pers. comm.). Inadequacies can be found here also in that the precision of the determination of RNA levels also possesses inherent limitations, namely detection of weakly expressed mRNAs, the difficulty of achieving equally high stringency across an entire high-density array and the potential for reduced specificity as the length of the cDNA probe increases. Nonetheless, this amazing technological achievement can be attributed to the chemical encoding of the information in nucleic acids. In these information-bearing molecules, what is important is the sequence, and not the chemical nature of the product. Chemically speaking, the gene or the mRNA corresponding to a protein is just a special kind of poly sugar-phosphate, whatever the final protein may be. Consequently, the methods developed for studying a given mRNA species or gene will possess universal application to almost any DNA or gene.

Thus, in the "brave new mRNA world", does this mean that transcriptomes represent the ultimate complement of genomics for functional genomics? The answer lies in the relative importance of the regulatory steps occurring between the DNA and the mRNA on one hand, and between the mRNA and the active protein on the other (Anderson and Anderson 1998), not to mention Metabolic Control Analysis and epigenetic phenomena. In most cases, the active product of a gene is a protein, and what is important for the cell is the level of activity of this protein, as determined by its molecular half-life and its influence on metabolic flux. However, if most of the regulatory events controlling the level of expression and activity of protein molecules are dictated during the production of the corresponding mRNA, then transcriptomics would suffice as an efficient short-cut to study most functional aspects of genomics. A primary objection to this statement lies in the importance of post-translational modifications for the modulation of protein activity. In signal transduction, many events occur via phosphorylation of proteins, i.e. without any modification of the levels of polypeptide expression. These events are not necessarily minor. Indeed, very important cellular decisions (e.g. differentiation, mitosis or apoptosis) can be taken from these simple phosphorylation events. Of course, these phosphorylation events end downstream by the activation of new genes, and are thus detectable at the level of the mRNA message. However, the site of action of these activated genes cannot be predicted from the genetic code. To establish whether proteomics is of worth when compared to transcriptomics, it is necessary to know whether the level of protein expression and protein activities can be predicted well from information obtained solely at the level of the mRNA template. This hypothesis can be tested by evaluating the correlation between mRNA levels and corresponding protein levels in cells. These experiments have been carried out on liver (Anderson and Seilhamer 1997; Anderson and Anderson 1998) and on yeast (Haynes et al. 1998). Each concluded with a correlation factor of ca. 0.4, indicating a very poor correlation. Importantly, however, this means that protein levels cannot be simply deduced from the levels of mRNA expression, nor can comparative analyses of protein expression be derived from a comparison of RNA levels. Worse still is the knowledge that a variation at the level of a given mRNA species cannot be transposed into an equivalent variation in the level of the corresponding proteins.

Cases are known where the extent of variation in RNA and protein are widely different (Rousseau et al. 1992), while examples are also known where the protein and RNA levels vary in opposite directions (Khochbin et al. 1991). These discrepancies can be attributed to the dramatic variation associated with the molecular half-life of mRNA species (Humphery-Smith 1999).

Given these limitations, proteomics may appear as the preferred tool for providing an understanding of how genomes function, as it analyses directly the end product of the genome (the output code as defined by Humphery-Smith and Blackstock 1997). Protein analysis is, however, very far from being trouble-free. Here, again, the reasons for the difficulties lie deep in the roots of biology and biochemistry. Nonetheless, it is a common-sense message that our knowledge of cellular molecular biology will not be complete without information pertaining to DNA, RNA and proteins. It must also be noted that any technological platform employed for holistic cellular analyses will have its inherent weaknesses, and thus, we must exploit a multitude of experimental procedures if we are to approach that "holy grail" of a complete understanding of biological processes within cells and tissues during health and disease.

In every large-scale analytical process, there are two independent phases, namely, separation of the biomolecules of interest and, subsequently, molecular characterization. In the analysis of nucleic acids, separation is most frequently achieved by cloning, PCR and/or preparation of high-density cDNA arrays. In the protein world, direct cloning or direct preparation of probes is not achieved as easily. The proteins have then to be chemically resolved by a separation method, and this is a major issue for proteomics. While nucleic acids are mainly information-driving molecules and are chemically homogeneous, proteins are the molecular machines of the cells and as such exist in varied forms corresponding to the functional needs of a cell. Proteins show widely different structures and therefore chemical properties. They also vary with respect to their intracellular abundance in different cell types and in health and disease. As a consequence, it is almost impossible to find an analytical technique suitable for all proteins and able to resolve and separate the potentially hundreds of thousands of protein forms (including post-translational variants) present in a particular cell or tissue. The only technique currently available with sufficient resolving power is two-dimensional gel electrophoresis, exploiting charge- and mass-driven separations. Owing to its (still increasing) resolving power, relative and increasing simplicity and range of application, two-dimensional gel electrophoresis has become the core technology of proteome research. As such, two chapters of this book will be devoted to this technique. The major problems associated with this technique are how to deal with the dynamic range of expression of proteins and how to deal with the chemical diversity of proteins. The latter results in an amazingly different range of molecular weights, isoelectric points and solubilities under experimental conditions. While enlarging the analysis window in isoelectric point and molecular weight is a pure electrophoretic problem, there is another set of problems linked to sensitivity at the level of protein detection and characterization and, possibly most importantly, protein solubility. In fact, protein solubility is limited under the conditions prevailing during two-dimensional gel electrophoresis, and proteins present above their solubility threshold will not be detected in the resulting maps. This problem can be overcome by:

(1) increasing protein solubility (to be dealt with in a later chapter in this book); and
(2) decreasing the detection threshold of proteins so that there will no longer be any solubility problem (to be covered in another chapter in this book).

Progress in detection methods cannot be extended too far without consequences. The main consequence is the "contemplative defect" which plagued large-scale protein analysis for almost two decades. Indeed, two-dimensional gel electrophoresis is an old separation method, as it was published in a rather mature form in 1975 (Klose 1975; O'Farrell 1975). At this time, DNA sequencing was in its infancy. The main problem was therefore not to separate proteins, which appeared as wonderfully well-resolved spots, but to know which protein was present within each spot on a 2-D electrophoresis gel. This was a major ordeal at that time, except perhaps for some serum proteins, where a good number of purified proteins or antibodies were available. This was largely because these molecules could be found in sufficiently high abundance in an easily studied tissue system, i.e. serum. Protein sequencing by Edman degradation had been feasible for a long time (e.g. Edman and Begg 1967) but required amounts of proteins out of the range of preparation via 2-D gel electrophoresis. Work with protein on a large-scale was therefore a rather contemplative and meditative work, because the characterization step, the second step in the large-scale analysis of macromolecules, was totally deficient for proteins. This situation changed dramatically in the late 1980s and in the 1990s, owing to the minaturization and increased precision of many analytical techniques applicable to the protein sciences and ranging from amino acid analysis and peptide sequencing through to mass spectrometry and highly sensitive immunodetection.

These techniques brought the detection threshold down to within the realm of picomole and even low femtomole sensitivity, compatible with what can be retrieved from within the 2-D gel environment. These techniques have subsequently and quite elegantly become the interface with gel electrophoresis. The importance of this revolution in characterization for proteomics is exemplified by the number of chapters devoted in this book to methods of protein characterization. This further highlights the need for proteomics to be seen as the combination of large-scale separation and large-scale characterization of proteins.

This revolution in the microcharacterization of proteins was made possible due to another revolution brought about by genetic engineering and large-scale DNA sequencing. In fact, it is usually no longer necessary to fully sequence a polypeptide separated as a spot on a 2-D gel in order to know the protein and the nature of its encoding gene. A partial characterization may be sufficient to make an univocal assignment among the protein sequences predicted from nucleic acids. This is of course especially true in organisms with completely sequenced genomes, and increasingly for human proteins in association with the information derived from extensive Expressed Sequence Tag (EST) cDNA libraries in both the public and private domain. In this respect, small protein sequence tags combined with amino acid composition or mass spectrometric data can provide assignments totally impossible without the wealth of data available in nucleic acid-derived databases. Although as much a part of proteomics as protein separation and micro-characterization, these databases and the improvements that have occurred in both the hardware and software underlying the computation sciences will not be dealt with specifically in this text. Nonetheless, they will be referred to regularly as an integral aspect of protein characterization. The computational aspects of proteomics, as well as dedicated subjects such as the study of post-translational modifications or specialized mass spectrometry, will be treated in subsequent texts within this series.

References

Anderson L, Seilhamer J (1997) A comparison of selected mRNA and protein abundances in human liver. Electrophoresis 18: 533–537

Anderson NL, Anderson NG (1998) Proteome and proteomics: new technologies, new concepts and new words.Electrophoresis 19 : 1853–1861

Blattner FR, Plunkett G 3rd, Bloch CA, Perna NT, Burland V, Riley M, Collado-Vides J, Glasner JD, Rode CK, Mayhew GF, Gregor J, Davis NW, Kirkpatrick HA, Goeden MA, Rose DJ, Mau B, Shao Y (1997) The complete genome sequence of *Escherichia coli* K-12. Science 277: 1453–1474

Bult CJ, White O, Olsen GJ, Zhou L, Fleischmann RD, Sutton GG, Blake JA, FitzGerald LM, Clayton RA, Gocayne JD, Kerlavage AR, Dougherty BA, Tomb JF, Adams MD, Reich CI, Overbeek R, Kirkness EF, Weinstock KG, Merrick JM, Glodek A, Scott JL, Geoghagen NSM, Venter JC (1996) Complete genome sequence of the methanogenic archaeon, *Methanococcus jannaschii*. Science 1996 273: 1058–1073

Cole ST, Brosch R, Parkhill J, Garnier T, Churcher C, Harris D, Gordon SV, Eiglmeier K, Gas S, Barry CE III, Tekaia F, Badcock K, Basham D, Brown D, Chillingworth T, Connor R, Davies R, Devlin K, Feltwell T, Gentles S, Hamlin N, Holroyd S, Hornsby T, Jagels K, Krogh A, McLean J, Moule S, Murphy L, Oliver K, Osborne J, Quail MA, Rajandream MA, Rogers J, Rutter S, Seeger K, Skelton J, Squares R, Squares S, Sulston JE, taylor K, Whitehead S, Barrell BG (1998) Deciphering the biology of Mycobacterium tuberculosis from the complete genome sequence. Nature 393: 537–544

Deckert G, Warren PV, Gaasterland T, Young WG, Lenox AL, Graham DE, Overbeek R, Snead MA, Keller M, Aujay M, Huber R, Feldman RA, Short JM, Olsen GJ, Swanson RV (1998) The complete genome of the hyperthermophilic bacterium *Aquifex aeolicus*. Nature 392: 353–358

DeRisi JL, Iyer VR, Brown PO (1997) Exploring the metabolic and genetic control of gene expression on a genomic scale. Science 278: 680–686

Edman P, Begg G (1967) A protein sequenator. Eur J Biochem 1: 80–91

Fleischmann RD, Adams MD, White O, Clayton RA, Kirkness EF, Kerlavage AR, Bult CJ, Tomb JF, Dougherty BA, Merrick JM, McKenney K, Sutton G, FitzHugh W, Fields C, Gocayne JD, Scott J, Shirley R, Liu L, Glodex A, Kelley JM, Weidmen JF, Phillips CA, Spriggs T, Hedblom E, Cotton MD, Utterback TR, Hanna MC, Nguyen DT, Saudek DM, Brandon RC, Fritchmann JL, Fuhrmann J, Geoghagen NSM, Gnehm CL, McDonald LA, Small KV, Fraser CM, Smith HO, Venter JC (1995) Whole-genome random sequencing and assembly of *Haemophilus influenzae* Rd.
Science 269: 496–512

Fraser CM, Gocayne JD, White O, Adams MD, Clayton RA, Fleischmann RD, Bult CJ, Kerlavage AR, Sutton G, Kelley JM, Fritchman JL, Weidman JM, Small KV, Sandusky M, Fuhrmann J, Nguyen D, Utterback TR, Saudek DM, Phillips CA, Merrick JM, Tomb JF, Dougherty BA, Bott KF, Hu PC, Lucier TS, Peterson JD, Smith HO, Hutchison A 3rd, Venter JC (1995) The minimal gene complement of *Mycoplasma genitalium*. Science 270: 397–403

Fraser CM, Casjens S, Huang WM, Sutton GG, Clayton R, Lathigra R, White O, Ketchum KA, Dodson R, Hickey EK, Gwinn M, Dougherty B, Tomb JF, Fleischmann RD, Richardson D, Peterson J, Kerlavage AR, Quackenbush J, Salzberg S, Hanson M, van Vugt R, Palmer N, Adams MD, Gocayne J, Weidman JM, Utterback TR, Watthey L, McDonald L, Artiach P, Bowman C, Garland S, Fujii C, Cotton MD, Horst K, Roberts K, Hatch B, Smith HO, Venter JC (1997) Genomic sequence of a Lyme disease spirochaete, *Borrelia burgdorferi*. Nature 390: 580–586

Fraser CM, Norris SJ, Weinstock GM, White O, Sutton GG, Dodson R, Gwinn M, Hickey EK, Clayton R, Ketchum KA, Sodergren E, Hardham JM, McLeod MP, Salzberg S, Peterson J, Khalak H, Richardson D, Howell JK, Chidambaram M, Utterback T, McDonald L, Artiach P, Bowman C, Cotton MD, Fujii C, Garland S, Hatch B, Horst K, Roberts K, Sandusky M, Weidman J, Smith HO,Venter JC

(1998) Complete genome sequence of *Treponema pallidum*, the syphilis spirochete. Science 281: 375–88

Goffeau A, Barrell BG, Bussey H, Davis RW, Dujon B, Feldmann H, Galibert F, Hoheisel JD, Jacq C, Johnston M, Louis EJ, Mewes HW, Murakami Y, Philippsen P, Tettelin H, Oliver SG (1996) Life with 6000 genes. Science 274: 546, 563–567

Haynes PA, Gygi SP, Figeys D, Aebersold R (1998) Proteome analysis: biological assay or data archive? Electrophoresis 19: 1862–1871

Himmelreich R, Hilbert H, Plagens H, Pirkl E, Li BC, Herrmann R (1996)Complete sequence analysis of the genome of the bacterium *Mycoplasma pneumoniae*. Nucleic Acids Res. 24: 4420–4449

Humphery-Smith, I. Replication-induced protein synthesis and its importance to proteomics. TIBTECH (In press)

Humphery-Smith I, Blackstock W (1997) Proteome analysis: genomics via the output rather than the input code. J Protein Chem. 16: 537–544

Humphery-Smith I, Cordwell SJ, Blackstock WP (1997) Proteome research: complementarity and limitations with respect to the RNA and DNA worlds. Electrophoresis 18: 1217–1242

James P (1997) Of genomes and proteomes. Biochem Biophys Res Commun 231: 1–6

Kaneko T, Sato S, Kotani H, Tanaka A, Asamizu E, Nakamura Y, Miyajima N, Hirosawa M, Sugiura M, Sasamoto S, Kimura T, Hosouchi T, Matsuno A, Muraki A, Nakazaki N, Naruo K, Okumura S, Shimpo S, Takeuchi C, Wada T, Watanabe A, Yamada M, Yasuda M, Tabata S (1996) Sequence analysis of the genome of the unicellular cyanobacterium *Synechocystis* sp. strain PCC6803. II. Sequence determination of the entire genome and assignment of potential protein-coding regions. DNA Res 3: 109–136

Kawarabayasi Y, Sawada M, Horikawa H, Haikawa Y, Hino Y, Yamamoto S, Sekine M, Baba S, Kosugi H, Hosoyama A, Nagai Y, Sakai M, Ogura K, Otsuka R, Nakazawa H, Takamiya M, Ohfuku Y, Funahashi T, Tanaka T, Kudoh Y, Yamazaki J, Kushida N, Oguchi A, Aoki K, Kikuchi H (1998) Complete sequence and gene organization of the genome of a hyper-thermophilic archaebacterium, *Pyrococcus horikoshii* OT3. DNA Res 5: 55–76

Khochbin S, Gorka C, Lawrence JJ (1991) Multiple control level governing H10 mRNA and protein accumulation. FEBS Lett 283: 65–67

Klenk HP, Clayton RA, Tomb JF, White O, Nelson KE, Ketchum KA, Dodson RJ, Gwinn M, Hickey EK, Peterson JD, Richardson DL, Kerlavage AR, Graham DE, Kyrpides NC, Fleischmann RD, Quackenbush J, Lee NH, Sutton GG, Gill S, Kirkness EF, Dougherty BA, McKenney K, Adams MD, Loftus B, Peterson S, Reich CI, McNeil LK, Badger JH, Glodek A, Zhou L, Overbeek R, Gocayne JD, Weidman JF, McDonald L, Utterback TR, Cotton MD, Spriggs T, Artiach P, Kaine BP, Sykes SM, Sadow PW, D'Andrea KP, Bowman C, Fujii C, Garland SA, Mason TM, Olsen GJ, Fraser CM, Smith HO, Woese CR, Venter JC (1997) The complete genome sequence of the hyperthermophilic, sulphate-reducing archaeon *Archaeoglobus fulgidus*. Nature 390: 364–370

Klose J (1975) Protein mapping by combined isoelectric focusing and electrophoresis of mouse tissues. A novel approach to testing for induced point mutations in mammals. Humangenetik 26: 231–243

Kunst F, Ogasawara N, Moszer I, Albertini AM, Alloni G, Azevedo V, Bertero MG, Bessieres P, Bolotin A, Borchert S, Borriss R, Boursier L, Brans A, Braun M, Brignell SC, Bron S, Brouillet S, Bruschi CV, Caldwell B, Capuano V, Carter NM, Choi SK, Codani JJ, Connerton IF, Cummings NJ, Daniel RA, Denizot F, Devine KM, D^{3}sterh[divide]ft A, Ehrlich SD, Emmeron PT, Entian KD, Errington J, Fabret C, Ferrari E, Foulger D, Fritz C, Fujita M, Fujita Y, Fuma S, Galizzi A, Galleron N, Ghim SY, Glaser P, Goffeau A, Golightly EJ, Grandi G, Giuseppi G, Guy BJ, Haga K, Haiech J, Harwood CR, H[Uacute]naut A, Hilbert H, Holsappel S, Hosono S, Hullo MF, Itaya M, Jones L, Joris B, Karamata D, Kasahara Y, Klaerr-Blanchard M, Klein C, Kobayashi Y, Koetter P, Koningstein G, Krogh S, Kumano M, Kurita K, Lapidus A, Lardinois S, Lauber J, Lazarevic V, Lee SM, Levine A, Liu H, Masuda S, Mau[Ugrave]l C, M[Uacute]digue C, Medina N, Mellado RP, Mizuno M, Moestl D, Nakai S, Noback M, Noone D, O'Reilly M, Ogawa K, Ogiwara A, Oudega B, Park SH, Parro V, Pohl TM, Portetelle D, Porwollik S, Prescott AM, Presecan E, Pujic P, Purnelle B, Rapoport G, Rey M, Reynolds S, Rieger M, Rivolta C, Rocha E, Roche B, Rose M, Sadaie Y, Sato T, Scanlan E, Schleich S, Schroeter R, Scoffone F, Sekiguchi J, Sekowska A, Seror SJ, Seror P, Shin BS, Soldo B, Sorokin A, Tacconi E, Takagi Y, Takahashi H, Takemaru K, Takeuchi M, Tamakoshi A, Tanaka T, Terpstra P, Tognoni A, Tosato A, Uchimaya S, Vandenbol M, Vannier F, Vassarotti A, Viari A, Wambutt R, Wedler E, Weitzenegger T, Winters P, Wipat A, Yamamoto H, Yamane K, Yasumoto K, Yata K, Yoshida K, Yoshikawa HF, Zumstein E, Yoshikawa H, Danchin A (1997) The complete genome sequence of the gram-positive bacterium *Bacillus subtilis*. Nature 390: 249–256

Nowak R (1995) Bacterial genome sequence bagged. Science 269: 468–470

O'Farrell PH (1975) High resolution two-dimensional electrophoresis of proteins. J Biol Chem 250: 4007–4021

Rousseau D, Khochbin S, Gorka C, Lawrence JJ (1992) Induction of H1(0)-gene expression in B16 murine melanoma cells. Eur J Biochem 208: 775–779

Rudd KE, Humphery-Smith I, Wasinger VC, Bairoch A (1998) Low molecular weight proteins: a challenge for post-genomic research. Electrophoresis 19: 536–544

Sanger F, Coulson AR, Hong GF, Hill DF, Petersen GB (1982) Nucleotide sequence of bacteriophage lambda DNA. J Mol Biol 162: 729–773

Scherrer K (1974) Control of gene expression in animal cells: the cascade regulation hypothesis revisited. Ad. Exp Med Biol 44: 169–219

Smith DR, Doucette-Stamm LA, Deloughery C, Lee H, Dubois J, Aldredge T, Bashirzadeh R, Blakely D, Cook R, Gilbert K, Harrison D, Hoang L, Keagle P, Lumm W, Pothier B, Qiu D, Spadafora R, Vicaire R, Wang Y, Wierzbowski J, Gibson R, Jiwani N, Caruso A, Bush D, Safer H, Patwell D, Prabhakar S, McDougall S, Shimer G, Goyal A, Pietrokovski S, Church GM, Daniels CJ, Mao JI, Rice P, N[divide]lling J, Reeve JN (1997) Complete genome sequence of *Methanobacterium thermoautotrophicum* deltaH: functional analysis and comparative genomics. J Bacteriol 179: 7135–7155

The C. elegans sequencing consortium
Genome sequence of the nematode *C. elegans*: a platform for investigating biology
Science 1998 282: 2012–2018

Tomb JF, White O, Kerlavage AR, Clayton RA, Sutton GG, Fleischmann RD, Ketchum KA, Klenk HP, Gill S, Dougherty BA, Nelson K, Quackenbush J, Zhou L, Kirkness EF, Peterson S, Loftus B, Richardson D, Dodson R, Khalak HG, Glodek A, McKenney K, Fitzegerald LM, Lee N, Adams MD, Hickey EK, Berg DE, Gocayne JD, Utterback TR, Peterson JD, Kelley JM, Cotton MD, Weidman JM, Fujii C, Bowman C, Watthey L, Wallin E, Hayes WS, Borodovsky M, Karp PD, Smith HO, Fraser CM, Venter JC (1997) The complete genome sequence of the gastric pathogen *Helicobacter pylori*. Nature 388: 539–547

Wasinger V, Humphery-Smith I (1998)
Small genes/gene-products of *Escherichia coli* K-12.
FEMS Microbiology Letters 169: 375–382

Wasinger VC, Cordwell SJ, Cerpa-Poljak A, Yan JX, Gooley AA, Wilkins MR, Duncan MW, Harris R, Williams KL, Humphery-Smith I (1995) Progress with gene-product mapping of the Mollicutes: *Mycoplasma genitalium*. Electrophoresis 16: 1090–1094

Wilkins MR, Williams KL, Appel RD, Hochstrasser DF (1997) Proteome Research: New Frontiers in Functional Genomics Springer, Berlin Heidelberg New York

Wodicka L, Dong H, Mittmann M, Ho MH, Lockhart DJ (1997) Genome-wide expression monitoring in *Saccharomyces cerevisiae*. Nat Biotechnol15: 1359–1367

CHAPTER 2

Solubilization of Proteins in Two-Dimensional Electrophoresis

T. Rabilloud[1] and M. Chevallet[1]

1 Introduction

The solubilization process for 2-D electrophoresis has to achieve several parallel goals:

1. *Breaking Macromolecular Interactions in Order to Yield Separate Polypeptide Chains.* This includes breaking disulfide bonds and disrupting all non-covalent interactions, both between proteins and between proteins and non-proteinaceous compounds such as lipids or nucleic acids.

2. *Preventing Any Artefactual Modification of Polypeptides in the Solubilization Medium.* Ideally, the perfect solubilization medium should freeze all the extracted polypeptides in their exact state prior to solubilization, both in terms of amino acid composition and in terms of post-translational modifications. This means that all the enzymes able to modify the proteins must be quickly and irreversibly inactivated. Such enzymes include of course proteases, which are the most difficult to inactivate, but also phosphatases, glycosidases, etc. In parallel, the solubilization protocol should not expose the polypeptides to conditions in which chemical modifications (e.g. deamidation of Asn and Gln, cleavage of Asp-Pro bonds) may occur.

3. *Allowing Removal of Substances that May Interfere with Two-Dimensional Electrophoresis.* In 2-D, proteins are the analytes. Thus, anything in the cell but proteins can be considered as an interfering substance. Many cellular compounds (e.g. coenzymes, hormones, simple sugars) do not interact with proteins or do not interfere with the electrophoretic process. However, many compounds bind to proteins and/or interfere with 2-D, and must be eliminated prior to electrophoresis if their amount exceeds a critical interference threshold. Such compounds mainly include salts, lipids, polysaccharides (including cell walls) and nucleic acids.

4. *Keeping Proteins in Solution During the Two-Dimensional Electrophoresis Process.* Although solubilization *stricto sensu* stops at the point where the sample is loaded onto the first dimension gel, its scope can be extended to the 2-D process,

[1] CEA- Laboratoire de Bioénergétique Cellulaire et Pathologique, EA 2019, DBMS/BECP CEA-Grenoble, 17 rue des martyrs, F-38054 GRENOBLE CEDEX 9, France.

as proteins must be kept soluble till the end of the second dimension. The second dimension is generally a sodium dodecyl sulfate (SDS) gel, and very few problems are encountered once the proteins have entered the SDS PAGE gel. The main problem in SDS gels is overloading of the major proteins when micro preparative 2-D is carried out, and nothing but scaling-up the SDS gel (its thickness and its other dimensions) can counteract overloading a SDS gel. However, severe problems can be encountered in the isoelectric focusing (IEF) step at three stages:

(a) During the initial solubilization of the sample, important interactions between proteins of widely different pI and/or between proteins and interfering compounds (e.g. nucleic acids) may happen. This may lead to poor solubilization of some components.
(b) During the entry of the sample in the focusing gel, there is a stacking effect due to the transition between a liquid phase and a gel phase with a higher friction coefficient. This stacking increases the concentration of proteins and may give rise to precipitation events.
(c) At, or very close to, the isoelectric point, the solubility of the proteins comes to a minimum. This can be explained by the fact that the net charge comes close to zero, with a concomitant reduction of the electrostatic repulsion between polypeptides. This can also result in protein precipitation or adsorption to the IEF matrix.

Solubilization of proteins for IEF is submitted to several constraints due to the separation mode. The two main constraints are the use of a low ionic strength, so that ionic interactions are difficult to break, and the respect of the natural charge of the proteins in the gel. This means that one of the best means to keep molecules apart, i.e. electrostatic repulsion, cannot be used in IEF gels, while it is one of the bases for the efficiency and popularity of SDS gels. Consequently, protein precipitation is frequently encountered in IEF, especially at high concentrations or for sparingly soluble proteins. The problem is of course more severe at or close to the isoelectric point (pI), when the net charge reaches to a minimum.

2 Rationale of Solubilization-Breaking Molecular Interactions

Apart from disulfide bridges, the main forces holding proteins together and allowing binding to other compounds are non-covalent interactions. Covalent bonds are encountered mainly between proteins and some coenzymes. The non-covalent interactions are mainly ionic bonds, hydrogen bonds and hydrophobic interactions. The basis for hydrophobic interactions is in fact the presence of water. In this very peculiar (hydrogen-bonded, highly polar) solvent, the exposure of non-polar groups to the solvent is thermodynamically not favoured compared to the grouping of these apolar groups together. Indeed, although the van der Waals forces give an equivalent contribution in both configurations, the other forces (mainly hydrogen bonds) are maximized in the latter configuration and disturbed in the former (solvent destruction). Thus, the energy balance is clearly in favour of the collapse of the apolar groups together (Tanford 1980). This explains why hexane and water are not miscible, and also that the lateral chain of apolar amino acids (L, V, I, F, W, Y) pack together and form the hydrophobic cores of the proteins (Dill 1985). These hydrophobic interactions are also responsible for some protein-protein interactions and for the binding of lipids and other small apolar molecules to proteins.

The constraints for a good solubilization medium for 2-D electrophoresis are therefore to be able to break ionic bonds, hydrogen bonds, hydrophobic interactions and disulfide bridges under conditions compatible with IEF, i.e. with very low amounts of salt or other charged compounds (e.g. ionic detergents).

2.1 Disruption of Disulfide Bridges

Breaking of disulfide bridges is usually achieved by adding to the solubilization medium an excess of a thiol compound. Mercaptoethanol was used in the first 2-D protocols (O'Farrell 1975), but its use does have drawbacks. A portion of the mercaptoethanol will ionise at basic pH (pK $\approx$ 9) enter the basic part of the IEF gel and ruin the pH gradient in its alkaline part because of its buffering power (Righetti et al. 1982). Although its pK is around 8, dithiothreitol (DTT) is much less prone to this drawback, as it is used at much lower concentrations (usually 50 mM instead of the 700 mM present in 5 % mercaptoethanol). However, DTT is still not the perfect reducing agent. Some proteins of very high cysteine content or with cysteines of very high reactivity are not fully reduced by DTT. In these cases, phosphines are very often an effective answer. First, the reaction is stoechiometric, which allows us in turn to use very low concentrations of the reducing agent (a few millimolar). Second, these reagents are not as sensitive as thiols to dissolved oxygen. The most powerful compound is tributylphosphine, which was the first phosphine used for disulfide reduction in biochemistry (Ruegg and Rüdinger 1977). However, the reagent is volatile, toxic, has a rather unpleasant odour, and needs an organic solvent to make it water-miscible. In the first uses of the reagent, propanol was used as a carrier solvent at rather high concentrations (50 %) (Ruegg and Rüdinger 1977). It was, however, found that dimethylsulfoxide (DMSO) or dimethylformamide (DMF) are suitable carrier solvents, which enable the reduction of proteins by 2 mM tributylphosphine (Kirley 1989). All these drawbacks have disappeared with the introduction of a water-soluble phosphine, tris (carboxyethyl) phosphine (available from Pierce), for which 1M aqueous stock solutions can be easily prepared and stored frozen in aliquots.

The use of phosphines in 2-D electrophoresis seems to be very promising. Some proteins seem much better solubilized with phosphines than with any other disulfide-breaking chemical (Herbert et al. 1998), so that improved 2-D patterns can be obtained. In addition, phosphines are compatible with acrylamide polymerization, so that they can be used to cast reducing IEF gels when the classical tube format with carrier ampholytes is used (T. Rabilloud, unpubl. results).

2.2 Disruption of Non-covalent Interactions

The perfect way to disrupt all types of non covalent interactions would be the use of a charged compound that disrupts hydrophobic interactions by providing a hydrophobic microenvironment. The hydrophobic residues of the proteins would be dispersed in that environment and not clustered together. This is the description of SDS, and this explains why SDS has been often used in the first stages of solubilization. However, SDS is not compatible with IEF, and must be removed from the proteins during IEF.

The other way of breaking most non-covalent interactions is the use of a chaotrope. It must be kept in mind that all the non covalent forces keeping molecules together must be taken into account with a comparative view of the solvent. This means that the final energy of interaction depends on the interaction per se and on its effects on the solvent. If the solvent parameters are changed (dielectric constant, hydrogen bond formation, polarizability, etc.), all the resulting energies of interaction will change. Chaotropes, which alter all the solvent parameters, exert profound effects on all types of interactions. For example, by changing the hydrogen bond structure of the solvent, chaotropes disrupt hydrogen bonds but also decrease the energy penalty for exposure of apolar groups and therefore favour the dispersion of hydrophobic molecules and the unfolding of the hydrophobic cores of a protein (Herskovits et al. 1970). Unfolding the proteins will also greatly decrease ionic bonds between proteins, which are very often not very numerous and highly dependent on the correct positioning of the residues. As the gross structure of proteins is driven by hydrogen bonds and hydrophobic interactions, chaotropes dramatically decrease ionic interactions both by altering the dielectric constant of the solvent and by denaturing the proteins, so that the residues will no longer be positioned correctly.

Non-ionic chaotropes, such as those used in 2-D, are unable to disrupt ionic bonds when high charge densities are present (e.g. histones, nucleic acids) (Sanders et al. 1980). In this case, it is often quite advantageous to modify the pH and to take advantage of the fact that the ionizable groups in proteins are weak acids and bases. For example, increasing the pH to 10 or 11 will induce most proteins to behave as anions, so that ionic interactions present at pH 7 or lower turn into electrostatic repulsion between the molecules, thereby promoting solubilization. The use of a high pH often results in much improved solubilizations (Horst et al.1980).

For 2-D electrophoresis, the chaotrope of choice is urea. Although urea is less efficient than substituted ureas in breaking hydrophobic interactions (Herskovits et al. 1970), it is more efficient in breaking hydrogen bonds, so that its overall solubilization power is greater. However, denaturation by urea induces the exposure of the totality of the proteins' hydrophobic residues to the solvent. This increases in turn the potential for hydrophobic interactions, so that urea alone is often not sufficient to quench completely the hydrophobic interactions. A first improvement is obtained when using another ancillary chaotrope in addition to urea. Among the various chaotropes described for protein denaturation, thiourea is a chaotrope of choice. Thiourea has been shown to be a much stronger denaturant than urea itself (Gordon and Jencks 1963) on a molar basis. Thiourea alone is weakly soluble in water (ca. 1 M), so that it cannot be used as the sole chaotrope. However, thiourea is more soluble in concentrated urea solutions (Gordon and Jencks 1963). Consequently, urea-thiourea mixtures (typically 2 M thiourea and 5 to 8 M urea, depending on the detergent used) exhibit a superior solubilizing power (Molloy et al. 1998) and are able to increase dramatically the solubility of membrane or nuclear proteins in IPG gels as well as protein transfer to the second dimension SDS gel (Rabilloud et al. 1997).

Chaotrope mixtures are, however, not sufficient, especially when lipids are present in the sample. This explains why detergents, which can be viewed as specialized agents for hydrophobic interactions, are almost always included in the urea-based solubilization mixtures for 2-D electrophoresis. Detergents act on hydrophobic interactions by providing a stable dispersion of a hydrophobic medium in the aqueous medium, through the presence of micelles, for example. Therefore, the hydrophobic molecules (e.g. lipids) are no longer collapsed in the aqueous solvent but will disaggregate in the micelles, provided the amount of detergent is sufficient to ensure maximal dispersion of the hydrophobic molecules. Detergents have polar heads that are able to contract other types of noncovalent bonds (hydrogen bonds, salt bonds for charged heads, etc.). The action of detergents is the sum of the dispersive effect of the micelles on the hydrophobic part of the molecules and the effect of their polar heads on the other types of bonds. This explains why various detergents show very variable effects, from a weak and often incomplete delipidation (e.g. Tweens) to a very aggressive action where the exposure of the hydrophobic core in the detergent-containing solvent is no longer energetically unfavoured and leads to denaturation (e.g. SDS).

Of course, detergents used for IEF must bear no net electrical charge, and only non-ionic and zwitterionic detergents may be used. However, ionic detergents such as SDS may be used for the initial solubilization, prior to isoelectric focusing, in order to increase solubilization and facilitate the removal of interfering compounds. Low amounts of SDS can be tolerated in the subsequent IEF (Ames and Nikaido 1976) provided that high concentrations of urea (Weber and Kuter 1971) and non-ionic (Ames and Nikaido 1976) or zwitterionic detergents (Rémy and Ambard-Bretteville 1987) are present to ensure complete removal of the SDS from the proteins during IEF. Increasing the chaotropic nature of the solvent decreases all hydrophobic interactions and thus SDS-protein interactions. However, the use of SDS is limited by its electric charge. As any other salt, SDS will migrate to the ends of the pH gradient and distort it, distortion increasing with the quantity of SDS. Therefore, high amounts of SDS must be removed prior to IEF, by precipitation (Hari 1981) for example. Thus, it must be kept in mind that SDS will only be useful for solubilization and for sample entry and will not cure isoelectric precipitation problems.

The use of non ionic or zwitterionic detergents in the presence of urea presents some problems due to the presence of urea itself. In concentrated urea solutions, urea is not freely dispersed in water but can form organized channels (see March 1977). These channels can bind linear alkyl chains, but not branched or cyclic molecules, to form complexes of undefined stoichiometry called inclusion compounds. These complexes are much less soluble than the free solute, so that precipitation is often induced upon formation of the inclusion compounds, precipitation being stronger with increasing alkyl chain length and higher urea concentrations. Consequently, many non ionic or zwitterionic detergents with linear hydrophobic tails (Dunn and Burghes 1983; Rabilloud et al.1990) and some ionic ones (Willard et al. 1979) cannot be used in the presence of high concentrations of urea. This limits the choice of detergents mainly to those with non-linear alkyl tails (e.g. Tritons, Nonidet P40, CHAPS) or with short alkyl tails (e.g. octyl glucoside), which are unfortunately less efficient in quenching hydrophobic interactions. However, sulfobetaine detergents with long linear alkyl tails have received limited applications, as they require low concentrations of urea. However, good results have been obtained in certain cases for sparingly soluble proteins (Satta et al. 1984), although this type of protocol seems rather delicate owing to the need for a precise control of all parameters to prevent precipitation.

Apart from the problem of inclusion compounds, the most important problem linked with the use of urea is carbamylation. Urea in water exists in equilibrium with ammonium cyanate, the level of which increases with increasing temperature and pH (Hagel et al. 1971). Cyanate can react with amines to yield substituted ureas. In the case of proteins, this reaction takes place with the α-amino group of the N-terminus and the ε-amino groups of lysines. This reaction leads to artefactual charge heterogeneity, N-terminus blocking and adduct formation detectable in mass spectrometry. Carbamylation must therefore be completely avoided. This can be easily made with some simple precautions. The use of a pure grade of urea (p.a.) decreases the amount of cyanate present in the starting material. Avoidance of high temperatures (never heat urea-containing solutions above 37 °C) considerably decreases cyanate formation. In the same trend, urea-containing solutions should be stored frozen (–20 °C) to limit cyanate accumulation. Last but not least, a cyanate scavenger (primary amine) should be added to urea-containing solutions. In the case of isoelectric focusing, carrier ampholytes are perfectly suited for this task. If these precautions are correctly taken, proteins seem to withstand long exposures to urea without carbamylation (Bjellqvist et al. 1993).

3
Initial Solubilization

From the rationale described above, the simplest way to solubilize a sample, for example intact cells, is to dilute them with a solution containing chaotropic agents (chiefly urea), a suitable detergent (in most cases non-ionic or zwitterionic, or SDS if an extra initial solubilizing power is needed), and a buffer, usually carrier ampholytes, which avoids any interference with the subsequent IEF and protects from carbamylation. Such a simple protocol may work well in some cases, but two main problems can be encountered. These are keeping the primary structure and the post-translational modifications of the proteins intact, on the one hand, and removing interfering compounds after disrupting their interactions with proteins, on the other hand.

3.1 Keeping Proteins Intact

The problem of keeping proteins intact is most prominent for complex samples (e.g. whole cells or tissues) and arises mainly from the hydrolases present in such samples (phosphatases, glycosidases and especially proteases). Most of these hydrolases are stored in lysosomes, but are released upon the action of the detergent on the cell membranes. They are of course denatured by the chaotrope present in the solubilization medium, leading to their inactivation. This inactivation is generally quick enough for glycosidases and phosphatases, so that the post-translational status of the proteins is generally kept intact. This is, however, not the case for proteases. Evidence of proteolysis after solubilization in 9 M urea (Colas des Francs et al. 1985) or SDS (Granzier and Wang 1993) has been already described. This is probably due to the fact that many proteases are resistant proteins, as shown by their ability to work in dilute SDS, used for example in peptide mapping (Cleveland et al. 1977). This means that their kinetics of denaturation in urea- or SDS-based solutions can be slow enough to allow them to work for a non-negligible time, while most of the other cellular proteins are already denatured and therefore expose a maximal number of protease-sensitive sites. This seems true especially at acidic or neutral pH (Segers et al. 1986), and solubilization at alkaline pH seems to limit proteolysis (Hochstrasser et al. 1988). This problem of proteolysis in denaturing solutions will of course strongly depend on the concentration of proteases in the sample, and seems to be more important in plant tissues (Colas des Francs et al. 1985), where the vacuole can be viewed as a giant lysosome. This work also demonstrated that addition of protease inhibitors was of weak, if any, efficiency to solve this problem (Colas des Francs et al. 1985). In such difficult cases, solubilization at high pH is not sufficient and the only solution is to increase the denaturing power of the solubilization process and the kinetics of denaturation as much as possible. For plant samples, which are very rich in proteases, solubilization in boiling SDS has been proposed (Harrison and Black 1982). In this case, the thermal denaturation synergizes the SDS denaturation and affords a faster inactivation of proteases. Another solution is to homogenize the sample in dilute trichloroacetic acid (TCA) (Wu and Wang 1984) or in TCA in acetone (Damerval et al. 1986), which inactivates and precipitates almost instantaneously all the proteins, including proteases. Subsequent resolubilization of the protein precipitate in a urea-containing buffer (Damerval et al. 1986) does not seem to yield to any reactivation of the proteases. Such procedures are probably the only efficient ones for tissues very rich in proteolytic activities. However, it must be kept in mind that protocols involving protein precipitation of any type are likely to lead to protein losses, as some proteins will precipitate but not be resolubilized by any means. Here, again, the use of the most efficient solubilization cocktail (i.e. urea-thiourea-detergent) is likely to minimize this recovery problem.

This example of protease inactivation emphasizes the fact that a rapid penetration of the denaturing solution is a key parameter for an efficient solubilization. Consequently, fresh tissues or cells are always much better solubilized than frozen or lyophilized samples (Franzen et al. 1995). Moreover, lyophilization is known to induce cross-linking of proteins via amide bonds, thereby leading to artefactual charge modification (Goodno et al. 1981). When cultured cells or cell organelles are used as samples, a very efficient protocol for solubilization is to prepare a concentrated cell or organelle suspension and to dilute it with a concentrated denaturing solution.

3.2 Disrupting Complexes and Removing Interfering Compounds

The other major problem in initial solubilization prior to 2-D electrophoresis is the maximal disruption of supramolecular complexes and the removal of any substance interfering with the electrophoretic process. These interfering substances are mainly salts, lipids and polysaccharides including nucleic acids.

3.2.1 Salts

Generally speaking, salts do not interfere through their binding to protein; they interfere directly with the electrophoresis process. Salts migrate through the pH gradients (producing Joule heat) and accumulate at both ends of the electrophoresis support. This accumulation builds very high conductivity zones, whose size will of course depend on the quantity of salts present in the sample. Because of the high conductivity, the voltage drop and thus the electric field are very low in

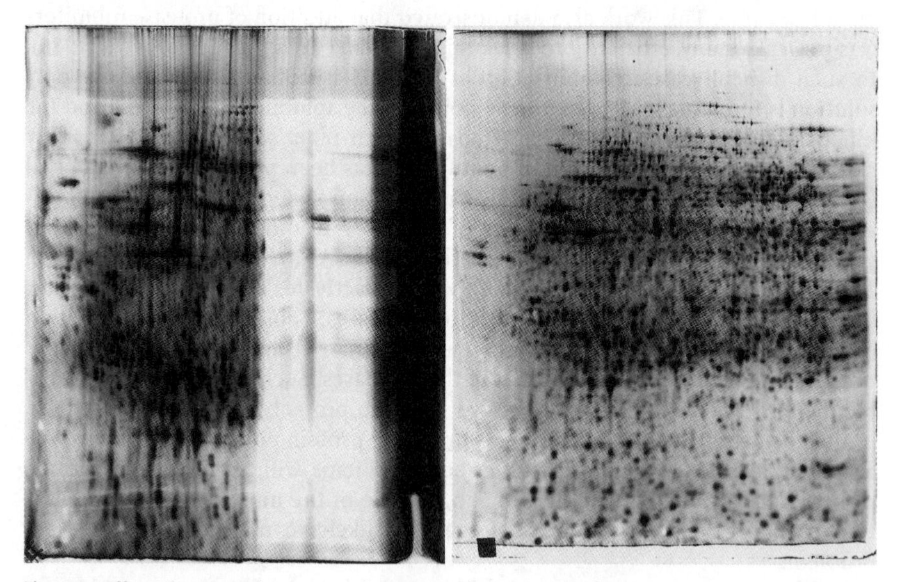

Fig. 2.1. Effect of excess salt on 2-D gels. 100 μg of mitochondrial proteins were separated by 2-D electrophoresis. The sample was kept pure (right), or polluted by sodium chloride (left). Note the typical empty and dark streaky zones where sodium accumulates. No effect can be seen at the other electrode (left side of the gel) because of chlorine formation and evacuation at the electrode. Salt accumulation zones can be seen after focusing as bulges in the IEF gel

these zones. Consequently, proteins cannot focus in these zones and appear as streaks (Fig. 2.1). These high conductivity zones frequently appear as bulges in the first-dimension gels. Bulging is induced by the accumulation of the hydration water surrounding the salt molecules.

When necessary, salt removal is carried out either by dialysis or by precipitation of proteins (e.g. by TCA or organic solvents). The classical drawback of these approaches is loss of proteins, by sticking to the dialysis membrane or by passing through it in the former method, by being soluble in the precipitation medium or irreversibly precipitated in the latter method. Samples rich in salts generally come from biological fluids or chromatographic eluates. However, whole cell extracts can also give this problem, especially at high concentrations. When moderately interfering amounts of salts are encountered, casting pH plateaus at both ends of the pH gradient is a very simple and efficient remedy, available when the 2-D electrophoresis protocol uses home-made immobilized pH gradients (IPGs) (Rabilloud et al. 1994). This strategy avoids the losses incurred by dialysis or precipitation. Salts present in the sample just collect in the plateau zones first and do not invade the useful pH gradient zone. The plateau zones are removed prior to the second dimension, so that any interference of the accumulated salts is avoided.

3.2.2 Lipids

Lipids give problems when supramolecular assemblies of lipids (membrane and derivatives thereof) are present. As a basic rule, the presence of detergents is the solution of choice to disrupt the membranes, solubilize the lipids, delipidate and solubilize the proteins bound to those membranes or vesicles. Consequently, many reviews have been written on the properties and uses of detergents and the reader is referred to some of them (Helenius et al. 1979; Hjelmeland and Chrambach, 1981; Hjelmeland, 1986), as well as to some comparative work on the efficiency of various detergents on membranes (Navarrete and Serrano 1983). However, detergents act by diluting the lipids into the micelles. A problem will therefore arise for high lipid levels. In this case, there will be soon an inadequation between the amount of lipids present in the sample and the amount of detergent which can be used in the desired sample volume (insufficient detergency). Two solutions may be envisioned. The first is to scale up the separation, and therefore dilute the sample, so that a correct detergency can be achieved. This is limited by the volumes which can be loaded on the electrophoretic gels. The second solution is to carry out a chemical delipidation on the sample prior to resolubilization of the proteins in the presence of detergents. Delipidation is achieved by extraction of the biological material with organic solvents (Van Renswoude and Kempf 1984), generally a mixture (Radin 1981), and often containing chlorinated solvents (Wessel and Flügge 1984). However, more conventional protein precipitation protocols, for example with ethanol or acetone, often provide a partial but useful delipidation (Penefsky and Tzagoloff 1971; Menke and Koenig 1980).

Generally speaking, these media based on organic solvents remove the excess of lipids very efficiently. However, severe losses in proteins may be experienced, either because some proteins are soluble in organic solvents (Radin, 1981), or because the precipitated proteins may not resolubilize. Special attention must be paid to the final removal of the organic solvents prior to resolubilization. If the solvent is not efficiently removed, emulsion problems or precipitation by the remaining solvent may arise. If the precipitated protein pellet is dried too extensively in order to remove the solvent completely, a tight and dry pellet impossible to resolubilize even in media of high denaturing and solubilizing power appears, with extremely severe losses. This leads to the picture that achievement of a proper and reproducible delipidation-solubilization cycle is very difficult, as soon as delipidation by organic solvents is required. As a practical rule, the process becomes more and more difficult as the solvent used becomes less and less miscible with water. Consequently, partial delipidation with alcohols is often easier than with acetone, which is itself much easier than with chlorinated or ether-based solvents.

3.2.3 Polysaccharides (Including Nucleic Acids)

The problems encountered because of polysaccharides are of several types. First, they behave as polyanions and are therefore able to fix proteins through electrostatic interactions. Second, polysaccharides, and especially nucleic acids, also bind ampholytes to give complexes (Galante et al. 1976) which also bind proteins and focus to give completely artefactual results with much streaking (Heizmann et al. 1980), as shown in Fig. 2.2. Third, many polysaccharides (e.g. DNA) are very long molecules which are able to increase considerably the viscosity of the solutions and also to clog the small pores of the polyacrylamide gels used to separate the proteins.

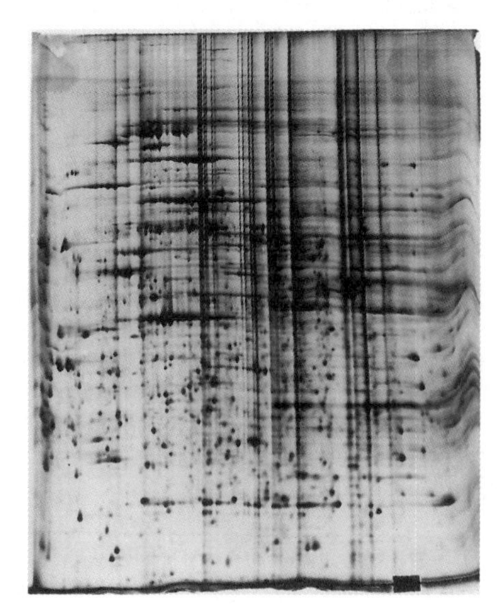

Fig. 2.2. Effect of nucleic acids on the resolution of 2-D electrophoresis. Transformed mouse pre-B lymphocytes were cultured in vitro, collected, washed with PBS and lysed in urea, detergent, reducer and ampholytes. The crude extract was then loaded onto a carrier ampholyte tube gel and submitted to 2-D electrophoresis. Proteins were detected with silver staining. Note the prominent vertical streaks due to DNA interference with 2-D electrophoresis

For these reasons, removal of polysaccharides is required, unless they are present at very low concentrations. This is especially important for nucleic acids. Because of their high charge density, nucleic acids give extremely severe artifacts in 2-D electrophoresis, by all the mechanisms described above. Moreover, nucleic acids are present at problematic concentrations in most total cell extracts, not to speak of nuclear extracts. The first removal method is digestion by nucleases, initially by a mixture of RNAses and DNAses (O'Farrell 1975). As with most of the enzyme-based removal methods, the main drawbacks are linked to the parallel action of the proteases contained in the sample, thereby degrading the proteins, and to the addition of extraneous proteins (the nucleases). In fact, the most efficient methods use centrifugation to get rid of the excess nucleic acids (Rabilloud et al. 1986). A last problem remains, however, in the fact that some proteins still stick to nucleic acids even in the presence of high concentrations of urea (Sanders et al. 1980). These proteins can be solubilized either with competing cations such as protamine (Sanders et al. 1980) or lecithins at acid pH (Willard et al. 1979). These methods are extremely efficient but introduce high amounts of charged compounds, so that only low sample amounts can be loaded. A more versatile strategy is to use a high pH during extraction, so that all the proteins will behave as anions and will be repelled from the anionic nucleic acids. To avoid overswelling of the nucleic acids, which decreases the subsequent removal by ultracentrifugation, this increase of pH can be mediated by the addition of a basic polyamine (e.g. spermine; Rabilloud et al. 1994) which will precipitate the nucleic acids. Alternatively, the alkaline pH can be obtained either by addition of a few millimoles of potassium carbonate to the urea–detergent–ampholytes solution (Horst et al. 1980), or by the use of alkaline ampholytes (Hochstrasser et al. 1988). In these protocols, ampholytes act as nucleic acid-precipitating agents. Of course, ultracentrifugation will remove all large polysaccharides, whether they are nucleic acids or not.

Ultracentrifugation-based methods could be seen as risky methods, since large proteins could also sediment during the removal of nucleic acids, especially if this removal has to be complete and include small nucleic acids. It must, however, be kept in mind that extraction is carried out with reagents which increase the density of the solvent (urea). This will lead to a decreased sedimentation of the proteins compared to the one of the polysaccharides, which have a very high buoyant density, so that minimal protein losses due to sedimentation have been reported (Shirey and Huang 1969, Chaudhury 1973).

3.2.4 Other Compounds

Apart from these major classes of interfering compounds, other interfering compounds can be found, mainly in extracts from plants. These include lignins, polyphenols, tannins, alkaloids, pigments, etc. (Gegenheimer 1990). For example, polyphenols bind protein by hydrogen bonds as long as they are in the reduced state, while oxidized polyphenols bind protein by covalent bonds. The use of polyvinylpyrrolidone (PVP) to trap polyphenolic compounds is common practice (Cremer and Van de Walle 1985; Gegenheimer 1990). Otherwise, extraction with organic solvents such as acetone removes efficiently all these interfering compounds present in plant material (Hari 1981; Damerval et al. 1986). The problems specifically encountered for solubilization of plant proteins and the solutions which have been proposed for carrying out correct solubilization of these plant proteins have been previously reviewed (Damerval et al. 1988).

3.3 Conclusion

As a conclusion, two solubilization protocols are given in Protocols 1 and 2. The first one is intended for animal and prokaryotic cells. It can be adapted to animal tissues if they are homogenized in the denaturing solution, or previously very finely ground under liquid nitrogen (to improve the penetration of the solution). The second protocol is intended for more difficult plant tissues. Both of them use a high pH to maximize extraction and minimize protease action, while the second protocol also uses SDS to resolubilize precipitated proteins. Other solubilization protocols and procedures can be found in Chapters 3 and 4, the former protocol being optimized for yeast cells.

Protocol 1 (Adapted from Rabilloud et al. 1994)

- *Alternate Extraction Solution* (1.2 × Concentrated):
  9.6 M urea (5.8 g/10 ml), 5 % CHAPS, 25 mM spermine base, 50 mM DTT

Note: This solution is prepared with gentle warming of the components (not above 37 °C). It can be stored frozen in aliquots at –20 °C. The spermine stock solution (1 M in water) is prepared by dissolving directly the contents of a bottle of spermine base in water. This stock solution is stored frozen in aliquots at –20 °C. Spermine base strongly absorbs carbon dioxide from air, so that it cannot be weighed accurately after some storage.

- *Alternate Extraction Solution* (1.2 × concentrated):
  8.4 M urea, 2.4M thiourea, 5 % CHAPS, 25 mM spermine base, 50 mM DTT

Note: This solution is prepared with gentle warming of the components (not above 37 °C). It can be stored frozen in aliquots at –20 °C. The partial specific volume of urea is 0.75 ml/g, while the partial specific volume of thiourea and CHAPS is 1 ml/g. This means that 1 g of dissolved urea accounts for a volume of 0.75 ml, while 1 g of dissolved thiourea or CHAPS accounts for 1 ml.

- *Sample Preparation* : A concentrated suspension of cells or subcellular fraction of interest is made in 10 mM Tris pH 7.5, 1 mM ethylene diamine tetraacetic acid (EDTA), 0.25 M sucrose. One volume of this suspension is placed in a polyallomer micro ultracentrifuge tube (e.g. TL100). Four volumes of concentrated extraction solution is added and immediately mixed by placing a piece of parafilm over the tube and inverting it several times. Most often, the solution become instantly very viscous (nucleic acids). Extraction is carried out at room temperature for 30 to 60 min with occasional mixing. The sample is ultracentrifuged (250 000 *g* for 1 h). A translucent pellet of nucleic acids should be obtained. The protein-containing supernatant is kept. Protein concentration can be determined by Coomassie Blue binding assay (Bradford type). After determination, carrier ampholytes (usually a wide pH range) are added at a final concentration of 0.4%. The sample can now be used or stored frozen.

Note: the ultracentrifuge tube must absolutely be made of polyallomer. Polycarbonate or cellulose ester tubes do not withstand the high pH of the extraction solution.

Protocol 2 (Adapted from Damerval et al. 1986)

- *Grinding Solution* (to be made fresh): 10% trichloroacetic acid and 0.07% mercaptoethanol in acetone.
- *Rinsing Solution*: 0.07% mercaptoethanol in acetone
- *Resolubilization Solution* (to be made fresh): 9.5 M urea, 5 mM potassium carbonate, 0.4% SDS, 0.5% DTT, 1% Ampholytes, 6% CHAPS

- *Sample Preparation*: The sample is dry-crushed in a liquid nitrogen-cooled mortar. The resulting powder is suspended in the grinding solution and the proteins are precipitated for 45 min at –20 °C. The suspension is centrifuged at 35 000 *g* for 15 min (4 °C). The pellet is resuspended in rinsing solution and extracted for 1 h at –20 °C. The suspension is centrifuged at 35 000 *g* for 15 min (4 °C). The supernatant is discarded and the pellet dried under vacuum. It is then resolubilized with the urea–carbonate solution (50 µl/mg of dry pellet) for 1 h at room temperature. The extract is centrifuged (ultracentrifugation at 200 000 *g* for 30 min is recommended). The supernatant is used immediately or stored frozen.

An interesting approach, especially for proteomics, is to use a serial extraction of complex samples. This has been first introduced by Klose and Zeindl (1984), but strong cross-contamination of the fractions was observed. This approach has been further refined recently by Molloy et al. (1998), using E. coli cells. In this case, considerable enrichment in membrane proteins was observed in the fractions extracted with the most denaturing solutions. This protocol is given below.

Protocol 3 (Adapted from Molloy et al. 1998)

- *First Extraction Solution*: 40mM Tris base in water.
- *Second Extraction Solution* : 8M urea, 4% CHAPS, 100mM DTT, 40 mM Tris base, 0.4% carrier ampholytes.
- *Third Extraction Solution*: 5M urea, 2M Thiourea, 2% CHAPS, 2% SB3–10 (decyl dimethyl ammonio propane sulfonate), 2mM Tributylphosphine (from a 100mM stock in tetramethylurea or DMF), 40 mM Tris base, 0.4% carrier ampholytes
- *Fourth Extraction Solution*: 1% SDS, 50mM DTT and 25% glycerol in 0.4 M Tris-HCl pH 8.8

- *Sample Preparation*: 15mg of lyophilized *E. coli* are reconstituted in 5 ml of first extraction solution. 150 Units of Benzonase are added and the mixture is vortexed for 5 min. It is then placed in an ultrasonic bath (bath temperature 10 °C) for 10 min. The sample is then re-vortexed and re-sonicated for an additional 5 and 10 min, respectively. The resulting suspension is centrifuged at 12 000 *g* for 10 min. The supernatant is the first soluble extract. The pellet is then

extracted with 5 ml of the second extraction solution by vortexing and sonication as described (sonication bath temperature: 20 °C). After centrifugation (12 000 *g* for 10 min), the supernatant is recovered as the second soluble extract. The pellet (≈ 10 % of the starting material) is re-extracted as described with 1 ml of third extraction solution, leading to the first membrane extract and a pellet, which is boiled for 5 min with 100 μl of fourth extraction solution. Centrifugation of this final extract (12 000 *g* for 10 min) leads to the final insoluble pellet and the second membrane extract. The proportion of *E. coli* proteins extracted in the different solutions is as follows:
First soluble extract: ≈ 40 %
Second soluble extract: ≈ 50 %
First membrane extract: ≈ 9 %
Second membrane extract: ≈ 1 %
The last two extracts are highly enriched in membrane proteins , while the first two extracts mainly contain soluble proteins (Molloy et al. 1998).

4 Solubility During IEF

Additional solubility problems often arise during the IEF at sample entry and solubility at the isoelectric point.

4.1 Solubility During Sample Entry

Sample entry is often quite critical. In most 2-D systems, sample entry in the IEF gel corresponds to a transition between a liquid phase (the sample) and a gel phase of higher friction coefficient. This induces a stacking of the proteins at the sample-gel boundary, which results in a very high concentration of proteins at the application point. These concentrations may exceed the solubility threshold of some proteins, thereby inducing precipitation and sometimes clogging of the gel, with poor penetration of the bulk of proteins. Such a phenomenon is of course more prominent when high amounts of proteins are loaded onto the IEF gel. The sole simple but highly efficient remedy to this problem is to include the sample in the IEF gel. This process abolishes the liquid-gel transition and decreases the overall protein concentration, as the volume of the IEF gel is generally much higher than the one of the sample. This decrease in protein concentration is generally highly beneficial for all solubility problems.

This process is, however, rather difficult for tube gels in carrier ampholyte-based IEF. The main difficulty arises from the fact that the thiol compounds used to reduce disulfide bonds during sample preparation are strong inhibitors of acrylamide polymerization, so that conventional samples cannot be used as such. Alkylation of cysteines and of the thiol reagent after reduction could be an answer, but many neutral alkylating agents (e.g. iodoacetamide, ethyl maleimide) also inhibit acrylamide polymerization. Owing to this situation, most workers describing inclusion of the sample within the IEF gel have worked with non-reduced samples (Chambers et al. 1985; Semple-Rowland et al. 1991). Although this presence of disulfide bridges is not optimal, inclusion of the sample within the gel has proven of great but neglected interest (Chambers et al. 1985; Semple-Rowland et al. 1991). It must, however, be pointed out that it is now possible to carry out acrylamide polymerization in an environment where disulfide bridges are reduced. The key is to use 2 mM tributylphosphine as reducing agent. Tetramethylurea is used as a carrier solvent to prepare a 100 mM tributylphosphine stock solution. These conditions ensure total reduction of disulfides and are totally compatible with acrylamide polymerization with the standard Temed/persulfate initiator (T. Rabilloud, unpubl. results). This modification should help the experimentators trying sample inclusion within the IEF gel when high amounts of proteins are to be separated by 2-D.

The process of sample inclusion within the IEF gel is, however, much simpler for IPG gels. In this case, rehydration of the dried IPG gel in a solution containing the protein sample is quite convenient and efficient, provided that the gel has a sufficiently open structure to be able to absorb proteins efficiently (Rabilloud et al. 1994). Coupled with the intrinsic high capacity of IPG gels, this procedure enables us to easily separate milligram amounts of protein (Rabilloud et al. 1994, Sanchez et al. 1997).

4.2 Solubility at the Isoelectric Point

Solubility at the isoelectric point is usually the second critical point for IEF. The isoelectric point is the pH of minimal solubility, mainly because the protein molecules have no net electrical charge. This abolishes the electrostatic repulsion between protein molecules, which maximizes in turn protein aggregation and precipitation.

The horizontal comet shapes frequently encountered for major proteins and for sparingly soluble proteins often arise from such a near-isoelectric precipitation. Such isoelectric precipitates are usually easily dissolved by the SDS solution used for the transfer of the IEF gel onto the SDS gel, so that the problem is limited to a loss of resolution, which, however, precludes the separation of large amounts of proteins.

The problem is, however, more severe for hydrophobic proteins. In this case, a strong precipitation of the isoelectric protein seems to occur, which is not reversed by incubation of the IEF gel in the SDS solution. The result is severe quantitative losses, which seem to increase with the hydrophobicity of the protein and the amount loaded (Adessi et al. 1997). This problem has been first described for IPG gels, but poor analysis of hydrophobic proteins seems to be the rule, whatever system is used for IEF. Less isoelectric precipitation seems to occur in carrier-ampholyte gels, probably because the hydrophobic proteins have precipitated at the entry point before. The sole solution to this serious solubility problem is to increase the chaotropicity of the medium used for IEF, by using both urea and thiourea as chaotropes (Pasquali et al. 1997; Rabilloud et al. 1997; Rabilloud et al. 1998), as exemplified in Fig. 2.3. The benefits of using thiourea-urea mixtures to increase protein solubility can be transposed to conventional, carrier ampholyte-based focusing in tube gels with minor adaptations. Thiourea strongly

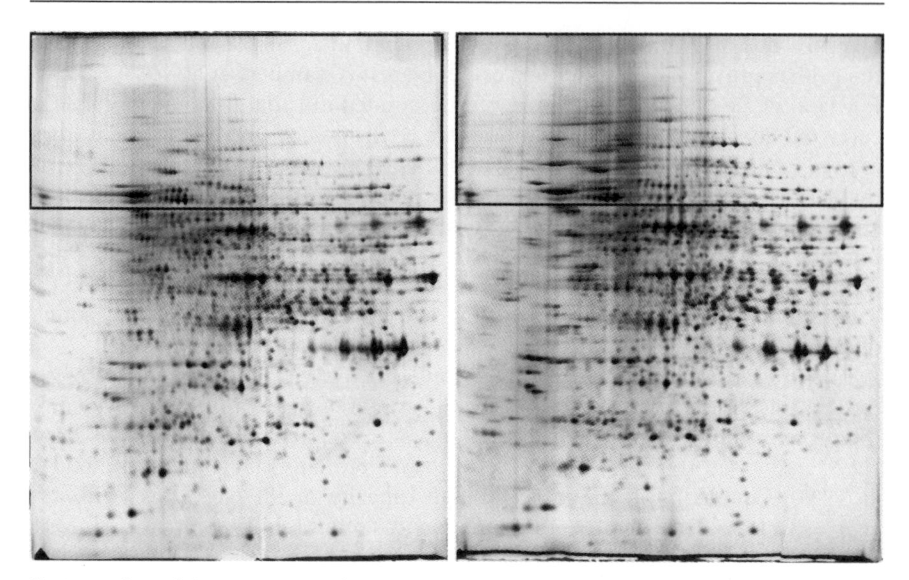

Fig. 2.3. Effect of thiourea on 2-D electrophoresis. 200 μg of a total extract from *S. cerevisiae* was loaded on 2-D gels (first dimension IPG pH 4 to 8). Detection by silver staining. Left: focusing carried out in 8 M urea, 4 %CHAPS; right: focusing carried out in 8 M urea, 1 M thiourea, 4 %CHAPS. The *rectangle* points to the high molecular weight zone where new and/or more abundant spots can be seen with thiourea

inhibits acrylamide polymerization with the standard Temed/persulfate system. However, photopolymerization with 30 μM methylene blue, 50 μM sodium toluene sulfinate and 25 μM diphenyl iodonium chloride (Lyubimova et al. 1993) enables acrylamide polymerization in the presence of 2M thiourea without any deleterious effect in the subsequent 2-D (Rabilloud 1998) so that higher amounts of proteins can be loaded without loss of resolution (Rabilloud 1998).

5 Conclusions: Current Limits and How to Push Them

Although this chapter has mainly dealt with the general aspects of solubilization, the main concluding remark is that there is no universal solubilization protocol. Standard urea-reducer-detergent mixtures usually achieve disruption of disulfide bonds and non-covalent interactions. Consequently, the key issues for a correct solubilization is the removal of interfering compounds, blocking of protease action, and disruption of infrequent interactions (e.g. severe ionic bonds). These problems will strongly depend on the type of sample used, the proteins of interest and the amount to be separated, so that the optimal solubilization protocol can vary greatly from one sample to another (Rabilloud 1996).

However, the most frequent bottleneck for the efficient 2-D separation of as many proteins and as much protein as possible does not usually lie in the initial solubilization but in keeping the solubility along the IEF step. In this field, the key feature is the disruption of hydrophobic interactions, which are probably responsible for most, if not all, of the precipitation phenomena encountered during IEF. This means that improving solubility during denaturing IEF will focus on the quest of ever more powerful chaotropes and detergents. In this respect, the use of thiourea may prove to be one of the keys to increasing the solubility of proteins in 2-D electrophoresis. However, chaotropes alone are not able to keep proteins in solution in IEF, especially at the pI. This may be due to the fact that the electric field used during IEF presses each protein molecule together with others, thereby excluding the solvent to some extent and thus maximizing protein-protein interactions. This is of course especially true at the pI, where there are no electrostatic repulsions to counteract this field effect. Consequently, the use of the correct detergent in conjunction with chaotropes is of paramount importance. One of the the roles of detergents in IEF is to coat the proteins and make them more hydrophilic, thereby preventing isoelectric precipitation. In this respect, using very efficient chaotropes has a drawback, as these chaotropes will also reduce the hydrophobic interactions between detergents and proteins. This reduces in turn the coating of the proteins by the detergent and thus solubility. This effect is shown in Fig. 2.4. Consequently, a good detergent will be able to coat proteins in highly chaotropic solutions, and thus have a large hydrophobic portion to be able to interact with proteins. However, the hydrophobic portion must not be too large. A large hydrophobic part decreases the critical micelle concentration (CMC), and thus the concentration of detergent available. Moreover, detergents

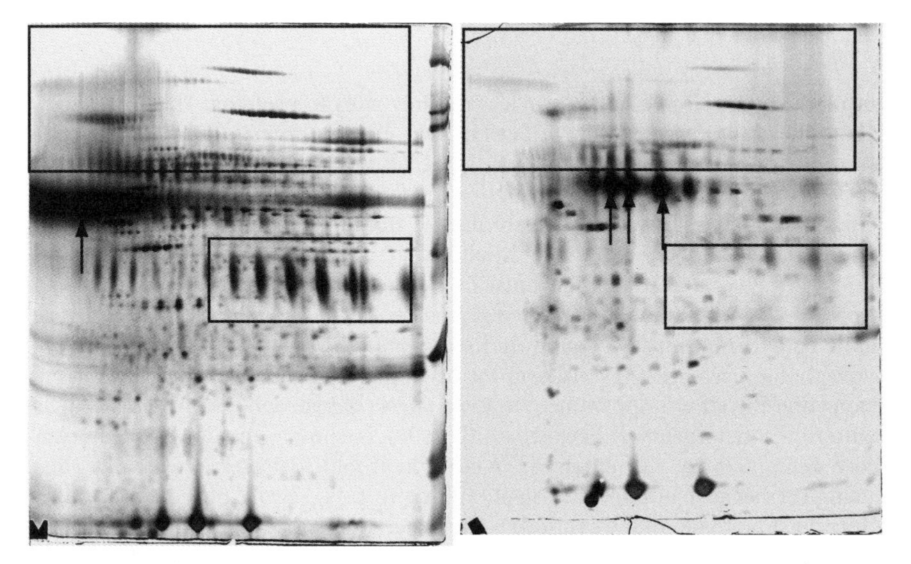

Fig. 2.4. Effect of excess chaotrope on 2-D electrophoresis. 500 µg of a membrane preparation from bovine neutrophils was loaded on 2-D gels (first dimension IPG pH 4 to 8). Detection by silver staining. Left: focusing carried out in 8 M urea, 1 M thiourea, 4 %CHAPS; right: focusing carried out in 7 M urea, 2 M thiourea, 4 %CHAPS. The *rectangles* point to zones where spots are decreased or lost in the presence of excess thiourea. This effect is attributed to the decrease of protein-detergent interactions by the excessive chaotropic power of the medium. Note, however, that high concentrations of chaotropes are required for proper focusing of other proteins (*arrows*)

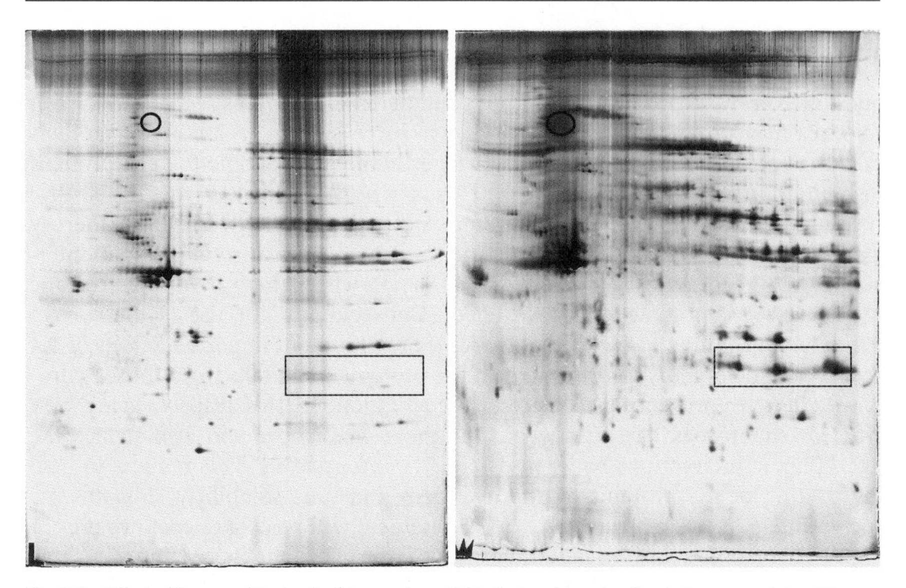

Fig. 2.5. Effects of new zwitterionic detergents on 2-D electrophoresis of membrane proteins. Human erythrocytes membranes (corresponding to 350 μg proteins) are dissolved in 7 M urea, 2 M thiourea, 0.4% ampholytes, 10 mM DTT, 0.2%Triton X-100 containing either 4%CHAPS (left) or 2% 3-[3-tetradecanoylamidopropyl] dimethyl ammonio propane sulfonate (ASB 14) (right). 2-D electrophoresis is carried out with IPG and spots are detected with silver staining. The *ellipsoid* points to band 3 (97 kDa, 11 transmembrane helixes) while the *rectangle* points to stomatin (30 kDa, 1 transmembrane domain)

with large hydrophobic parts have less binding sites to proteins. This problem is exemplified by the use of alkyl sulfates in zone electrophoresis. Decyl sulfate does not bind strongly enough to proteins. Undecyl and dodecyl sulfates exhibit correct and uniform binding, while higher homologues (tetradecyl and hexadecyl sulfate) are less efficient due to more limited binding (Lopez et al. 1991).

In designing a detergent, the hydrophilic part must be as hydrophilic as possible. In IEF where the detergent must be neutral, the hydrophilic part will be a strong dipole (e.g. sulfobetaines) or a glycoside. However, as shown before, a very subtle equilibrium must be found in the hydrophobic part. Detergents with large hydrophobic parts are very efficient for hydrophobic proteins, which have many strong binding sites. Improvement in the analysis of hydrophobic proteins by new zwitterionic detergents (Chevallet et al. 1998) is shown on Fig. 2.5. However, these detergents are not efficient for more hydrophilic proteins, which are lost during IEF because of poor detergent binding (e.g. Fig. 2.3). Consequently, much effort must be made in the synthesis of new detergents and in testing them alone or in cocktails to achieve an as universal solubilization as possible, as SDS does. This might, however, not be possible for IEF, so that the use of different gel systems, optimized for different classes of proteins, might be necessary to cover as many proteins as possible.

Last but not least, it is almost sure that cells contain proteins which are so hydrophobic that they will not be amenable to analysis by IEF. In this case, 2-D electrophoresis systems relying on zone electrophoresis only will be the sole possibility to analyze very hydrophobic proteins (e.g. Booth 1977; MacFarlane 1989). The main drawback of these systems is that the separation is mainly diagonal, due to a major sieving effect in both dimensions (Booth 1977; MacFarlane 1989). Here again, the synthesis of new detergents able of maximizing "off diagonal effects" by differential binding to proteins may be a solution to improve such systems.

References

Adessi C, Miege C, Albrieux C, Rabilloud T (1997) Two-dimensional electrophoresis of membrane proteins: a current challenge for immobilized pH gradients. Electrophoresis 18: 127–135

Ames GFL, Nikaido K (1976) Two-dimensional gel electrophoresis of membrane proteins. Biochemistry 15: 616–623

Bjellqvist B, Sanchez JC, Pasquali C, Ravier F, Paquet N, Frutiger S, Hughes GJ, Hochstrasser DF (1993) Micropreparative two-dimensional electrophoresis allowing the separation of samples containing milligram amounts of proteins. Electrophoresis 14: 1375–1378

Booth AG (1977) A novel system for the two-dimensional electrophoresis of membrane proteins. Biochem J 163: 165–168.

Chambers JAA, Degli Innocenti F, Hinkelammert K, Russo VEA (1985) Factors affecting the range of pH gradients in the isoelectric focusing dimension of two-dimensional gel electrophoresis: the effect of reservoir electrolytes and loading procedures. Electrophoresis 6: 339–348

Chaudhury S (1973) Fractionation of chromatin nonhistone proteins. Biochim Biophys Acta 322: 155–165

Chevallet M, Santoni V, Poinas A, Rouquié D, Fuchs A, Kieffer S, Rossignol M, Lunardi J, Garin J, Rabilloud T (1998) New zwitterionic detergents improve the analysis of membrane proteins by two-dimensional electrophoresis. Electrophoresis 19, 1901–1909

Clare Mills EN, Freedman RB (1983) Two-dimensional electrophoresis of membrane proteins. Factors affecting resolution of rat-liver microsomal proteins. Biochim Biophys Acta 734: 160–167

Cleveland DW, Fischer SG, Kirschner MW, Laemmli UK (1977) Peptide mapping by limited proteolysis in sodium dodecyl sulfate and analysis by gel electrophoresis. J Biol Chem 252: 1102–1106

Colas des Francs C, Thiellement H, De Vienne D (1985) Analysis of leaf proteins by two-dimensional electrophoresis. Plant Physiol 78: 178–182

Cremer F, Van de Walle C (1985) Method for extraction of proteins from green plant tissues for two-dimensional polyacrylamide gel electrophoresis. Anal Biochem 147: 22–26

Damerval C, De Vienne D, Zivy M, Thiellement H (1986) Technical improvements in two-dimensional electrophoresis increase the level of genetic variation detected in wheat-seedling proteins. Electrophoresis 7: 52–54

Damerval C, Zivy M, Granier F, De Vienne D (1988) Two-dimensional electrophoresis in plant biology. In: Chrambach A, Dunn MJ, Radola BJ (eds) Advances in Electrophoresis, vol 2. VCH Weinheim, pp 263–340

Dill KA (1985) Theory for the folding and stability of globular proteins. Biochemistry 24: 1501–1509

Dunn MJ, Burghes AHM (1983) High resolution two-dimensional polyacrylamide electrophoresis. I. Methodological procedures. Electrophoresis 4: 97–116

Franzen B, Hirano T, Okuzawa K, Uryu K, Alaiya AA, Linder S, Auer G (1995) Sample preparation of human tumors prior to two-dimensional electrophoresis of proteins. Electrophoresis 16: 1087–1089

Galante E, Caravaggio T, Righetti PG (1976) Binding of ampholine to transfer RNA. Biochim Biophys Acta 442: 309–315

Gegenheimer P (1990) Preparation of extracts from plants. Methods Enzymol 182: 174–193

Goodno CC, Swaisgood HE, Catignani GL (1981) A fluorimetric assay for available lysine in proteins. Anal Biochem 115: 203–211

Gordon JA, Jencks WP (1963) The relationship of structure to the effectiveness of denaturing agents for proteins. Biochemistry 2: 47–57.

Granzier HLM, Wang K (1993) Gel electrophoresis of giant proteins: solubilization and silverstaining of titin and nebulin from single muscle fiber segments. Electrophoresis 14: 56–64

Gyenes T, Gyenes E (1987) Effect of "stacking" on the resolving power of ultrathin-layer twodimensional gel electrophoresis. Anal Biochem 165: 155–160

Hagel P, Gerding JJT, Fieggen W, Bloemendal H (1971) Cyanate formation in solutions of urea. I. Calculation of cyanate concentrations at different temperature and pH. Biochim. Biophys. Acta 243: 366–373

Hari V (1981) A method for the two-dimensional electrophoresis of leaf proteins. Anal Biochem 113: 332–335
Harrison PA, Black CC (1982) Two-dimensional electrophorestic mapping of proteins of bundle sheath and mesophyll cells of the C4 grass *Digitaria sanguinalis*(L) Scop.(crabgrass). Plant Physiol 70: 1359–1366
Heizmann CW, Arnold EM, Kuenzle CC (1980) Fluctuations of non-histone chromosomal proteins in differentiating brain cortex and cerebellar neurons. J Biol Chem 255: 11504–11511
Helenius A, McCaslin DR, Fries E ,Tanford C (1979) Properties of detergents. Methods Enzymol 56: 734–749
Herbert B, Molloy MP, Gooley AA, Walsh BJ, Bryson WG, Williams KL (1998) Improved protein solubility in two-dimensional electrophoresis using tri butyl phosphine as a reducing agent. Electrophoresis 19: 845–851
Herskovits TT, Jaillet H, Gadegbeku B (1970) On the structural stability and solvent denaturation of proteins. II. Denaturation by the ureas. J Biol Chem 245: 4544–4550
Hjelmeland LM (1986) The design and synthesis of detergents for membrane biochemistry. Methods Enzymol 124: 135–164
Hjelmeland LM, Chrambach A (1981) Electrophoresis and electrofocusing in detergent-containing media: a discussion of basic concepts. Electrophoresis 2: 1–11
Hochstrasser DF, Harrington MG, Hochstrasser AC, Miller MJ, Merril CR (1988) Methods for increasing the resolution of two-dimensional protein electrophoresis. Anal Biochem 173: 424–435
Horst MN, Basha MM, Baumbach GA, Mansfield EH, Roberts RM (1980) Alkaline urea solubilization, two-dimensional electrophoresis and lectin staining of mammalian cell plasma membrane and plant seed proteins. Anal Biochem 102: 399–408.
Kirley TL (1989) Reduction and fluorescent labelling of cyst(e)ine-containing proteins for subsequent structural analyses. Anal Biochem 180: 231–236
Klose J, Zeindl E (1984) An attempt to resolve all the various proteins in a single human cell type by two-dimensional electrophoresis: I. Extraction of all cell proteins. Clin Chem 30: 2014–2020.
Lopez MF, Patton WF, Utterback BF, Chung-Welch N, Barry P, Skea WM, Cambria RP (1991) Effect of various detergents on protein migration in the second dimension of two-dimensional gels. Anal Biochem. 199: 35–44
Lyubimova T, Caglio S, Gelfi C, Righetti PG, Rabilloud T (1993) Photopolymerization of polyacrylamide gels with methylene blue. Electrophoresis 14: 40–50
MacFarlane D (1989) Two dimensional benzyldimethyl-n-hexadecylammonium chloride-sodium dodecyl sulfate preparative polyacrylamide gel electrophoresis: a high capacity high resolution technique for the purification of proteins from complex mixtures. Anal Biochem 176: 457–463
March J (1977) Advanced organic chemistry, 2nd edn, McGraw-Hill, London, pp83–84
Menke W, Koenig F (1980) Isolation of thylakoid proteins. Methods Enzymol 69: 446–452
Molloy MP, Herbert B, Walsh BJ, Tyler MI, Traini M, Sanchez JC, Hochstrasser DF, Williams KL, Gooley AA (1998) Extraction of membrane proteins by differential solubilization for separation using two-dimensional electrophoresis. Electrophoresis 19: 837–844
Navarrete R, Serrano R (1983) Solubilization of yeast plasma membranes and mitochondria by different types of non-denaturing detergents. Biochim Biophys Acta 728: 403–408
O'Farrell PH (1975) High resolution two-dimensional electrophoresis of proteins. J Biol Chem 250: 4007–4021
Pasquali C, Fialka I, Huber LA (1997) Preparative two-dimensional gel electrophoresis of membrane proteins. Electrophoresis 18: 2573–2581
Penefsky HS, Tzagoloff A (1971) Extraction of water soluble enzymes and proteins from membranes. Methods Enzymol 22: 204–219
Rabilloud T (1996) Solubilization of proteins for electrophoretic analyses. Electrophoresis 17: 813–829
Rabilloud T (1998) Use of thiourea to increase the solubility of membrane proteins in two-dimensional electrophoresis. Electrophoresis 19: 758–760
Rabilloud T, Hubert M, Tarroux P (1986) Procedures for two-dimensional electrophoretic analysis of nuclear proteins. J. Chromatogr 351: 77–89
Rabilloud T, Gianazza E, Catto N, Righetti PG (1990) Amidosulfobetaines, a family of detergents with improved solubilization properties: application for isoelectric focusing under denaturing conditions. Anal Biochem 185: 94–102
Rabilloud T, Valette C, Lawrence JJ (1994) Sample application by in-gel rehydration improves the resolution of two-dimensional electrophoresis with immobilized pH gradients in the first dimension Electrophoresis 15: 1552–1558
Rabilloud T, Adessi C, Giraudel A, Lunardi J (1997) Improvement of the solubilization of proteins in two-dimensional electrophoresis with immobilized pH gradients. Electrophoresis 18: 307–316

Radin NS (1981) Extraction of tissue lipids with a solvent of low toxicity. Methods Enzymol 72: 5–7
Remy R, Ambard-Bretteville F (1987) Two-dimensional electrophoresis in the analysis and preparation of cell organelle polypeptides. Methods Enzymol 148: 623–632
Righetti PG, Tudor G, Gianazza E (1982) Effect of 2-mercaptoethanol on pH gradients in isoelectric focusing. J Biochem Biophys Methods 6: 219–227
Ruegg UT, Rüdinger J (1977) Reductive cleavage of cystine disulfides with tributylphosphine. Methods Enzymol 47: 111–116
Sanchez JC, Rouge V, Pisteur M, Ravier F, Tonella L, Moosmayer M, Wilkins MR, Hochstrasser DF (1997) Improved and simplified in-gel sample application using reswelling of dry immobilized pH gradients. Electrophoresis 18: 324–327
Sanders MM, Groppi VE, Browning ET (1980) Resolution of basic cellular proteins including histone variants by two-dimensional gel electrophoresis: evaluation of lysine to arginine ratios and phosphorylation. Anal Biochem 103: 157–165
Satta D, Schapira G, Chafey P, Righetti PG, Wahrmann JP (1984) Solubilization of plasma membranes in anionic, non-ionic and zwitterionic surfactants for iso-dalt analysis: a critical evaluation. J Chromatogr 299: 57–72
Segers J, Rabaey M, DeBruyne G, Bracke M, Mareel M (1986) Protein degradation during different extraction procedures in malignant cells. In: Dunn MJ (ed) Electrophoresis '86. VCH, Weinheim, pp 642–645
Semple-Rowland SL, Adamus G, Cohen RJ, Ulshafer RJ (1991) A reliable two-dimensional gel electrophoresis procedure for separating neural proteins. Electrophoresis 12: 307–312
Shirey T, Huang RCC (1969) Use of sodium dodecyl sulfate, alone, to separate chromatin proteins from deoxyribonucleoprotein of *Arbacia punctulata* sperm chromatin.Biochemistry 8: 4138–4148
Tanford C (1980) The hydrophobic effect, 2nd edn, Wiley, New York
Van Renswoude J, Kempf C (1984) Purification of integral membrane proteins. Methods Enzymol 104: 329–339
Weber K, Kuter DJ (1971) Reversible denaturation of enzymes by sodium dodecyl sulfate. J Biol Chem 246: 4504–4509
Wessel D, Flügge UI (1984) A method for the quantitative recovery of protein in dilute solution in the presence of detergents and lipids. Anal Biochem 138: 141–143
Willard KE, Giometti C, Anderson NL, O'Connor TE, Anderson NG (1979) Analytical techniques for cell fractions. XXVI. A two-dimensional electrophoretic analysis of basic proteins using phosphatidyl choline/urea solubilization. Anal Biochem 100: 289–298
Wilson D, Hall ME, Stone GC, Rubin RW (1977) Some improvements in two-dimensional gel electrophoresis of proteins. Protein mapping of eukaryotic tissue extracts. Anal Biochem 83: 33–44
Wu FS, Wang MY (1984) Extraction of proteins for sodium dodecyl sulfate-polyacrylamide gel electrophoresis from protease-rich plant tissues. Anal Biochem 139: 100–103.

CHAPTER 3

Two-Dimensional Electrophoresis with Carrier Ampholytes

C. Monribot[1] and H. Boucherie[1]

1 Introduction

Two-dimensional (2-D) gel electrophoresis, which was originally described by O'Farrell (1975), separates proteins in the first dimension according to their isoelectric point, and in the second dimension according to their molecular weight. It offers the opportunity of separating several hundred proteins from a total cellular extract. In combination with the recent development of methods of protein identification based on microsequencing, amino acid composition and mass spectrometry, the technique of 2-D gel electrophoresis now provides an invaluable tool for proteomic studies.

There are presently two different ways of separating proteins in the first dimension on the basis of their isoelectric point. According to the first one, proteins are separated in a pH gradient generated by applying an electric field to a gel containing a mixture of free carrier ampholytes (An der Lan and Chrambach, 1985). Carrier ampholytes are low molecular mass components with both amino and carboxyl groups. According to the second way, the pH gradient is generated by a different type of chemicals, the immobilines (Bjellqvist et al. 1982). The immobilines are acrylamide derivatives carrying amino or carboxyl groups. These immobilines are copolymerized with the acrylamide gel matrix such that an immobilized pH gradient is generated.

The relative advantages and drawbacks of 2-D gel electrophoresis using carrier ampholytes or immobilines have been discussed in several reports (Corbett et al. 1994; Blomberg et al. 1995; Klose and Kobalz 1995; Lopez and Patton 1997). If it is clear that each method has its own advantages and drawbacks, it is also clear that both of them can yield reproducible gels for intra- and interlaboratory studies (Blomberg et al. 1995; Lopez and Patton 1997). This chapter is devoted to 2-D gel electrophoresis based on the use of carrier ampholytes, while Chapter 4 is devoted to 2-D electrophoresis with immobilized pH gradients. We will first report general aspects of 2-D gel electrophoresis with carrier ampholytes. Then we will describe step by step the different procedures that lead from protein sample preparation to visualization of proteins on a 2-D pattern.

[1] Institut de Biochimie et Génétique Cellulaires, UPR CNR S9026, 1, rue Camille Saint-Saëns, 33700 Bordeaux Cedex, France.

1.1
Principles of Two-Dimensional Gel Electrophoresis

Two-Dimensional gel electrophoresis of proteins is carried out under denaturing conditions. In order to separate proteins in the first dimension, proteins are solubilized in the presence of urea which essentially works by disrupting hydrogen bonds. This denaturant has the advantage that it does not affect the intrinsic charge of proteins so that it allows us to separate proteins only on the basis of their charge. When loaded on a pH gradient of adequate porosity, proteins will migrate until they have no net charge, ie when they reach the pH of the gradient corresponding to their isoelectric point (pI).

After separation of proteins according to their charge, proteins are separated in a second dimension in the presence of sodium dodecylsulfate (SDS). SDS is an anionic detergent that binds to proteins according to a constant weight ratio (1.4 g SDS per gram of protein) independent of the nature of the protein (Reynolds and Tanford 1970). The intrinsic charges of polypeptides are negligible compared to the negative charges provided by SDS, so that SDS-polypeptide complexes have essentially identical charge densities. Under this condition, proteins migrate in polyacrylamide gels strictly according to their size (Weber and Osborn 1969).

1.2
General Aspects of Two-Dimensional Gel Electrophoresis with Carrier Ampholytes

1.2.1
Sample Preparation

The method of sample preparation varies greatly depending on the cell type or the tissue from which proteins are extracted. It must, however, satisfy four main rules:
(1) efficient protein solubilization,
(2) avoidance of proteolysis and of other protein modification that would result in artefactual spots,
(3) avoidance of interfering substances such as nucleic acids, lipids, particulate material or salts that would alter protein migration and
(4) compatibility with the first dimension electrophoresis.

Usually protein solubilization is achieved by using a sample buffer containing high (8–9.5 M) urea concentration. To further improve the solubilizing effect of urea, this denaturant is used in combination with a nonionic detergent such as NP-40 or Triton X-100 or a zwitterionic detergent like CHAPS. Ionic detergents such as SDS can be used in low amounts, provided nonionic or zwitterionic detergent are present in a large excess to ensure their complete removal from proteins prior to electrophoresis. Solubilization of proteins also requires disruption of disulfide bonds. This is obtained either by use of β-mercaptoethanol or by use of dithiothreitol.

It must be emphasized that any step carried out to inactivate proteases or to remove interfering substances brings some risk of altering the 2-D protein pattern. For example, use of protease inhibitors may induce some modification of protein charge (Dunn 1993). TCA precipitation in order to remove interfering material can be followed by a poor resolubilization of some proteins. Finally, it is strongly advised to extract proteins as quickly as possible, in order to avoid the action of enzymes susceptible to altering protein migration, and to avoid any step which is not absolutely necessary.

This section was intentionally limited to general considerations. For more details, the reader should consult Chapter 2. The reader may also consult Dunbar (1987), Rickwood et al. (1990) and Rabilloud (1996).

1.2.2
The Different Two-Dimensional Gel Electrophoresis Techniques

There are three major 2-D electrophoresis techniques based on the use of carrier ampholytes. They essentially differ by the separation of proteins in the first dimension.

Standard Two-Dimensional Gel Electrophoresis: IEF/SDS Gel Electrophoresis. This 2-D method is based on the use of isoelectric focusing (IEF) for separating proteins in the first dimension. It corresponds to the technique originally devised by O'Farrell in 1975. It is generally used to resolve proteins with a pH ranging from 4 to 7. It can also be used for narrow pH intervals. According to this technique, proteins are loaded on the basic side of the gel and migration is towards the anode. Polypeptides migrate through the pH gradient determined by ampholytes until they have no net charge. This takes several hours under the electrophoretic conditions generally used. Although size may affect the rate at which polypeptides migrate through the gel, their final position is determined only by their isoelectric point. When combined with SDS polyacrylamide gel electrophoresis (SDS-PAGE) in the second dimension, IEF allows us to visualize 1000 to 2000 proteins from the total cellular protein extract.

It must be emphasized that under the IEF/SDS-PAGE conditions, proteins are separated according to two parameters, isoelectric point and molecular weight, that can be deduced from their amino acid composition. Migration of a known protein can thus be predicted from its sequence. Generally, predictability of protein migration is ± 0.15 pH units for migration on the IEF gel and ± 5000 daltons for migration on the SDS gel (Boucherie et al. 1995; Garrels et al. 1997; Link et al. 1997). The difference between the expected migration of an identified protein and its observed 2-D gel location may be used for investigating post-translational modifications. Conversely, migration of an unknown protein provides valuable information regarding its identity.

NEPHGE/SDS Gel Electrophoresis. Nonequilibrium pH gradient electrophoresis (NEPHGE) separates basic proteins (O'Farrell et al. 1977). The proteins are applied to the acidic end of the first-dimensional gel and separated in a basic pH gradient. Electrophoresis is towards the cathode. Under these conditions, if the gel is allowed to run to equilibrium, there is a collapse of the basic end of the pH gradient. Most of the basic proteins would then migrate out of the gel. To avoid the run off of basic proteins the gel is not run to equilibrium. Accordingly, optimum resolution of proteins is obtained after relatively short periods of electrophoresis. The pH range of the ampholytes varies depending upon the range of proteins to be resolved. Large ampholyte ranges (pH 3–10) are generally used. In this case, following migration, acidic proteins are compressed at the anodic end of the gel and basic proteins are well resolved. Narrow basic pH ranges (pH 7–10) can also be used for detailed studies of basic proteins.

Migration of the proteins in NEPHGE gels is dependent upon both their electrophoretic mobility and their isoelectric point. Using NEPHGE/SDS PAGE gels, several hundred proteins not resolved by IEF can be visualized.

Giant Gels. As shown above, the complete 2-D separation of a complex protein sample using carrier ampholytes requires two separate 2-D runs, one based on an IEF separation and the other on an NEPHGE separation, and results in two partially overlapping protein patterns. Since 1975, Klose et al. (Klose 1975; Klose and Feller 1981; Klose and Kobalz 1995) combining several refinements of the 2-D technique, have devised a 2-D method that offers a good separation of both acidic and basic proteins. As for NEPHGE separation, proteins are loaded on the anodic side of the gel. A wide pH range is used in combination with the use of very long first-dimensional gels (46 cm). Focusing is stopped before the basic proteins reach the cathodic end of the gel. This technique provides the most powerful resolving 2-D method so far reported in the literature. More than 10 000 polypeptide spots can be resolved on these giant gels (46 × 30 cm). However, this technique, given its sophistication, is not accessible to the beginner and remains limited to a few experienced laboratories. It has been described in detail by Klose and Kobaltz (1995), as well as the special equipment required.

1.2.3 Two-Dimensional Gel Electrophoresis: The Choices

The beginner will have to face several important technical choices at each step of 2-D gel electrophoresis before deciding on a definitive strategy for running 2-D gels. Each of these choices will have an impact on the final quality of 2-D separation. The following section deals with the main choices that will be encountered.

Equipment. The first dimension based on the use of carrier ampholytes is generally carried out in vertical cylindrical gels and the second dimension is most often run on vertical slab gels. Many different commercialized apparatus for both the first and the second dimension of these types are available, e. g. the Multi Cell (Bio-Rad), Iso-Dalt 2-D gel electrophoresis system (Amersham Pharmacia-Biotech) and Esa 2-D system (Esa Inc.). However, first dimension and second dimension apparatus are very conventional and can be easily constructed in the laboratory.

Alternatively, the first dimension can be run on horizontal slab gels; the strips corresponding to each sample run are sliced, and applied to the surface of a horizontal slab gel for running the second dimension (Multiphor II horizontal electrophoresis apparatus; Amersham Pharmacia-Biotech).

First Dimension. The length of the first-dimensional gel routinely used can vary from 13 to 25 cm. It should be kept in mind that the number of spots to be detected is highly dependent on the size of the gel. The diameter of the cylindrical gels is also important: depending on its size (more or less than 1 mm), equilibration prior to the second dimension may or may not be required (see below). Finally, it must be taken into account that the acrylamide concentration of first dimensional gels must be low enough in order to render protein migration independent of the size of the protein. However, it should be sufficient to allow easy manipulation of the gels. This is generally obtained by acrylamide concentration around 3.5 to 4.5 %.

Equilibration Versus Nonequilibration. To facilitate transfer of proteins from the first-dimensional gel to the second dimension, it is generally recommended to equilibrate gels in the presence of SDS. The main objective of equilibration is to coat proteins with SDS before starting migration in the second dimension. Then, all molecules of the same protein will run out of the first-dimensional gel at the same time and proteins will be separated as round spots in the second dimension. In the absence of equilibration, the SDS is provided by the electrophoresis buffer of the upper chamber. At the onset of the second dimension, SDS enters the first-dimensional gel, binds to proteins, and subsequently SDS-coated proteins enter the second-dimensional gel. A progressive binding of SDS to proteins at this stage may occur, resulting in streaking.

However, it should be kept in mind that prolonged equilibration results in protein loss, particularly of low molecular weight species. The loss of proteins during equilibration can be as high as 15 to 25 % (Garrels 1989; Rickwood et al. 1990). It is thus better to avoid this step when not necessary. This is the case when first-dimensional gels are 1 mm in diameter or less.

Second Dimension. In the second dimension, a uniform separating gel (generally 10 or 11 % acrylamide) or a gradient separating gel (10–15 %) can be used. However, it is advisable to use gradient gels only when necessary. It increases the difficulty of obtaining reproducible gels as it is difficult to ensure that gradients are exactly the same over a long period of months. For proteomic studies, the aim of which is to obtain an overview of whole cell proteins, it should be kept in mind that the average size of proteins is 50 000. Gradient gels often allow a good resolution of proteins between 7000 and 30 000 whereas the vast majority of proteins (30 000 to 100 000) are compressed in the upper part of two-dimensional gels.

To avoid protein streaking in the second dimension, a stacking gel is often recommended. The addition of this stacking gel may provide a further source of nonreproducibility. When the diameter of first-dimensional gels is small enough, no stacking gel is required.

Precast Gels. Precast gels for the first dimension with carrier ampholytes and for the second dimension can be purchased from Esa Inc. (Chelmsford, Massachusetts) and from Amersham Pharmacia-Biotech (Uppsala, Sweden).

1.2.4 Post-Two-Dimensional Gel Electrophoresis Procedures

There are many different ways to detect proteins after two-dimensional gel separation, e. g. staining, autoradiography, fluorography, fluorescence. Special mention should be given to the use of the phosphor screen technology. This technique presently offers the best way to quantify spots, provided radioactive labelling of protein is possible. It is the most sensitive technique presently known for the detection of radioactive proteins. Also, phosphor imaging plates have a linear dynamic range for radioactivity detection which covers five orders of magnitude (Johnston et al, 1990). This allows us to easily quantify all polypeptide spots detectable on a gel in only one gel exposure. In comparison, X-ray films for autoradiography are 10 to 250 times less sensitive. In addition, they require several time exposures as the linear dynamic range of their response to radioactivity covers only two orders of magnitude. For similar reasons, quantification of proteins after silver staining requires the running of several gels with different loading of protein sample. Phosphor screens allowed us to detect ^{14}C, ^{35}S, ^{32}P and ^{3}H radioisotopes.

Identification of separated polypeptides can be carried out by microsequencing, mass spectrometry (Schevchenko et al. 1996) or amino acid composition (Garrels et al. 1994; Maillet et al. 1996; see also part II of the present book). When the amount of protein contained in the polypeptide spot to be identified is too low, it is necessary to concentrate the protein by using several spots excised from different gels. Two different elution-concentration gel systems can be used. One is based on the use of a vertical slab gel containing a large polyacrylamide stacking gel and a small resolving gel (Rasmussen et al. 1991). The other is based on the use of a Pasteur pipette containing only a small polyacrylamide stacking gel (Gevaert et al. 1996; Kristensen et al. 1997).

2 Sample Preparation

The sample preparation described here was originally devised for extracting proteins from yeast cells (Boucherie et al. 1995, 1996). Because yeast cells are surrounded by a cell wall, protein extraction cannot be simply carried out by an osmotic shock as it is, for example, with mammalian cells. Extensive breakage of yeast cells is required. This is obtained by vigorously shaking cells in the presence of glass beads. To avoid proteolysis, this step is performed on lyophilized cells, in the absence of buffer. Apart from preventing proteolysis, a further interest of this procedure is the possibility of sending cells for protein extraction by ordinary mail without the requirement of frozen ice. Another particularity of this sample preparation is the heating of the sample in the presence of SDS at the first step of protein solubilization. The rationale for this heating is not only to increase solubilization, but also to inactivate proteases or other enzymes that may alter protein size or charge.

No protease inhibitors are added to the sample according to our procedure (some inhibitors are known to induce charge alteration; Dunn 1993). In contrast, exogenous DNase and RNase are added in order to eliminate nucleic acids. Care should be taken to use protease-free DNase and RNase. The entire procedure, as applied to the extraction of proteins from yeast cells, is described in Protocol 1 (Sect. 2.4). This extraction procedure has been revealed to be also well adapted to protein extraction of bacterial cells. With modifications (see Sect. 2.4) it has been successfully used for extracting proteins from mammalian cells.

The final concentration of the protein sample is 0.03 M Tris-HCl pH 8.0, 9.5 M urea, 0.1 % SDS, 0.7 % CHAPS, 1.75 % β-mercaptoethanol, 0.2 % ampholytes. We observed that increasing the SDS concentration to 1 % or (and) the CHAPS concentration to 4 % does not improve protein solubilization.

2.1 Chemicals

CHAPS, 4.9 M, $MgCl_2$ solution, RNase A type XII-A from bovine pancreas, Trizma Base, Trizma hydrochloride (Sigma); dodecylsulfate sodium salt LAB (SDS), urea beads ultra pure, β-mercaptoethanol for molecular biology (Merck); Pharmalyte 3–10 (Amersham Pharmacia-Biotech); DNase I, RNase free, from bovine pancreas, 10 000 units/ml (Boehringer Mannheim).

2.2 Buffers

- *Extraction Buffer A* (0.1 M Tris-HCl pH 8.0, 0.3 % SDS). To prepare 100 ml of lysis buffer, dissolve 888 mg of Trizma hydrochloride, 530 mg of Trizma Base and 300 mg of SDS in deionized water and make up to 100 ml. Filter through a 0.45-μm pore size filter and store as 500-μl aliquots at -20 °C.
- *RNase A Solution* (0.05 M $MgCl_2$, 200 Kunitz units/ml RNAse A, 0.5 M Tris-HCl pH 7.0). To prepare 2.5 ml of RNase solution, dissolve 5 mg RNase A in 1.7 ml of 0.75 M Tris-HCl pH 7.0. Add 25.5 μl of 4.9 M $MgCl_2$ and 0.775 ml deionized water. Store as small aliquots (25 μl) at -20 °C.
- *Extraction Buffer B* (4.75 M urea, 4 % CHAPS, 1 % Pharmalyte 3–10, 5 % β-mercaptoethanol). To prepare 10 ml sample buffer, dissolve 2.85 g urea, 0.4 g CHAPS, 0.5 ml β-mercaptoethanol and 0.25 ml Pharmalyte 3–10 in 7 ml deionized water. Store as aliquots (150 μl) at -20 °C.

2.3 Equipment

MiniBeadBeater (Biospec Products); acid-washed glass beads (0.45 mm, glass beads; B. Braun).

2.4 Procedure

Protocol 1. Sample Preparation

1. The amount of cells to be used for protein extraction is calculated such that the final protein concentration of the sample ranges between 3 and 10 mg/ml. Washed cells are transferred into a "screw cap" microcentrifuge tube, resuspended in a small amount of ice-cold deionized water (50 to 100 μl) and frozen at -80 °C prior to being lyophilized. It is of importance not to lyophilize cells as a pellet. This would greatly impair the efficiency of cell breakage.
2. Lyophilize cells, taking care not to exceed too much the minimum time required for complete lyophilization. Overpassing this time results in a decrease in cell disruption efficiency.
3. Add acid-washed glass beads (0.45 mm in diameter) to the lyophilized cells. The volume of glass beads must be equivalent to the volume of resuspended cells prior to lyophilization.
4. Disrupt cells by shaking lyophilized cells in the presence of glass beads on a MiniBeadBeater. The tubes are agitated five times for 20 s, at 20-s intervals, leaving on ice between bursts of shaking.
5. Solubilize proteins by adding extraction buffer A previously kept at 100 °C. Briefly vortex the sample.
6. Maintain the sample for 10 s in a water bath at 100 °C.
7. Leave on ice for 1 min.
8. Add per microliter of extraction buffer A, 0.1 μl RNase A solution and 0.02 μl DNase I (10 000 units/ml) to yield a final concentration per milliliter of 20 000 units RNase A and 200 units of DNase I. Add β-mercaptoethanol to bring the final concentration to 2.5 % (v/v).
9. Incubate for 1 min at 4 °C.
10. Add per microliter of sample: 1.4 mg of urea and 0.5 μl of extraction buffer B. Mix by gently moving the tube upside-down several times.
11. Leave 5 min at room temperature.
12. Gently vortex the sample and centrifuge for 3 min at 13 000 rpm .

Note: the final volume of the sample will be 3.12 × the initial volume of extraction buffer A added at step 5. Steps 3 and 4 can be omitted when using mammalian cells and more generally cells which can be easily lysed by an osmotic shock. The samples can be used immediately or stored at -80 °C for several months. Samples should not be thawed and frozen again. Accordingly, prepare aliquots that will be thawed only once.

Protocol 2. Rapid Sample Preparation

When there is no risk of proteolysis, heating of extraction buffer A is not necessary, and steps 5 and 6 can be omitted. In practice, after step 4, proteins are solublized in a mixture of extraction buffer A, RNase and DNase solutions and β-mercaptoethanol, prepared according to the ratios reported in step 8. Continue then as described in step 9.

2.5 Measurement of Sample Protein Concentration

The amount of protein loaded on the first dimension varies depending on the technique to be used for protein visualization. It is better to load as small an amount of protein as possible in order to prevent alteration of the pH gradient of the first dimension. When yeast proteins are radioactively labelled (^{14}C or ^{35}S), the equivalent of 5×10^6 cpm is loaded. This generally corresponds to a 10-μl sample containing 30 μg of proteins. Using phophor screen technologies, proteins can be visualized after a one-night exposure. When yeast proteins have to be revealed by silver staining 300 μg is loaded. Usually, 30 μl of a sample with a protein concentration of 10 mg/ml is loaded.

2.5.1 Assay for Protein Radioactivity in Samples

Pipette 2 μl of protein sample on a glass microfibre filter (GF/C, Whatman). Let filters dry at room temperature for 15 min. Soak filters twice for 10 min in 5 % TCA containing 1 g/l of the same amino acid as the one used for labelling proteins. Then place dried filters in a counting vial. Counting is done in 5 ml of liquid scintillaton (Ready value; Beckman).

2.5.2 Measurement of Protein Concentration

Protein samples prepared according to our procedure can be assessed for protein concentration using the modified Bradford assay of Ramagli and Rodriguez (1985). The modification consists of an acidification of the samples prior to determining protein concentration.

3 First Dimension: Standard Isoelectric Focusing

The pH range of protein separation according to this standard isoelectric focusing is from 4.5 to 7. Isoelectric focusing is carried out on gel rods which are 24 cm long and 1 mm in diameter. Given the small diameter of these gels, equilibration is not required prior to submitting proteins to the second dimension. An example of 2-D protein separation obtained by using this type of isoelectric focusing in combination with standard 2-D gel electrophoresis (described in Sect. 4) is shown in Fig. 3.1.

Acrylamide concentration: 3.6 % T, 4.7 % C
Gel composition: 3.4 % acrylamide, 0.17 % bisacrylamide, 9.5 M urea, 3.6 % CHAPS, 4 % ampholytes
Gel size: 24 cm long, 1 mm in diameter
pH gradient: 4.5–7

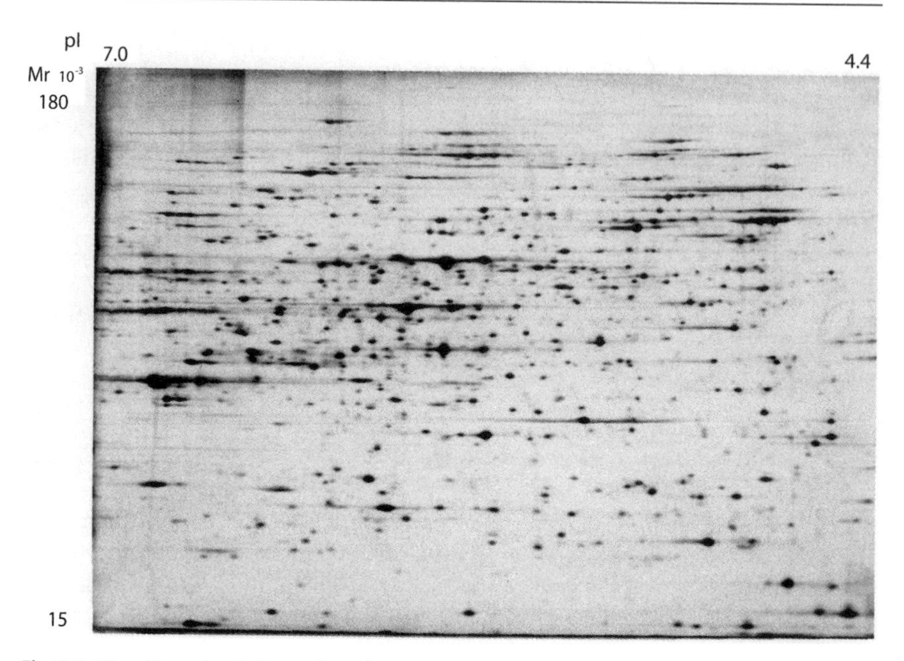

Fig. 3.1. Two-dimensional electrophoresis using standard isoelectric focusing. First dimension IEF. Second dimension SDS-PAGE (11 % acrylamide). Silver stain. Sample: *Saccharomyces cerevisiae*

3.1 Chemicals

N, N'-methylenebisacrylamide ultrapure, electrophoretic grade, ammonium persulfate (Boehringer Mannheim); CHAPS (Sigma); urea beads ultra pure (Merck); acrylamide IEF PlusOne, Pharmalyte 3–10, Pharmalyte 5–6, Pharmalyte 5–8 (Amersham Pharmacia-Biotech).

3.2 Reagent Solutions

- *IEF Acrylamide Solution* (29.8 % T, 4.7 % C). 28.4 % acrylamide (w/v) and 1.4 % (w/v) N, N'-methylenebisacrylamide. To prepare 25 ml of the solution, dissolve 7.1 g acrylamide and 0.35 g bisacrylamide in 20 ml deionized water. Make up to 25 ml, filter through a 0.45-μm pore size filter and store at 4 °C for no more than 2 weeks. Keep in a brown bottle to protect from light.
- *CHAPS Solution.* 10 % (w/v) in deionized water. Dissolve 2.5 g of CHAPS in 20 ml deionized water and make up to 25 ml. Filter through a 0.45-μm pore size filter and keep at 4 °C for no more than 1 month.
- *Ammonium Persulfate Solution* (10 % APS). To prepare 2 ml of solution, dissolve 200 mg ammonium persulfate in 2 ml deionized water. This solution should be prepared just before use.

- *Cathodic Solution* (0.1M NaOH). To prepare 1 l of cathodic solution, dissolve 4 g NaOH in 1 l deionized water. Deaerate under vacuum for 30 min while continuously stirring. The cathodic solution is prepared just before use.
- *Anodic Solution* (0.08 M H_3PO_4). To prepare 1 liter of anodic solution dissolve 5.5 ml of 85 % phosphoric acid in 1 l of deionized water. Prepare just before use.
- *Overlay Solution* (2.37 M urea, 2 % CHAPS, 0.5 % Pharmalyte 3–10). To prepare 10 ml of overlay solution, dissolve 1.4 g urea and 0.2 g of CHAPS in deionized water, add 125 µl Pharmalyte 3–10 and make up to 10 ml. Store as aliquots (200 µl) at -20 °C.
- *Sample Buffer* (0.03 M Tris-HCl pH 8.0, 9.5 M urea, 0.1 % SDS, 0.7 % CHAPS, 1.75 % β-mercaptoethanol, 0.2 % ampholytes). To prepare sample buffer, mix the various components used for sample preparation in the same proportion as described in Section 2.4, only omitting RNase and DNase solutions.

3.3 Equipment

We use glass tubes specially designed for isoelectric focusing (see Fig. 3.2; available on special request from Atelier Jean Premont, 36 av de Labarde, 33300 Bordeaux, France). The part of the tube in which the gel is polymerized is 24 cm long and has an inner diameter of 1 mm. It is surmounted by a 2.2-cm glass tube which has the same external diameter (8 mm) and an inner diameter of 2.5 mm. This part of the tube is used as a funnel in which the sample is loaded. A 10-ml pipette-pump is required to fill the tubes with the gel solution (Poly Labo).

In our laboratory the isoelectric focusing gel tubes are mounted into a homemade vertical electrophoresis apparatus with an upper and a lower electrode chamber. This apparatus can hold 12 gel rods. Silicone rubber grommets convenient for holding the tubes can be purchased from BioRad (Cat no. 165–1985). However, any commercialized apparatus for isoelectric focusing capable of accomodating glass tubes which are 26 cm long can be used.
The isoelectric focusing gels are polymerized and run in an incubator maintained at 26 °C.

3.4 Preparation of Isoelectric Focusing Gels

Recipe for 12 tubes. The first-dimensional gels are prepared the day before isoelectric focusing. We observe a better reproducibility under this condition.
1. Keep the glass tubes for at least 1 hour at 26 °C.
2. Prepare the 10 % ammonium persulfate solution.

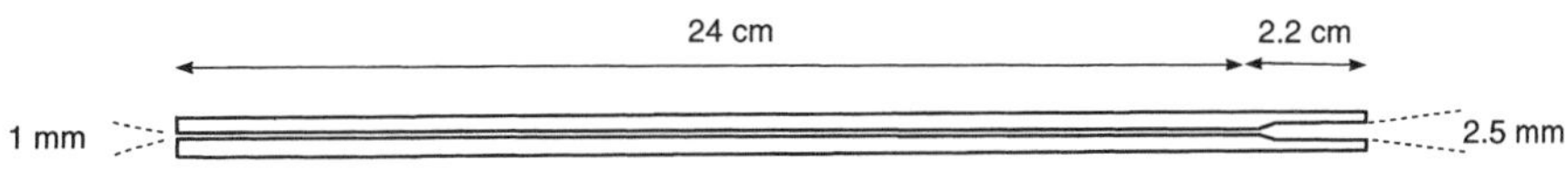

Fig. 3.2. Schematic representation of glass tube for IEF gels

3. Prepare the first-dimensional acrylamide gel solution in a corex tube by adding the components in the following order:
   - 4 g of urea
   - 0.85 ml of IEF acrylamide solution
   - 2.6 ml of 10 % CHAPS
   - 166 μl of Pharmalyte 3–10
   - 166 μl of Pharmalyte 5–6
   - 332 μl of Pharmalyte 5–8.
4. Dissolve urea by gentle mixing. Warm the urea solution by keeping the corex tube in the palm of the hand. Do not heat the urea solution!
5. Deaerate the solution under vacuum for 3 min.
6. Initiate polymerization by adding 20 μl of 10 % APS freshly prepared.
7. Swirl gently the corex tube in the hand. Take care not to re-introduce oxygen into the solution.
8. Glass tubes can be filled simultaneously with the first-dimensional gel acrylamide solution, using a gel-casting apparatus such as the one described by Garrels (1983) or Dunbar (1987). According to this procedure, the gel solution is displaced upwards into the glass tubes by overlaying the gel solution with water. Alternatively, glass tubes can be filled individually by using a pipette-pump. For this purpose, insert the upper end of the glass tube ("loading funnel") into the pipette-pump. Pump the gel solution until it enters a few millimeters into the "loading funnel". Maintain the tube horizontally and remove the pipette-pump. Immediately press the thumb on the upper end of the tube in order to prevent leakage of the gel solution. Eliminate excess acrylamide solution by carefully releasing the pressure of the thumb (acrylamide solution must be only 1 mm inside the loading funnel) and lay the tube horizontally for polymerization. This procedure is routinely used in our laboratory.
9. Leave the gels to polymerize overnight at 26 °C.

Note: TEMED is not required for polymerization.

3.5
Running Standard Isoelectric Focusing Gels

According to this procedure the samples are loaded at the basic end of the gels. A prefocusing step is performed prior to loading the samples.

1. Place the isoelectric focusing tubes into the electrophoresis stand. Fill the lower chamber with 0.08 M H_3PO_4. If necessary remove air trapped at the bottom end of the tubes.
2. Overlay the gels with 15 μl of sample buffer. Fill up the tubes with the 0.1-M NaOH solution previously deaerated.
3. Fill the upper chamber with the 0.1-M NaOH solution, taking care not to disturb the sample buffer layer on the top of the gels.
4. Prefocus the gels as follows:
   - 500 V for 30 min
   - 1000 V for 45 min
   - 1500 V for 15 min.

5. Remove the upper electrode buffer and the sample buffer.
6. Load the protein sample. Cover with 15 µl of overlay solution. Fill up the tubes and the upper chamber with fresh upper electrode buffer.
7. Focusing is carried out at:
   - 500 V for 15 min
   - 1000 V for 30 min
   - 1600 V for 21 h.
8. After focusing, the gels are immediately extruded from the glass tubes. To extrude the gels, use a 2.5-ml syringe filled with water and fitted with a yellow pipette tip. Insert the tip into the top end of the glass tube and push out the gel by pressure on the syringe. The gels are directly extruded onto a piece of Parafilm and are kept at −80 °C. A corner of the Parafilm is cut, indicating the basic side of the gel.

The prefocusing and the focusing are carried out in an incubator maintained at 26 °C.

4 Second Dimension: Standard Slab Gel Electrophoresis

The second dimension is run on a vertical slab gel. This slab gel (90 cm large) allows us to run three second-dimensional gels in parallel, which maximizes comparability of two-dimensional gels. There is no stacking gel and the first-dimensional gels are layed on the top of the slab gel without any sealing with agarose. Another particularity of this second dimension procedure is that the gel composition does not contain SDS as usual. The only SDS present in the gel during the second dimension electrophoresis comes from the electrophoresis buffer of the upper chamber. It enters the gel when the current is applied. We found that the absence of SDS in the slab gel has a "stacking" effect on proteins when they leave the first-dimensional gel and enter into the slab gel. This second-dimensional gel resolves proteins with molecular weight that range between 180 000 and 17 000.

Acrylamide concentration: 11 % T, 3.3 % C
Gel composition: 0.36 M Tris-HCl pH 8.5, 10.6 % acrylamide, 0.35 % bisacrylamide
Gel size: 85 cm long, 20 cm high, 1 mm thick
Molecular weight resolution: 17 000 to 180 000

4.1 Chemicals

Acrylamide, N, N'-methylenebisacrylamide, ammonium persulfate (APS), tetramethylethylenediamine (TEMED) (Boehringer Mannheim); Trizma base, Trizma hydrochloride (Sigma); glycine for electrophoresis, dodecylsulfate sodium salt LAB (SDS) (Merck).

4.2
Solutions for Second-Dimensional Gels

- *Slab Gel Acrylamide Solution* (30,1 % T, 3.3 % C). 29.1 % acrylamide (w/v) and 0.99 % N, N'-methylenebisacrylamide. To prepare 1 l of acrylamide solution, dissolve 291 g acrylamide and 9.9 g bisacrylamide in 800 ml deionized water. Fill up to 1000 ml. Store at 4 °C, for no more than 2 weeks. Keep in a brown bottle to protect from light.
- *Slab Gel Buffer* (1.5 M Tris-HCl pH 8.5). Dissolve 130 g Trizma base and 66.3 g Trizma hydrochloride in 800 ml deionized water. Fill up to 1 l with deionized water. Store at 4 °C.
- *Ammonium Persulfate Solution.* 10 % in deionized water. To prepare 2 ml of solution, dissolve 200 mg ammonium persulfate in 2 ml deionized water. This solution should be prepared just before use.
- *Electrode Buffer* (192 mM glycine, 25 mM Trizma base, 0.2 % SDS). To make 4000 ml electrode buffer add 12.12 g Trizma base, 57.6 g glycine and 8 g SDS. Dissolve in deionized water and fill up to 4000 ml. Prepare before use.

4.3
Equipment

The polymerization cassette consists of two glass plates, a back plate and a front plate (Fig. 3.3a, b) each 925 mm long, 215 mm high and 4 mm thick. The cassette allows us to cast a wide slab gel corresponding to three second-dimensional gels to be run in parallel. The front plate has a notch 15 mm deep and 845 mm long. A PVC U-frame 1 mm thick (Fig. 3.3c) is used to delimit the gel. PVC strips 1 mm thick are used as spacers (Fig. 3.3d). Plexiglas spacers (1 mm thick) are used to fix the distance between the glass plates.

The apparatus for running slab gels is a modification of the one described by Garrels (1983; Fig. 3.4; available on special request from Deco Volume, 30 rue Denfert Rochereau, 33400 Talence, France). It is a vertical system that allows us to run only one wide slab gel at the same time. Usually, two apparatuses are run in parallel.

4.4
Preparation of the Casting Cassette

1. Before use, wash the glass plates with deionized water and carefully air-dry.
2. Place the U-frame and the PVC spacers on the back plate (Fig. 3.3c, d). Make sure that the U-frame and the spacers do not overlap.
3. Place the front plate on the U-frame (Fig. 3.3e)
4. Hold the plates together with clamps (Fig. 3.3f)
5. Slightly insert Plexiglas spacers between the plates in order to maintain the distance between the plates (Fig. 3.3g, h). Press the two glass plates against these spacers by means of plastic U-clamps.

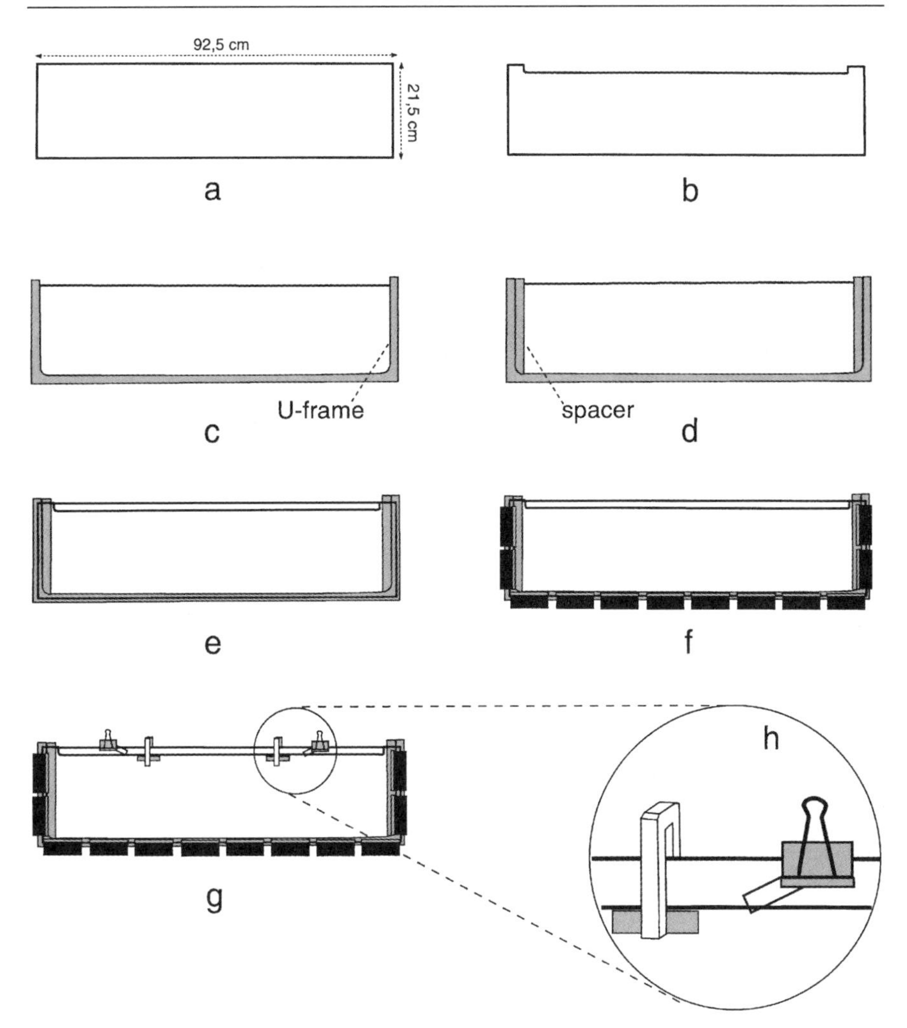

Fig. 3.3. Preparation of the casting cassette for second-dimensional gels. *a* back plate; *b* front plate; *c-h* preparation of the casting cassette (see text)

4.5 Preparation of Slab Gels

1. In a 500-ml vacuum flask add successively:
   - 51 ml of slab gel buffer
   - 76.5 ml of slab gel acrylamide solution
   - 81 ml of deionized water
   - Deaerate under vacuum for 3 min
   - Add 1 ml of 10 % APS freshly prepared
   - Mix gently
   - Add 136 μl of TEMED

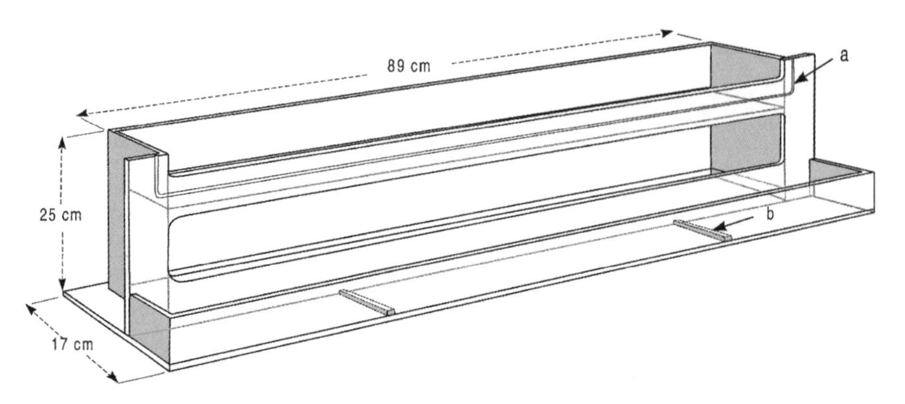

Fig. 3.4. Schematic representation of slab gel electrophoresis apparatus. The gel cassette is clamped against the front of the upper electrode chamber. A silicon sponge cord is placed in a notch (*a*) to seal the upper chamber against the front glass plate. Spacers (*b*) maintain the glass plates off the bottom of the lower electrode chamber

- Transfer to a 500-ml beaker. Take care not to re-introduce oxygen into the solution.

2. Pump the gel solution with a 50-ml syringe and pour it down inside the cassette, by making the solution run down alternately on the left and right side of the plates (never in the middle of the cassette). Add the gel solution up to a level 2 or 3 mm below the bottom of the notch.
3. Gently overlay the gel solution with deionized water.
4. Allow the slab gel to polymerize for at least 1 h at room temperature.

Note: there is no SDS in the gel composition.

4.6 Running Two-Dimensional Gels

Three two-dimensional gels are run simultaneously on each of the slab gels.

1. Prepare the electrophoresis buffer (192 mM glycine, 25 mM Trizma base, 0.2 % SDS).
2. Remove the U-frame of the casting cassette. Remove any residual liquid on the top of the gel by blotting or aspiration. Rinse the gel surface with deionized water and dry before loading isoelectric focusing gels.
3. Lay the SDS gel cassette at 45° with the front plate upward in order to facilitate the application of the first-dimensional gels.
4. Transfer the first dimensional gel from the Parafilm to the back glass (internal side) of the casting cassette, taking care that the first-dimensional gel is parallel to the top of the slab gel.
5. With a blunt-ended spatula push the first-dimensional gel between the glass plates, seating it carefully on the top of the slab gel. A few drops of running buffer will help to slide the gel between the glass plates. Be sure that no air bubbles are trapped between the IEF gel and the surface of the acrylamide slab gel.

6. Do the same with the two other first-dimensional gels.
7. Insert the cassette into the electrophoresis apparatus.
8. Carefully fill the space of the cassette above the rod gels with electrophoresis buffer using a pipette. Take care not to displace the first-dimensional gels.
9. Fill the upper and higher chamber with electrophoresis buffer.
10. Take care that no air bubbles are trapped between the lower surface of the slab gel and the two glass plates of the cassette. If so, remove the bubbles by a stream of buffer from a bent needle connected to a 50-ml syringe.
11. Run the slab gel at room temperature as follows:
    - 5 W for 15 min
    - 25 W until the Bromophenol blue tracking dye reaches the bottom of the gel. The running time is about 6 h.
12. After running, open the cassette with a spatula. Cut the slab gel into three individual gels, each corresponding to one of the first-dimensional gel loaded. Cut the lower corner of the gel corresponding to the basic side of the first dimensional gel to indicate its orientation. Then process the gels to detect polypeptides.

Note: as already mentioned in the previous isoelectric focusing section, equilibration of first-dimensional gels prior to loading on top of the second-dimensional gels is omitted.

5 Modifications of the Standard Two-Dimensional Gel Method

5.1 Extension of the pH Gradient and of the Molecular Weight Scale

The extension of the pH gradient and of the molecular weight scale described here allows us to visualize, in addition to the proteins separated by the standard 2-D gel system, some acidic and basic proteins as well as low molecular weight proteins. For this purpose, a modification in the ampholyte composition of the first-dimensional gel is used to improve the resolution of basic and acidic proteins. An increase in the acrylamide concentration of the slab gels allows us to visualize low molecular weight proteins. In return, proteins separated by the standard 2-D gel method are restricted to an area of gel smaller than on standard gels. Protein separation ranges from pH 3.4 to 7.2 in the first dimension and from molecular weight 180 000 to 8000 in the second dimension. An example of gels obtained according to this procedure is shown in Fig. 3.5.

Preparation of a protein sample for loading on the first-dimensional gels is the same as for standard IEF gels.

First dimension:
Acrylamide concentration: 3.6 % T, 4.7 % C
Gel composition: 3.4 % acrylamide, 0.17 % bisacrylamide, 9.5 M urea, 3.6 % CHAPS, 4 % ampholytes
Gel size: 24 cm long, 1 mm in diameter
pH gradient: 3.4–7.2

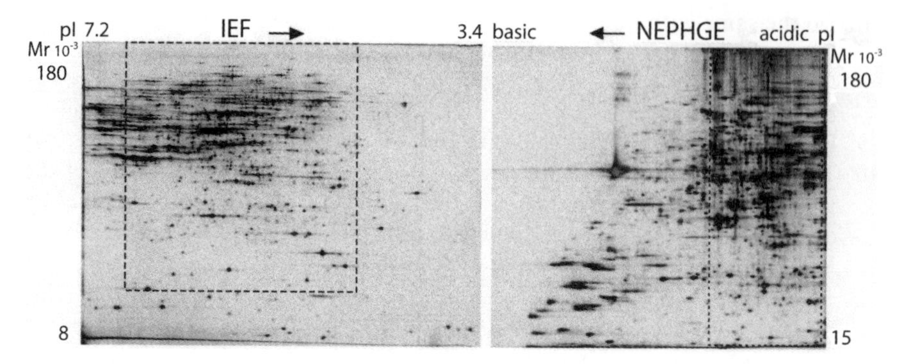

Fig. 3.5. Two-dimensional electrophoresis using modifications of standard 2-D gel electrophoresis. *Left* Extension of the pH gradient and of the molecular weight scale. First dimension modified IEF procedure (see Sect. 5.1). Second dimension SDS-PAGE (13 % acrylamide). *Right* Separation of basic proteins. First dimension NEPHGE (see Sect. 5.2). Second dimension SDS-PAGE (11 % acrylamide). Silver stain. Sample: *Saccharomyces cerevisiae*. *Delimited area* indicates proteins which are resolved by standard IEF/SDS PAGE electrophoresis

Second dimension:
Acrylamide concentration: 13 % T, 3.3 % C
Gel composition: 12.53 % acrylamide, 0.45 % bisacrylamide
Running buffer: 0.192 M glycine, 25 mM Trizma base, 0.2 % SDS
Gel size: 85 cm large, 20 cm high, 1 mm thick
Molecular weight separation: 8000 to 180 000

5.1.1 Chemical

Pharmalyte 2–4, Pharmalyte 7–9 (Amersham Pharmacia-Biotech). Other chemicals are as for IEF gels (see Sect. 3.1).

5.1.2 Procedure

First Dimension. Reagent solutions for preparing first-dimensional gels are as for IEF gels (see Sect. 3.2). To prepare the first-dimensional acrylamide gel solution for 12 gels, add the components in a corex tube in the following order:
- 4 g of urea
- 0.85 ml of IEF acrylamide solution
- 2.6 ml of 10 % CHAPS
- 249 μl of Pharmalyte 3–10
- 166 μl of Pharmalyte 2–4
- 124 μl of Pharmalyte 5–8
- 124 μl of Pharmalyte 7–9.

All other steps, for preparation of the first-dimensional gel and running the first dimension, are as for IEF gels (see Sects. 3.4 and 3.5).

Second Dimension. Chemicals and reagent solutions for preparing the second dimensional gels are as for standard slab gels (see Sects. 4.1 and 4.2). To prepare the slab gel solution, add successively in a 500-ml vacuum flask:
- 51 ml of slab gel buffer
- 90.4 ml of slab gel acrylamide solution
- 67 ml of deionized water
- Deaerate under vacuum for 3 min
- Add 1 ml of 10 % APS freshly prepared
- Mix gently
- Add 136 μl of TEMED
- Transfer to a 500-ml beaker.

All other steps, for preparation of the second dimensional gel and running the second dimension, are as for standard slab gels (see Sects. 4.5 and 4.6).

5.2 Separation of Basic Proteins by NEPHGE

Prepartion of a protein sample for loading NEPHGE gels is the same as for standard IEF gels. In contrast to IEF gels, the sample is loaded at the acidic end of the gel, and the gel is run without prefocalization. The second dimension is the same as for standard two-dimensional gel electrophoresis.

Acrylamide concentration: 3.6 % T, 5.4 % C
Gel composition: 3.4 % acrylamide, 0.19 % bisacrylamide, 9.35 M urea, 5.6 % CHAPS, 2 % ampholytes
Gel size: 24 cm long, 1 mm in diameter
pH gradient: 4–10

5.2.1 Chemicals

Acrylamide IEF PlusOne, Pharmalyte 3–10, Ampholine 3.5–10 (Amersham Pharmacia-Biotech); N, N'-methylenebisacrylamide ultrapure, electrophoretic grade, ammonium persulfate, tetramethylethylenediamine (TEMED) (Boehringer Mannheim); CHAPS (Sigma); urea beads ultra pure (Merck).

5.2.2 Solutions

First dimension solutions (ammonium persulfate solution, anodic and cathodic solutions) except acrylamide solution are as described in Section 3.2. Second dimension solutions (slab gel acrylamide solution, gel buffer and APS) are as described in Section 4.2.

NEPHGE Acrylamide Solution (30 % T, 5.4 % C). 28.38 % Acrylamide (w/v) and 1.62 % (w/v) N, N'-methylenebisacrylamide. To prepare 50 ml of the solution, dissolve 14.19 g of acrylamide and 0.81 g of bisacrylamide in deionized water. Filter and store at 4 °C for no more than 2 weeks. Keep in a brown bottle to protect from light.

5.2.3 Procedure

NEPHGE Gel Preparation. Recipe for 12 tubes. The gels are prepared the day before use.

1. Prepare the NEPHGE solution in a corex tube by adding in the following order:
   - 5.5 g of urea
   - 1.16 ml of NEPHGE acrylamide solution
   - 0.55 g of CHAPS
   - 250 μl of Ampholines 3.5–10
   - 250 μl of Pharmalytes 3–10
   - 4.17 ml of deionized water.
2. Dissolve urea by gentle mixing.
3. Deaerate for 3 min under vacuum.
4. Initiate polymerization by adding 20 μl of 10 % APS freshly prepared and 12.75 μl of TEMED.
5. Mix by gentle agitation.
6. Fill glass tubes with the first-dimensional gel acrylamide solution by using a pipette-pump.
7. Leave the tubes to polymerize horizontally at 26 °C for one night.

Note: in contrast to IEF gels, TEMED is added for polymerization of NEPHGE gels because polymerization is less efficient at alkaline pH. The alkaline pH results from the high amount of basic carrier ampholytes present in the NEPHGE gel composition.

Running NEPHGE Gels

1. The first-dimensional tubes are mounted into a vertical electrophoresis stand with an upper and a lower electrode chamber. The upper chamber is filled with 0.08 M H_3PO_4 (anodic solution) and the lower chamber is filled with 0.1 M NaOH (cathodic solution). The NaOH solution is deaerated for 30 min before use.
2. Samples are loaded at the acidic end of the gels.
3. Electrophoresis is carried out at 500 V for 30 min and then 1400 V for 6 h.
4. After focusing, the gels are processed as standard IEF gels (see Sects. 3.5)

The focusing is carried out in an incubator maintained at 26 °C.

Note: the gel is run from the acidic end to the basic end, i.e. in the opposite direction compared to the typical IEF gel. The H_3PO_4 buffer is placed in the upper chamber and the NaOH solution in the lower. The connections of the power supply are reversed.

6
Visualization of Separated Proteins

Depending on experiments, proteins can be visualized by dye or silver staining, autoradiography and fluorography. All these techniques have been widely described previously, and are described in detail in Chapter 5 of this book. The reader should refer to this chapter for description of detection protocols.

Acknowledgments. We would like to thank those who throughout the years, with their expertise, technical skill or "mistakes", have contributed to the development of 2-D gel electrophoresis in our laboratory. Particular thanks are extended to Marie France Peypouquet, Nelly Bataillé, Didier Thoraval, Manuel Sabiani, Sandrine Piot, Christelle Deletoile, Christophe Lehman and Richard Joubert.

References

An der Lan B, Chrambach A (1985) Analytical and preparative gel electrofocusing. In: Hames BD, Rickwood D (eds) Gel electrophoresis of proteins. A practical approach. IRL Press, Oxford

Ansorge W (1982) Fast visualization of protein bands by impregnation in potassium permanganate and silver nitrate. In: Stathakos D, (ed) Electrophoresis, 82. Walter de Gruyter, Berlin, pp 235–242

Bjelqvist B, Ek K, Righetti PG, Gianazza E, Görg A, Westermeier R, Postel W (1982) Isoelectric focusing in immobilised pH gradients: principle, methodology and some applications. J Biochem Biophys Methods 6: 317–339

Blomberg A, Blomberg L, Fey SJ, Mose-Larsen P, Larsen PM, Roepstorff P, Degand P, Boutry M, Posch A, Görg A (1995) Interlaboratory reproducibility of yeast protein patterns analyzed by immobilized pH gradient two-dimensional gel electrophoresis. Electrophoresis 16: 1935–1945

Boucherie H, Dujardin G, Kermorgant M, Monribot C, Slonimski P, Perrot M (1995) Two-dimensional protein map of *Saccharomyces cerevisiae*: construction of a gene-protein index. Yeast 11: 601–613

Boucherie H, Sagliocco F, Joubert R, Maillet I, Labarre J, Perrot M (1996) Two-dimensional gel protein database of *Saccharomyces cerevisiae*. Electrophoresis 17: 1683–1699

Corbett J, Dunn MJ, Posch A, Görg A (1994) Positional reproducibility of protein spots in two-dimensional gel electrophoresis using immobilized pH gradient isoelectric focusing in the first dimension: an interlaboratory comparison. Electrophoresis 15: 1205–1211

Dunbar BS (1987) Sample preparation for electrophoresis; Method for isoelectric focusing. In: Dunbar BS (ed) Two dimensional electrophoresis and immunological techniques. Plenum Press, New York

Dunn MJ (1993) Gel electrophoresis of proteins. BioS Scientific Publishers Alden Press, Oxford

Garrels JI (1983) Quantitative two-dimensional gel electrophoresis of proteins. Methods Enzymol 100: 411–423

Garrels JI (1989) The QUEST system for quantitative analysis of two-dimensional gels. J Biol Chem 264: 5269–5282

Garrels JI, Futcher B, Kobayashi R, Latter GI, Schwender B, Volpe T, Warner JR, McLaughlin CS (1994) Protein identification for a *Saccharomyces cerevisiae* protein database. Electrophoresis 11: 1466–1490

Garrels JI, McLaughlin CS, Warner JR, Futcher B, Latter GI, Kobayashi R, Schwender B, Volpe T, Anderson DS, Mesquita-Fuentes R, Payne WE (1997) Proteome studies of *Saccharomyces cerevisiae*: identification and characterization of abundant proteins. Electrophoresis 18: 1347–1360

Gevaert K, Verschelde J-L, Puype M, Van Damme J, Goethals M, De Boeck S, Vandekerckhove S (1996) Structural analysis and identification of gel-purified proteins, available in the femtomole range, using a novel computer program for peptide sequence assignment, by matrix-assisted laser desorption ionization-reflectron time-of-flight-mass spectrometry. Electrophoresis 17: 918–924

Johnston RF, Pickett SC, Barker DL (1990) Autoradiography using storage phosphor technology. Electrophoresis 11: 355–360

Klose J (1975) Protein mapping by combined isoelectric focusing and electrophoresis in mouse tissues. A novel approach to testing for induced point mutation in mammals. Humangenetik 26: 231–243

Klose J, Feller M (1981) Two-dimensional electrophoresis of membrane and cytosol proteins of mouse liver and brain. Electrophoresis 2: 12–24

Klose J, Kobalz U (1995) Two-dimensional electrophoresis of proteins: an updated protocol and implications for a functional analysis of the genome. Electrophoresis 16: 1034–1059
Kristensen DB, Inamatsu M, Yoshizato K (1997) Elution concentration of proteins cut from two-dimensional polyacrylamide gels using Pasteur pipettes. Electrophoresis 18: 2078–2084
Link AJ, Robison K, Church GM (1997) Comparing the predicted and observed properties of proteins encoded in the genome of *Escherichia coli* K-12. Electrophoresis 18: 1259–1313
Lopez MF, Patton WF (1997) Reproducibility of polypeptide spot positions in two-dimensional gels run using carrier ampholytes in the isoelectric focusing dimension. Electrophoresis 18: 338–343
Maillet I, Lagniel G, Perrot M, Boucherie H, Labarre J (1996) Rapid identification of yeast proteins on two-dimensional gels. J Biol Chem 271: 10263–10270
O'Farrell PH (1975) High resolution two-dimensional electrophoresis of proteins. J Biol Chem 250: 4007–4021
O'Farrell PZ, Goodman HM, O'Farrell PH (1977) High resolution two-dimensional electrophoresis of basic as well as acidic proteins. Cell 12: 1133–1142
Rabilloud T (1996) Solubilization of proteins for electrophoretic analyses. Electrophoresis 17: 813–829
Ramagli LS, Rodriguez LV (1985) Quantitation of microgram amounts of protein in two-dimensional polyacrylamide gel electrophoresis sample buffer. Electrophoresis 6: 559–563
Rasmussen H-H, Van Damme J, Puype M, Celis E-J, Vandekerkhove J (1991) Microsequencing of proteins recorded in human two-dimensional gel protein databases. Electrophoresis 12: 873–882
Reynolds JA, Tanford C (1970) Binding of dodecyl sulfate to proteins at high binding ratios. Possible implications for the state of proteins in biological membranes. Proc Natl Acad Sci USA 66: 1002–1007
Rickwood D, Alec J, Chambers A, Spragg SP (1990) Two-dimensional gel electrophoresis. In: Hames BD, Rickwood D (eds) Gel electrophoresis of proteins. A practical approach. IRL Press, Oxford
Schevchenko A, Jensen ON, Podtelejnikov AV, Sagliocco F, Wilm M, Vorme O, Mortensen P, Schevchenko A, Boucherie H, Mann M (1996) Linking genome and proteome by mass spectrometry: large scale identification of yeast proteins from two-dimensional gels. Proc Natl Acad Sci USA 93: 14440–14445
Weber K, Osborn M (1969) The reliability of molecular weight determinations by dodecyl sulfate polyacrylamide gel electrophoresis. J Biol Chem 244: 4406–4412

Appendix: Problems and Troubleshooting

Two-dimensional gel electrophoresis involves many different steps. Each of them is sensitive and may have a dramatic consequence on the final quality of the 2-D protein pattern. This section deals with some of the major problems that may be encountered by the novice starting to practise this methodology. They are categorized according to the step from which they originate. For general electrophoresis troubleshooting, refer to the troubleshooting section of Chapter 4 (Appendices A and B).

Equipment and Chemicals

- Gel plates and glass tubes need to be perfectly clean. After use they must be washed in sulfochromic acid or in strong detergent. Then, they must be extensively rinsed with tapwater and finally with deionized water. Gel plates are dried with filter paper (Kimwipes) and glass tubes are dried with air pressure. Glass tubes used for NEPHGE gels are periodically coated with Repel-Silane (Amersham Pharmacia-Biotech) in order to facilitate the extrusion of first-dimensional gels.
- First dimension and second dimension apparatuses are extensively rinsed with tapwater and, finally, with deionized water, after each run. They are carefully wiped with filter paper (Kimwipe) and periodically cleaned with a detergent.
- Carrier ampholytes of a given pH range are complex mixtures of a large number of different molecules. Their composition may vary from batch to batch. To overcome this problem it is advisable to use a combination of carrier ampholytes corresponding to different ranges of pH (as we do for IEF) or a mixture of carrier ampholytes of the same pH range but different trade marks (as we do for NEPHGE gels).
- The reproducibility and quality of gels are highly dependent upon the quality of reagents and water. Changing the origin of reagents may strongly affect protein separation. Concerning water, our laboratory is equipped with a Milli-Q Water system (Millipore). We found that it is important to avoid the oxidation or hydration of some reagents which may happen when the container has been opened for too long a time. Ammonium persulfate, β-mercaptoethanol and TEMED are particularly sensitive and have to be discarded once the vial has been opened for 2 months.

Sample Preparation

- Abnormal row of spots with the same molecular weight: proteins are carbamylated by isocyanate. Use very pure urea, freshly prepared urea solutions and avoid high temperatures.
- Abnormal proportion of small molecular weight proteins (less than 25 000): protease activity during protein extraction or in the sample. Cells were not completely lyophilized prior to protein extraction. Work rapidly until urea is added to the sample. Some samples may contain robust proteases. If so, use protease inhibitors, but keep in mind that the use of some inhibitors may introduce artefacts such as multiple spots.
- Streaking in the first dimension, mainly at the top of the gel: presence of particulate material in the sample (centrifuge), high concentration of salts (desalt the sample by dialysis) or of nucleic acids (increase the time of incubation in the presence of RNase and DNase). An inadequate solubilization of proteins may be also responsible for streaking: this may result from a too small amount of solubilization buffer added after cell breakage or from an insufficient length of time of solubilization, which in both cases will result in aggregates.

Focusing

- Urea precipitates in the tube after polymerization: precipitate will disappear during focusing.
- Gel breakage during focusing: sample contains high salt concentration or protein concentration is too high.
- All proteins are incorrectly focused. If proteins stay at the top of the first-dimensional gel: air bubble in the sample on top of the gel or air bubble at the bottom end. If proteins are spread all along the first dimension but incompletely focused: lengthen focusing time.
- Partial loss of the basic part of the gradient: (1) protein overloading: apply less protein. If not overloaded: (2) gradient drift. This happens because carrier ampholytes at the basic end of the gel are continuously passing into the reservoir buffer: focus for a shorter time.
- Acidic proteins are incorrectly focused. These proteins have to migrate through the whole length of the gel before they reach their isoelectric point under standard IEF conditions: they have still not reached their isoelectric point. Focus for a longer time.
- High molecular weight proteins or a few proteins are incorrectly focused: (1) the migration of large proteins may be slowed down because of the pore size of the focusing gel; (2) the entry of proteins into the gel may be delayed because of aggregation, interaction with other proteins or with nucleic acids on top of the gel. Lengthen the focusing time.
- Urea precipitates in first-dimensional gels once focusing is arrested, or before extruding the gel from the glass tube: this precipitate increases the risk of breaking the gel during extrusion; warm the gel in order to resolublize urea.
- No proteins in the 2-D gel: the polarity of the electrode did not match the electrode solutions. The acidic solution should be connected to the anode (+ elec-

trode), the basic solution to the cathode (- electrode). This is true for both standard IEF (cathode at the top) and NEPHGE (anode at the top).

Second Dimension

- Individual vertical streaks starting from the top of the second-dimensional gel, or sometimes in the middle of the gel: results from dust particles which have fallen on the gel surface or in the gel solution. Protect the surface of the gel and solutions, and work in a dust-free area. Carefully rinse gloves with water before manipulation.
- The front of migration is crooked: air bubbles between the bottom of the gel and the running buffer.
- Low molecular weight proteins are duplicated: first-dimensional gel was not in contact with the second-dimensional gel when the second dimension started.

CHAPTER 4

Two-Dimensional Electrophoresis with Immobilized pH Gradients

A. GÖRG[1] and W. WEISS[1]

1 Introduction

Two-dimensional polyacrylamide gel electrophoresis (2-D PAGE) (O'Farrell 1975), in which proteins are separated according to charge (pI) by isoelectric focusing (IEF) in the first dimension and according to size (M_r) by SDS-PAGE in the second dimension, has a unique capacity for the resolution of complex mixtures of proteins, permiting the simultaneous analysis of hundreds or even thousands of gene products. However, the exchange of 2-D gel data between laboratories has been a major problem because of spatial irreproducibility between 2-D gels generated by the conventional method of 2-D PAGE using carrier ampholyte (CA) IEF. Equilibrium CA-IEF cannot be achieved because of pH gradient instability with prolonged focusing time, as the pH gradient moves towards the cathode ('cathodic drift') and flattens in the centre ('plateau phenomenon'). Consequently, time-dependent protein patterns are obtained. In addition, reproducibility of pH gradient profiles is limited by the batch-to-batch variability of CA preparations.

Finally, the problems of pH gradient instability and reproducibility were overcome by the introduction of immobilized pH gradients (IPG) for IEF (Bjellqvist et al. 1982). IPGs are based on the principle that the pH gradient is generated by a limited number (6–8) of well-defined chemicals (the 'Immobilines') which are co-polymerized with the acrylamide matrix. Thus, cathodic drift is eliminated, reproducibility enhanced and pattern matching and interlaboratory comparisons simplified. IPGs allow the generation of pH gradients of any desired range (broad, narrow or ultra-narrow) between pH 3 and 12. Since sample loading capacity of IPGs is also much higher than with CA-IEF, especially in combination with narrow (1 pH unit) or ultra-narrow (0.1 pH unit) IPGs, 2-D PAGE with IPGs is the method of choice for micropreparative separation and spot identification.

[1] Technische Universität München, Department of Food Technology, 85350 Freising-Weihenstephan, Germany.

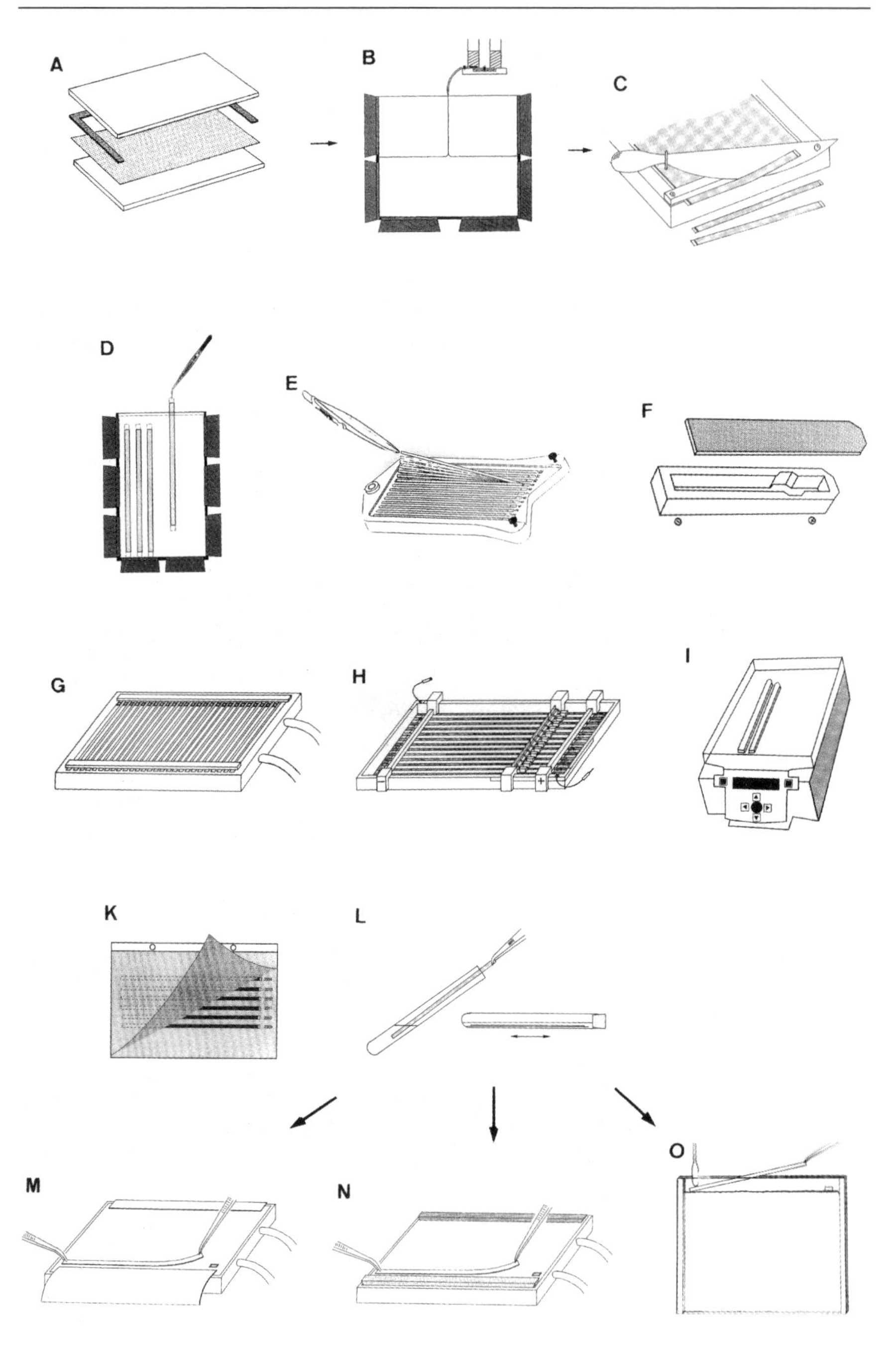
A
B
C
D
E
F
G
H
I
K
L
M
N
O

1.1 Two-Dimensional Electrophoresis with IPGs (IPG-Dalt)

After experiencing initial problems in handling IPGs, a basic protocol for horizontal and vertical two-dimensional electrophoresis with IPGs in the first dimension *(IPG-Dalt)* was established (Görg et al. 1988). Since that time, the protocol has not been changed essentially (Figs. 4.1–4.7). Compared to classical 2-D electrophoresis with carrier ampholytes (O'Farrell 1975), the employment of IPG-Dalt has produced significant improvements in 2-D electrophoretic separation, permitting higher resolution, especially with narrow-range IPGs (cf. Fig. 4.8) and reproducibility of 2-D patterns both within a laboratory and, more important, between laboratories (Corbett et al. 1994; Blomberg et al. 1995). Moreover, basic proteins (pI >7.5), normally lost by the cathodic drift of carrier ampholyte focusing or separated by NEPHGE (O'Farrell et al. 1977) with limited reproducibility, were perfectly separated under equilibrium conditions using IPGs 4–9, 4–10, and 6–10 for the separation of highly diverse samples such as plant, yeast, mouse liver, human heart, or myeloblast proteins (Görg et al. 1988; Görg 1991, 1993) (cf. Fig. 4.9). Recently, very alkaline IPGs up to pH 12 were successfully generated for the 2-D electrophoresis of ribosomal proteins and histones (Görg et al. 1997, 1998, 1999).

Due to these features together with the high loading capacity of IPG-Dalt for micro-preparative runs (up to 10 mg of a crude sample preparation can be applied onto a single 2-D gel) (Hanash et al. 1991; Bjellqvist et al. 1993; Posch et al. 1995), IPG-Dalt advanced to the core technology of proteome analysis (Wilkins et al. 1997), facilitating spot identification by peptide mass fingerprinting, MALDI or tandem mass spectrometry, amino acid composition analysis, and N-terminal and/or internal peptide microsequencing.

1.2 The Protocol of IPG-Dalt

The first dimension of IPG-Dalt, isoelectric focusing (IEF), is performed in individual IPG gel strips 3 mm wide and cast on GelBond PAGfilm (either ready-made Immobiline DryStrips or laboratory-made, obtained from washed and dried Immobiline gels cast by the gradient casting technique of Görg et al. (1980,

Fig. 4.1. Procedure of IPG-Dalt based on the protocol of Görg *et al.* (1988). *A* Assembly of the polymerization cassette for the preparation of IPG and SDS gels on plastic backings (Glass plates, GelBond PAGfilm and 0.5-mm-thick U-frame). *B* Casting of IPG- and/or pore-gradient gels. *C* Cutting of washed and dried IPG slab gels (or Immobiline DryPlates) into individual IPG strips. *D* Rehydration of individual IPG strips in a vertical rehydration cassette, in the reswelling tray (*E*), or in the IPG strip holder (IPGphor) (*F*). *G* IEF in individual IPG gel strips directly on the cooling plate of the IEF chamber, in the DryStrip kit (*H*), or on the IPGphor (*I*). *K* Storage of IPG strips after IEF. *L* Equilibration of IPG strips prior to SDS-PAGE. *M* Transfer of the equilibrated IPG strip onto the surface of the laboratory-made horizontal SDS gel along the cathodic electrode wick, or onto the surface of a ready-made horizontal SDS gel along the cathodic buffer strip (*N*). *O* Loading of the equilibrated IPG gel strip onto the surface of a vertical SDS gel

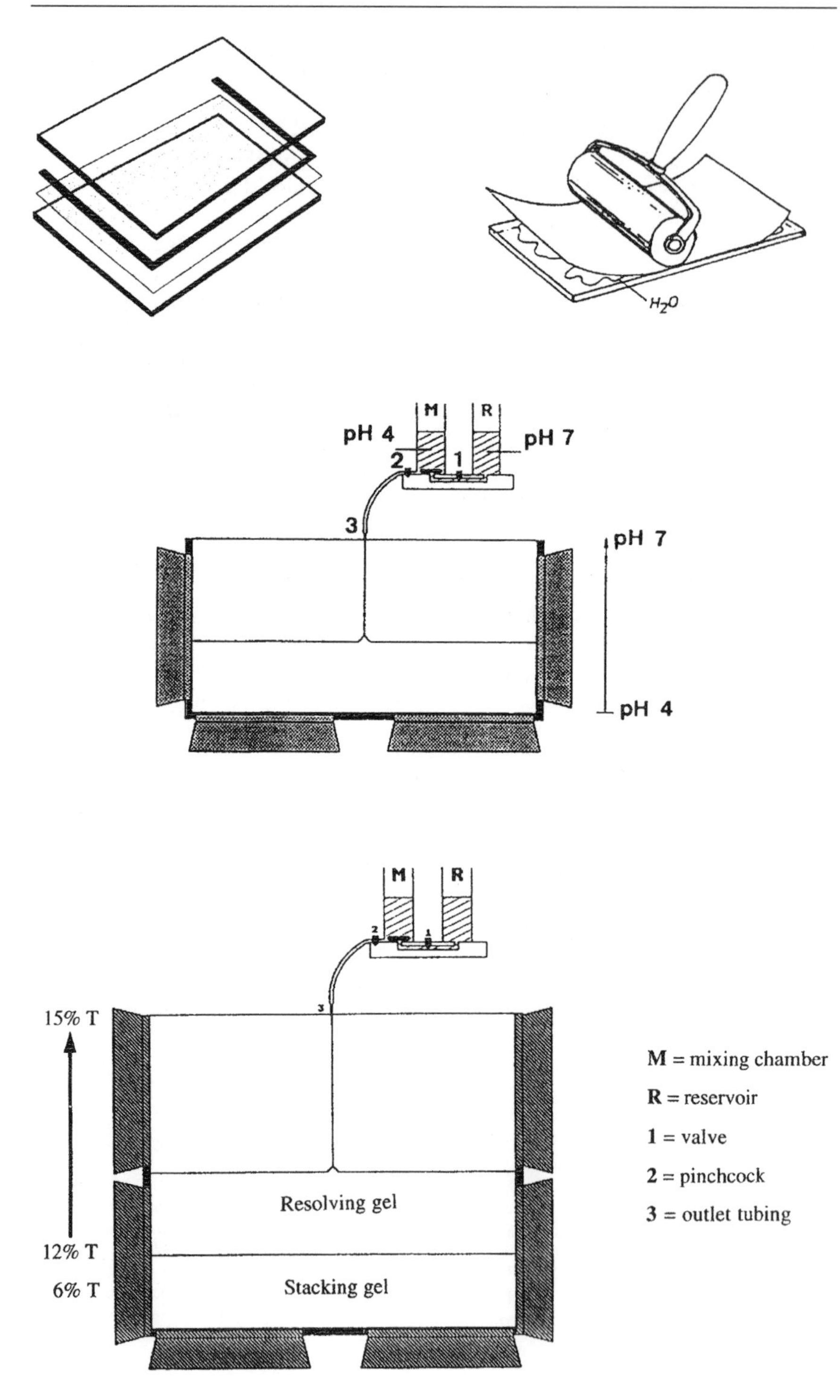
H_2O
M
R
pH 4
pH 7
2
1
3
pH 7
pH 4
M
R
2
1
3
15% T
12% T
6% T
Resolving gel
Stacking gel
M = mixing chamber
R = reservoir
1 = valve
2 = pinchcock
3 = outlet tubing

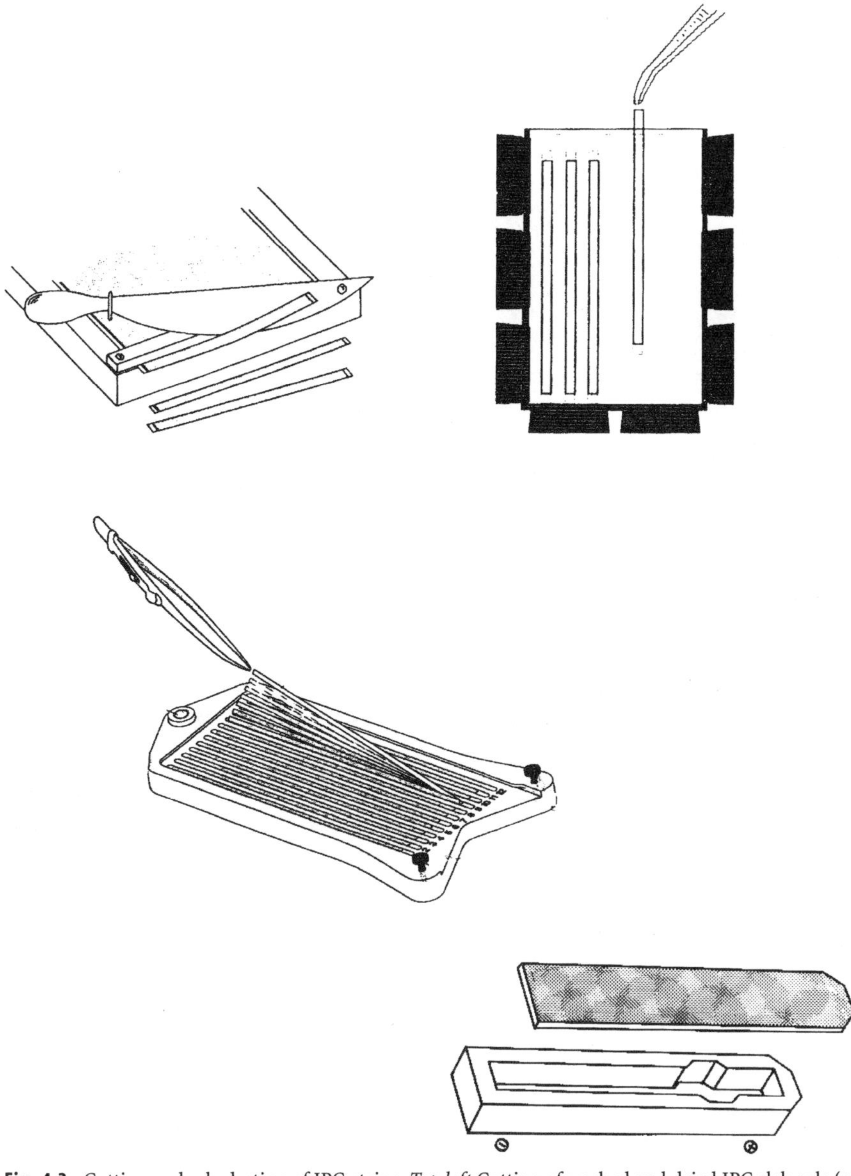

Fig. 4.3. Cutting and rehydration of IPG strips. *Top left* Cutting of washed and dried IPG slab gels (or Immobiline DryPlates) into individual IPG strips. *Top right* Rehydration of IPG strips in a vertical rehydration cassette. *Center* Rehydration of IPG strips in the reswelling tray. *Bottom* Rehydration of an IPG strip in the IPGphor strip holder

Fig. 4.2. Casting of IPG and SDS pore gradient gels on GelBond PAGfilm. *Top left* Assembly of the polymerization cassette for IPG and SDS gel casting on plastic backing (Glass plates, GelBond PAGfilm, U-frame 0.5 mm thick). *Top right* Application of the GelBondPAGfilm onto the glass plate. *Center* IPG gel casting. *Bottom* SDS pore gradient gel casting

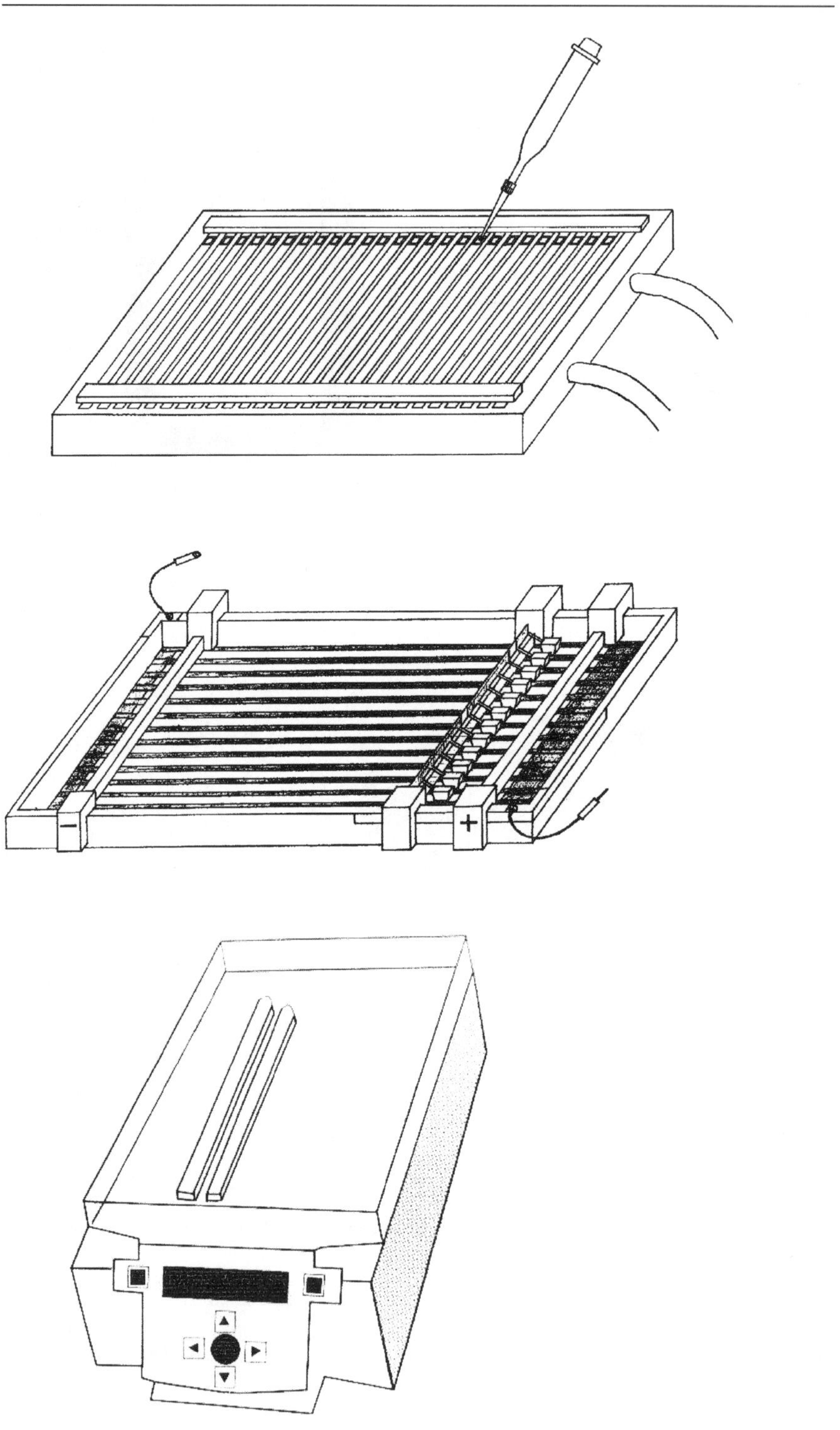

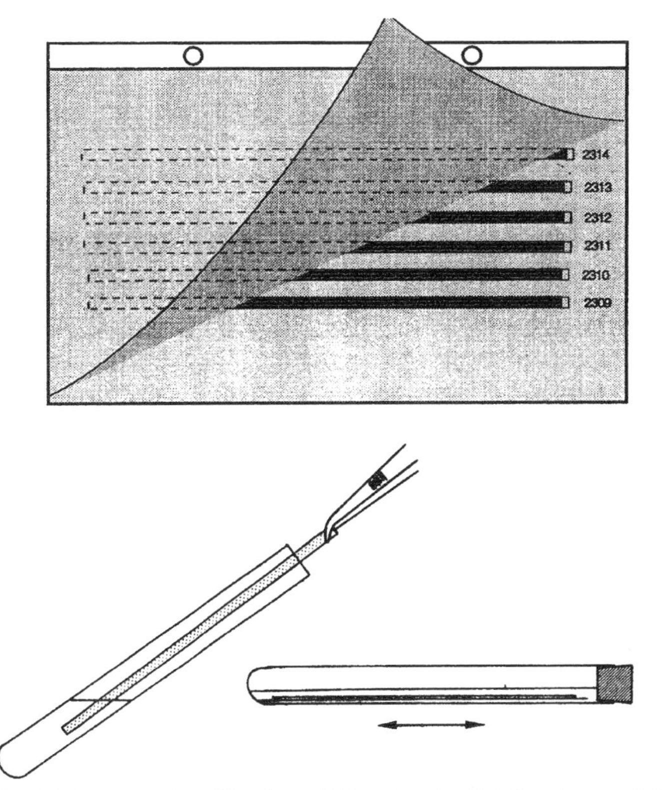

Fig. 4.5. Storage and equilibration of IPG strips after IEF. *Top* Storage of IPG strips between two sheets of plastic film. *Bottom* Equilibration of IPG strip prior to SDS-PAGE

1986). Prior to IEF, IPG dry strips are rehydrated to their original thickness of 0.5 mm with a solution containing 8 M urea, 0.5–2 % (non-ionic or zwitterionic) detergent, 0.2 % dithiothreitol (DTT) and 0.2 % carrier ampholytes. The rehydrated strips are then placed onto the cooling plate of an electro-focusing chamber and sample cups are placed onto the surface of the gel strips. Sample entry is critical, and best results are obtained using diluted samples dissolved in 9.5 M urea, 2–4 % non-ionic or zwitterionic detergent, 1 % DTT, and 0.2 % carrier ampholyte (O'Farrell 1975), or–in the case of very hydrophobic proteins–by a mixture of 2 M thiourea and 7 M urea instead of 9.5 M urea and/or other detergents (Rabilloud et al. 1997). For better sample entry, a low voltage gradient is applied across the gel for the first hour. Voltage is then increased to 3500 V (Multiphor) (Görg et al. 1988), or even up to 8000 V (IPGphor) until the steady state with constant focusing patterns is obtained.

◄

Fig. 4.4. IEF in individual IPG gel strips. *Top* Analytical IEF on the Multiphor cooling plate. *Center* Micropreparative IEF using Amersham Pharmacia's Immobiline DryStrip Kit. *Bottom* IEF on the IPGphor

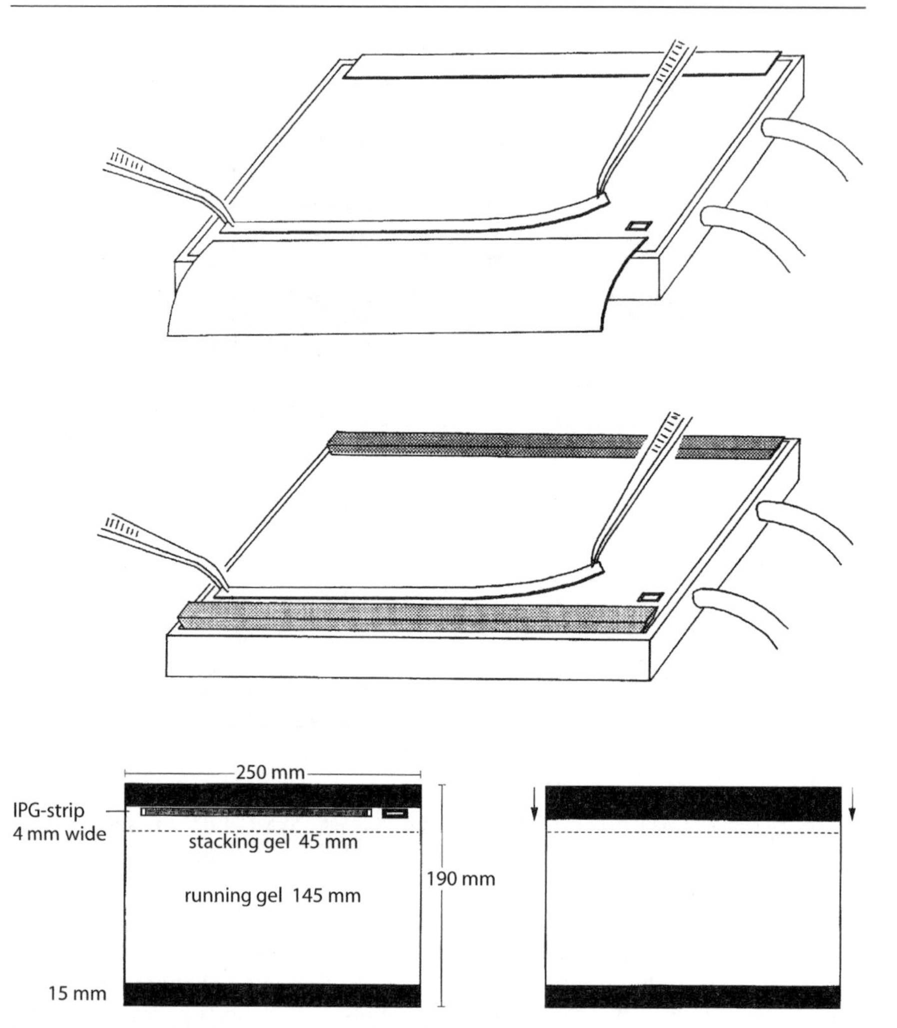

Fig. 4.6. Horizontal SDS-PAGE. *Top* Transfer of the equilibrated IPG gel strip onto the surface of the laboratory-made horizontal SDS gel alongside the cathodic electrode wick. *Center* Transfer of the equilibrated IPG gel strip onto the surface of a ready-made horizontal SDS gel alongside the cathodic polyacrylamide buffer strip. *Bottom* Removal of IPG strips after protein transfer from first to second dimension

As an alternative to cup-loading, samples can also be applied by in-gel rehydration (Rabilloud et al. 1994). The latter procedure is especially advantageous for high sample loads such as for micropreparative 2-D PAGE. An exciting new development for simplification of IPG-IEF is the introduction of an integrated system (IPGphor) where in-gel rehydration and IEF are performed in one step overnight, without personal assistance (Islam et al. 1998).

Whatever system is used for isolectric focusing, after IEF to the steady state, the IPG strips are equilibrated in the presence of SDS, DTT, urea, glycerol and

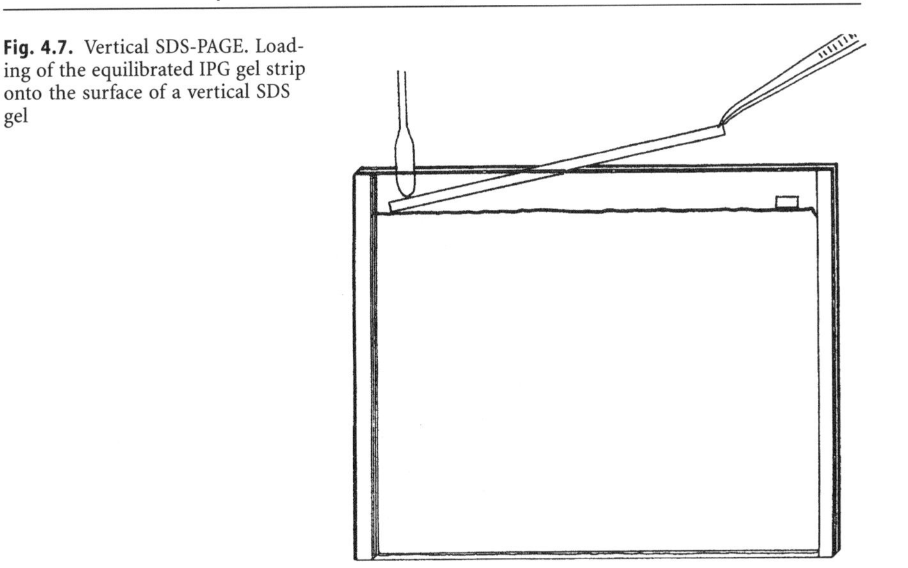

Fig. 4.7. Vertical SDS-PAGE. Loading of the equilibrated IPG gel strip onto the surface of a vertical SDS gel

iodoacetamide (IAA), and then placed onto the surface of a horizontal or on top of a vertical SDS gel (Görg et al. 1988). Alternative procedures using tributylphosphine instead of DTT and IAA have been successfully applied for hydrophobic proteins (e.g. wool filament proteins) (Herbert et al. 1998).

For horizontal setups, the laboratory-made or ready-made SDS-PAGE gel (ExcelGel SDS), cast on plastic backing, is placed onto the cooling plate of a horizontal electrophoresis system. The equilibrated IPG gel strip(s) is (are) transferred gel-side-down onto the surface of the SDS gel alongside the cathodic electrode wick. For vertical setups, the equilibrated IPG gel strips are loaded on top of vertical SDS polyacrylamide gels, with or without embedding in agarose. Vertical setups are especially useful for multiple runs (up to 20 at a time) (Anderson and Anderson 1978).

Upon completion of electrophoresis, the polypeptides are either stained with Coomassie Brilliant Blue or silver nitrate, or detected by fluorescence or autoradiography. Alternatively, proteins are transferred onto an immobilizing membrane and detected with specific reagents such as antibodies. For spot identification, the spots are excised from the gel or the blotting membrane and subjected directly or after (enzymatic or chemical) cleavage to Edman degradation, amino acid composition analysis, or mass spectrometry.

2 Sample Preparation

Pretreatment of samples for 2-D PAGE involves solubilization, denaturation and reduction to completely break up the interactions between the proteins (Rabilloud 1996, see also chapter 2 of this book). Although desirable, there is no single method of sample preparation that can be universally applied due to the diverse

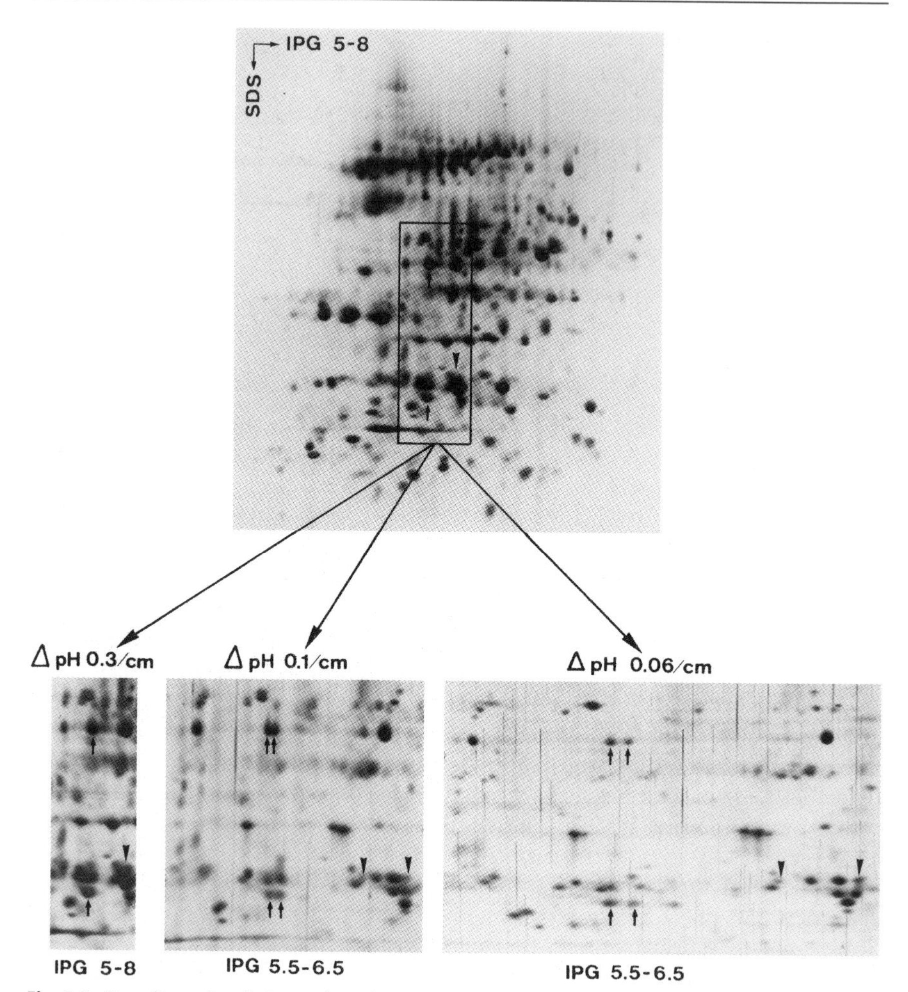

Fig. 4.8. Two-dimensional electrophoresis using narrow pH gradients (Blow-ups). First dimension: IPG 5–8 and 5.5–6.5 (separation distance: 110 mm and 160 mm, respectively). Second dimension: horizontal SDS-PAGE on laboratory-made gels (12–15 % T, 4 % C). Silver stain. Sample: pea seed proteins (Görg et al. 1988)

samples which are analyzed by 2-D gel electrophoresis. The ideal sample solubilization procedure for 2-D PAGE would result in the disruption of all non-covalently bound protein complexes and aggregates into a solution of individual polypeptides (Herbert et al. 1997). However, whatever method of sample preparation is chosen, it is most important to minimize protein modifications which might result in artefactual spots on the 2-D maps. Samples containing urea must not be heated as this may introduce considerable charge heterogeneity due to carbamylation of the proteins by isocyanate formed from the decomposition of urea. Generally speaking, samples should be subjected to as minimal handling as possible and kept cold at all times (Dunn 1993).

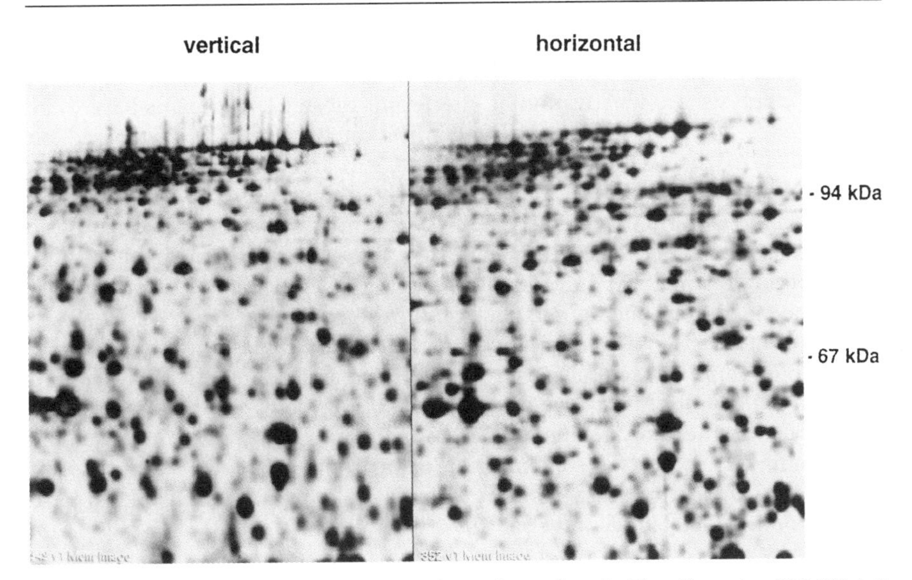

Fig. 4.9. Comparison of horizontal and vertical 2-D electrophoresis. First dimension: IPG-IEF 4–9 (separation distance 180 mm) (section). Second dimension: *Left* vertical SDS-PAGE (12 %T homogeneous); *right* horizontal SDS-PAGE (laboratory-made gel, 12 % T homogeneous). Silver stain. Sample: mouse liver proteins (Görg et al. 1995)

Another consideration is whether proteins from whole cells or from subcellular fractions only (e.g. mitochondria, ribosomes) (Corthals et al. 1997; Görg et al. 1997) should be analyzed, or which kind of solubility fraction (e.g. a total protein extract or selective extraction procedure such as water-soluble proteins or alcohol-extractable proteins) gives best results (Weiss et al. 1992, 1993).

Lysis of different types of cells and tissues may be achieved by homogenizers (e.g. Potter homogenizer), liquid nitrogen-cooled mortar and pestle technique, ultrasonic disintegrators, enzymatic lysis (e.g. with lysozyme), detergents (e.g. NP-40, Triton X-100, CHAPS, SDS), osmotic shock, repeated freezing and thawing, or a combination of these methods.

After cell lysis, it is necessary in most cases to inactivate interfering substances (e.g. plant phenols or nucleic acids) and to remove insoluble components by high-speed centrifugation. Plant phenols tend to adsorb proteins and cause disturbed protein patterns, horizontal and vertical streaks. Remedies are precipitation of proteins with trichloroacetic acid (TCA) (20 %) and removal of phenols by rinsing with ice-cold acetone/TCA, or to adsorb phenols to (insoluble) polyvinylpolypyrrolidone (PVPP). If the cells contain high levels of nucleic acids, the samples may be treated with a protease-free DNase/RNase mixture.

Proteases present within samples can produce artifactual spots on 2-D maps. It is often recommended to add protease inhibitors [e.g. phenylmethylsulfonyl fluoride (PMSF), Pefabloc, EDTA, pepstatin, or protease inhibitor cocktails], but it must be kept in mind that such reagents can also modify proteins and introduce charge artifacts (Dunn 1993). Other possibilities for protease inactivation are protein precipitation with ice-cold TCA or boiling with SDS sample buffer, but it should always be remembered that several proteases may retain their activity at these conditions.

After cell lysis and inactivation of interfering substances, proteins have to be solubilized. The most widely used solubilization procedure is that based on O'Farrell (1975), using a mixture of 2 % NP-40, 9 M urea, 1 % DTT and 0.8 % carrier ampholytes (Lysis buffer). Instead of NP-40, the non-ionic detergent Triton X-100 or the zwitterionic detergent CHAPS are often preferred. Although these methods give excellent results in many cases, not all protein complexes are fully disrupted by this mixture. In contrast, the anionic detergent SDS disrupts most non-covalent protein interactions, but cannot be applied in IEF gels. However, SDS can be used as a pre-solubilization procedure for samples prior to IEF, where the sample is initially solubilized in 1 % SDS and then diluted with a five-fold excess of Lysis buffer to displace the SDS from the proteins and replace it with a non-ionic or zwitterionic detergent, thereby maintaining the proteins in a soluble state (Dunn 1993). For the solubilization of hydrophobic proteins, mixtures of urea and thiourea and detergents other than NP-40 (such as sulfobetaines) have been recommended (Rabilloud et al. 1997, Chevallet et al. 1998).

Protein extracts should not be too diluted to avoid loss of protein due to adsorption to the wall of the vessel (glass or plastics). The minimum protein concentration should not be less than 0.1 mg/ml, and optimum concentration is 1–5 mg/ml. If samples are rather diluted and contain relatively high concentrations of salts which can interfere with IEF, samples may be desalted. Alternatively, proteins can be precipitated with ice-cold TCA/acetone to remove salts. Diluted samples with a low salt concentration may also be applied directly without further treatment, if the dried IPG strips are reswollen in sample solution. In this case, solid urea, CHAPS and DTT are added to the sample until the desired concentration is obtained (Rabilloud et al. 1994).

Short-time storage (several hours to overnight) of extracts is often possible in the refrigerator (4 °C). For a longer time, storage in a freezer or on dry ice at –78 °C is preferred. However, repeated freezing and thawing of the sample must be avoided. It is better to make portions (aliquots) and thaw only once.

For more details see Chapter 2 on sample solubilization by Rabilloud and Chevallet (this Vol.).

2.1 Chemicals

CHAPS, dithiothreitol (DTT) (Sigma), Serdolit MB-1 (Serva), urea, Pefabloc (Merck), Pharmalyte pH 3–10 (Amersham Pharmacia Biotech).

2.2 Buffers

- **Lysis Buffer:** 9.5 M urea, 2 % (w/v) CHAPS, 2 % (v/v) Pharmalyte pH 3–10, 1 % (w/v) DTT and 5 mM Pefabloc. To prepare 50 ml of lysis buffer, dissolve 30 g of urea in deionized water and make up to 50 ml. Add 0.5 g of Serdolit MB-1, stir for 10 min and filter. Add 1g of CHAPS, 0.5 g DTT, 1 ml Pharmalyte pH 3–10 and – immediately before use – 50 mg Pefabloc proteinase inhibitor to 48 ml of the urea solution. Lysis buffer should always be prepared freshly. Alternatively, make small aliquots (1 ml) and store at –78 °C for up to several months.

Note:
(1) Lysis buffer thawed once should not be refrozen!
(2) Never heat urea solutions above 37 °C in order to reduce the risk of protein carbamylation.

2.3 Procedure

In general, an adequate amount of sample (e.g. yeast or mammalian cells or ground plant seeds) is suspended in lysis buffer, disrupted by sonication in an ice bath (3 × 10 s) and centrifuged (60 min, 42 000 *g*, 15 °C):
(1) yeast cells (120 mg), myeloblasts ($5x10^8$ cells) or human or animal tissue such as liver or heart (50–100 mg), or ground seeds from legumes (20 mg) or cereals (50–100 mg) are homogenized in a liquid nitrogen cooled mortar. The powder is then suspended in lysis buffer (1 ml) so that the final protein concentration lies between 5 and 10 mg/ml.
(2) plant tissue proteins (*e.g.* leaf proteins) are treated with cold TCA/acetone (–20 °C) to remove phenolic compounds; leaves are crushed in a liquid nitrogen cooled mortar, and the powder is resuspended in a pre-cooled (–20 °C) solution of 10 % TCA in acetone with 0.07 % 2-mercaptoethanol. Proteins are allowed to precipitate overnight at –20 °C. After centrifugation the pellet is washed with ice-cold acetone containing 0.07 % 2-mercaptoethanol. The supernatant is discarded and the pellet dried in vacuo.

It is also possible to specifically extract certain protein fractions only, e.g. water-soluble (albumins) or alcohol-soluble proteins (gliadins) from cereals and to dilute these extracts with Lysis buffer prior to IPG-IEF (Weiss et al. 1992, 1993).

Protein extracts are either used immediately or are stored at –78 °C. For analytical runs, typically 20 µl of sample solution is applied to a single IPG gel strip, whereas for micropreparative runs up to several hundred microliters can be applied, portion by portion. The amount of protein to be loaded onto a single IPG gel strip (separation distance: 180 mm) varies between 50 and 100 µg for analytical, and 0.5 and 10 mg for micropreparative runs, respectively. Alternatively, the sample can be applied directly by in-gel rehydration.

3 First Dimension: IEF with IPGs

The first dimension of IPG-Dalt, isoelectric focusing (IEF), is performed in individual 3-mm-wide IPG gel strips cast on GelBond PAGfilm (either ready-made Immobiline DryStrips or laboratory-made). Prior to IEF, the IPG dry strips are rehydrated and placed onto the cooling plate of an electrofocusing chamber. The sample is applied either with sample cups or by in-gel rehydration. IPG-IEF can be simplified by use of an integrated system, the IPGphor, where rehydration with sample and IEF are performed automatically.

The standard protocol described here is valid for *broad* gradients in the pH range between 3 and 12 (e.g. IPG 4–7, 4–9, 6–10, 3–10, or 8–12, as well as IPG 3–12 and 4–12; Görg et al., 1998, 1999), and for narrow IPGs in the acidic and neutral range (e.g. IPG 5–6). For narrow pH gradients above pH 9 (*e.g.* IPG 9–12 or 10–12), a modified protocol has been developed which is described in detail elsewhere (Görg et al. 1997).

3.1 Apparatus and Chemicals

3.1.1 Equipment

Multiphor II horizontal electrophoresis apparatus, EPS 3500 XL power supply (3500 V), Multitemp II thermostatic circulator, IPGphor, Immobiline DryStrip Kit, gradient mixer (2 × 15 ml), glass plates with a 0.5-mm-thick U-frame (200 × 260 mm^2), plain glass plates (size 200 × 260 mm^2), reswelling cassette (125 × 260 mm^2), clamps, reswelling tray, GelBond PAGfilm (200 × 260 mm^2), roller, IEF sample applicator strip (Amersham Pharmacia Biotech), heating cabinet (Heraeus, Germany), laboratory shaker (rocking platform) (GFL, Germany), paper cutter (Dahle, Germany), Parafilm (roll, 50 cm × 15 m) (ANC, USA), Milli-Q System (Millipore).

3.1.2 Ready-Made IPG Gels

Immobiline DryPlate 4–7, Immobiline DryStrips 4–7, Immobiline DryStrips 3–10L, Immobiline DryStrips 3–10NL (Amersham Pharmacia Biotech).

3.1.3 Chemicals

Immobiline II chemicals, Pharmalyte (pH range 3–10), IPG buffers, repel silane (Amersham Pharmacia Biotech), acrylamide, bisacrylamide, ammonium persulfate, tetramethylethylenediamine (TEMED), Serdolit MB-1 mixed bed ion exchanger resin, silicone oil (Serva), CHAPS (Sigma), urea, glycerol (Merck).

3.2 Reagent Solutions

- **Acrylamide/Bisacrylamide Solution** (30 % T, 3 % C): 29.1 % (w/v) acrylamide and 0.9 % (w/v) N, N'-methylenebisacrylamide. To make 100 ml of the solution dissolve 29.1 g acrylamide and 0.9 g bisacrylamide in deionized water and fill up to 100 ml. Add 1 g Serdolit MB-1, stir for 10 min and filter. This solution can be stored for 1 week at 4 °C. However, for optimum results it is advisable to prepare it freshly the day you use it.

Table 4.1. Recipes for casting immobiline gels with pH gradients 4-7, 4-9, 6-10, 3-12

Linear pH gradient	pH 4-7		pH 4-9		pH 6-10		pH 3-12	
	Acidic solution pH 4	Basic solution pH 7	Acidic solution pH 4	Basic solution pH 9	Acidic solution pH 6	Basic solution pH 10	Acidic solution pH 3	Basic solution pH 12
Immobiline pk 1.0	–	–	–	–	–	–	1287 ml	–
Immobiline pK 3.6	578 l	302 µl	829 µl	147 µl	941 µl	100 µl	306 µl	–
Immobiline pk 4.6	110 µl	738 µl	235 µl	424 µl	–	–	414 µl	–
Immobiline pk 6.2	450 µl	151 µl	232 µl	360 µl	273 µl	333 µl	558 µl	336 µl
Immobiline pk 7.0	–	269 µl	22 µl	296 µl	243 µl	361 µl	496 µl	168 µl
Immobiline pk 8.5	–	–	250 µl	71 µl	260 µl	239 µl	112 µl	699 µl
Immobiline pk 9.3	–	876 µl	221 µl	663 µl	282 µl	326 µl	84 µl	157 µl
Immobiline pk 10.0	–	–	–	–	–	–	25 µl	342 µl
Immobiline pk >13	–	–	–	–	–	–	–	258 µl
Acrylamide/bisacrylamide	2.0 ml	2.0 ml	2.0 ml	2.0 m	2.0 ml	2.0 ml	2.25 ml	2.25 ml
Deionized water	8.9 ml	10.7 ml	8.3 ml	11.1 ml	8.1 ml	11.7 ml	6.45 ml	10.8 ml
Glycerol (100 %)	3.75 g	–	3.75 g	–	3.75 g	–	3.75 g	–
TEMED (100 %)	10.0 µl	10.0 µl	10.0 µl	10.0 µl	10.0 µl	10.0 µl	10.0 µl	10.0 µl
Persulfate (40 %)	15.0 µl	15.0 µl	15.0 µl	15.0 µl	15.0 µl	15.0 µl	15.0 µl	15.0 µl
Final volume	15.0 ml	15.0 ml	15.0 ml	15.0 ml	15.0 ml	15.0 ml	15.0 ml	15.0 ml

Note: For effective polymerization, acidic and basic solutions are adjusted to pH with 4 N sodium hydroxide and 4 N acetic acid, respectively, before adding the polymerization catalysts (TEMED and ammonium persulfate).

- **Ammonium Persulfate Solution:** 40 % (w/v) in deionized water. To prepare 1 ml of the solution, dissolve 0.4 g ammonium persulfate in 1 ml deionized water. This solution should be prepared freshly just before use.
- **Solutions for Casting Immobiline Gels:** to prepare 15 ml each of acidic and basic solutions mix chemicals and reagent solutions as described in Table 4.1. A huge selection of recipes for many types of narrow or broad pH gradients has been calculated (Righetti 1990). Table 4.1 describes our favourite pH gradients for 2-D electrophoresis.
- **Rehydration Solution:** 8 M urea, 0.5 % CHAPS, 15 mM DTT and 0.2 % ampholyte. To prepare 50 ml of the solution, dissolve 25 g urea in deionized water and complete to 50 ml. Add 0.5 g Serdolit MB-1, stir for 10 min and filter. To 48 ml of this solution add 0.25 g CHAPS, 100 mg DTT, and 0.25 ml IPG buffers (*alternatively:* Pharmalyte pH 3–10) and complete to 50 ml with deionized water. Rehydration solution should be prepared freshly the day you use it.

Note: do not heat urea solutions > 37 °C. Otherwise protein carbamylation may occur.

3.3 IPG Gel Casting

IPG gels are 0.5 mm thick and cast on GelBond PAGfilm. The mold consists of two glass plates, one covered with the GelBond PAGfilm, the other bearing the U-frame (0.5 mm thick) and is loaded in a vertical position and filled from the top with the help of a gradient mixer according to the casting procedure of Görg et al. (1980, 1986) for ultrathin pore gradient gels (Fig. 4.2).

Prior to use the glass plates are thoroughly washed with a mild detergent, rinsed with deionized water and air-dried. If new glass plates are used, pipette 1–2 ml of repel silane on the glass plate that bears the U-frame and distribute it evenly with a fuzz-free filter paper (Kimwipe). Let it dry for a few minutes, rinse again with water and let it air-dry. Repeat this procedure occasionally in order to prevent the gels from sticking to the glass plates.

Gel composition: 4 % T, 3 % C (4.5 % T, 3 % C for IPG 3–12, respectively)
Gel size: $180 \times 250 \times 0.5$ mm^3
pH gradient: lab-made pH gradients 4–7, 4–9, 6–10 and 3–12 are listed in Tab. 1; ready-made gels with pH gradients 4–7, 3–10L and 3–10NL are available from Amersham Pharmacia Biotech (Immobiline DryPlate or Immobiline DryStrip)

IPG slab gels with linear gradients pH 4–7, 4–9, 6–10 and 3–12 are cast according to Görg et al. (1986) with the recipes of Righetti (1990) and Görg et al. (1999). Two starter solutions (an acidic one and a basic one) are prepared as described in Table 4.1. For better polymerization, the acidic and basic solutions are adjusted to pH 7 with sodium hydroxide and acetic acid, respectively. The acidic, dense solution is pipetted into the mixing chamber and the basic, light solution into the reservoir of the gradient mixer (Fig. 4.2). When a pH plateau (2 cm wide) for the sample application area is desired, an extra portion of the dense solution is prepared and pipetted into the mold prior to pouring the gradient. After pouring the gradient into the precooled mold (refrigerator), the mold is kept at room temperature for 15 min to allow adequate levelling of the density gradient prior to polymerization for 1h at 50 °C. After polymerization, the mold is kept at room temperature for at least 15 min. Then the IPG gel is removed from the mold and extensively washed with deionized water, impregnated with 2 % glycerol, and dried at room temperature in a dust-free cabinet, covered with a plastic film and stored at –20 °C. The dried gels can be stored frozen for at least 1 year.

Instead of laboratory-made gels, ready-made gels (Immobiline DryPlate or Immobiline DryStrip) can be used.

Procedure

1. To assemble the polymerization cassette (Fig. 4.2), wet the plain glass plate (size 260 × 200 mm^2) with a few drops of water. Place the Gelbond PAGfilm, hydrophilic side upwards, on the wetted surface of the plain glass plate. The GelBond PAGfilm should overlap the upper edge of the glass plate for 1–2 mm to facilitate filling of the cassette. Expel excess water with a roller (Fig. 4.2). Place the glass plate which bears the U-frame (0.5 mm thick) on top of the GelBond PAGfilm and clamp the cassette together. Put it in the refrigerator for 30 min.
2. To cast the IPG gel, pipette 12.5 ml of the acidic, dense solution into the mixing chamber of the gradient mixer (Fig. 4.2). Outlet and valve connecting the mixing chamber and reservoir must be closed. Add 8 µl of TEMED and 12 µl of ammonium persulfate and mix. Open the connecting valve between the chambers for a second to release any air bubbles.
3. Pipette 12.5 ml of the basic, light solution into the reservoir of the gradient mixer. Add 8 µl of TEMED and 12 µl of ammonium persulfate and mix with a spatula.
4. Switch on the magnetic stirrer at a reproducible and rapid rate. However, avoid excessive vortex. Remove the polymerization cassette from the refrigerator and put it underneath the outlet of the gradient mixer. Open the valve
   (1) connecting the chambers and, immediately afterwards, the pinchcock
   (2) on the outlet tubing so that the gradient mixture is applied slowly, but steadily into the cassette from a height of about 5 cm just by gravity flow. Take care that the level in both chambers drops equally fast. Formation of the gradient is completed in 2–3 min.

Note: when one of the chambers is emptying faster than the other, there is a risk that the resulting pH gradient will not be linear. Check if there is an air bubble in the connecting line or whether the speed of the magnetic stirrer is inappropriate.

5. Keep the mold at room temperature for 15 min to allow adequate levelling of the density gradient. Then polymerize the gel for 1 h at 50 °C in a heating cabinet.
6. After polymerization at 50 °C, allow the cassette to cool down to room temperature (at least 15 min). Then insert a spatula between the glass plates and the gel, pry apart the glass plates and carefully remove the gel from the cassette. Wash the IPG gel for 1 h with 10-min changes of deionized water (500 ml each)

in a glass tray on a rocking platform. Equilibrate the gel in 2 % (w/v) glycerol for 30 min and dry it overnight at room temperature, using a fan, in a dust-free cabinet. Afterwards, protect the surface of the dry gel with a sheet of plastic film. The dried IPG gel can be stored in a sealed plastic bag at –20 °C for at least several months without loss of function. Dried IPG gels in several pH ranges are also commercially available (Immobiline DryPlate).

7. Prior to IEF, cut the dried IPG gels – or the ready-made Immobiline DryPlates – into individual 3- to 4-mm-wide strips with the help of a paper cutter (Fig. 4.3). Alternatively, ready-cut IPG strips (Immobiline DryStrip) can also be used.

Note: in order to ensure the reproducibility of the IPG gradient, the volume of the cassette should be constant. Therefore, it is recommended to check the volume of the cassette from time to time since it diminishes on ageing of the U-frame.

3.4 Running IPG Strips

Prior to IEF, the desired number of IPG gel strips are rehydrated to the original gel thickness (0.5 mm). The rehydrated strips are then placed on the cooling block of the electrofocusing chamber or in the grooves of the strip aligner of the tray of the Pharmacia DryStrip Kit, and samples are applied to silicone frames or sample cups placed directly on the surface of the IPG gel strips. Alternatively, the sample can be applied directly by in-gel rehydration using Amersham Pharmacia's reswelling tray or IPGphor strip holders.

3.4.1 IEF Following Sample Cup Loading

For analytical purposes, typically 50–100 μg of protein is applied to a single, 180-mm-long IPG gel strip, whereas for micropreparative purposes up to several milligrams of protein can be applied in the case that Pharmacia's Immobiline DryStrip Kit with sample cups is used. When IEF is performed using the DryStrip Kit, the IPG gel strips can be covered by a layer of silicone oil which is advantageous when running very basic pH gradients (e.g. IPG 10–12) (Görg et al. 1997) or narrow pH gradients with long focusing times for micropreparative IPG-Dalt (Bjellqvist et al. 1993).

3.4.1.1 Rehydration of IPG Strips

1. To rehydrate the IPG gel strips to their original thickness (0.5 mm), take two layers of Parafilm (roll, 50 cm wide) and cut out a U-frame which has the same shape and size as the glass plate's U-frame. Assemble the rehydration cassette in this way so that a 0.7-mm-thick U-frame (0.5 mm original thickness plus two layers of Parafilm, 0.1 mm each to compensate for the 0.2-mm-wide GelBond PAGfilm on which the IPG gel is polymerized) is obtained. If larger sam-

ple volumes (> 100 μl) are to be applied, it is advantageous to use one layer of parafilm only. Clamp the cassette together and fill in the rehydration solution either from the bottom with a syringe or from the top with a pipette.

2. Take an appropriate number of dry IPG gel strips and pull off their protective covers. Lower them carefully, but without delay, into the rehydration cassette, gel side towards the glass plate bearing the U-frame (Fig. 4.3). Allow the strips to rehydrate overnight at room temperature. Make sure that the cassette does not leak.
3. After the IPG gel strips have been rehydrated, pour the rehydration solution out of the reswelling cassette, remove the clamps and open the cassette with the help of a spatula. Using clean forceps, rinse the rehydrated IPG gel strips with deionized water for a second and place them, gel side up, on a sheet of water saturated filter paper. Wet a second sheet of filter paper with deionized water, blot it to remove excess water and put it onto the surface of the IPG gel strips. Blot them gently for a few seconds to remove excess rehydration solution in order to prevent urea crystallization on the surface of the gel during isoelectric focusing.

3.4.1.2 Analytical IEF for Multiple Runs

1. Moisten the flat-bed cooling block with 2–3 ml of kerosene and place the IPG gel strips *(up to 40)* side by side, 1–2 mm apart, on it (Fig. 4.4). The acidic end of the IPG gel strips must face towards the anode.
2. Cut two IEF electrode strips or paper strips prepared from 2 mm thick filter paper (*e.g.* MN 440, Macherey and Nagel, Germany) to a length corresponding to the width of all IPG gel strips lying on the cooling plate. Soak the electrode strips with deionized water (conductivity 2 μS; however, not Milli-Q quality) and remove excessive moisture by blotting with filter paper.
3. Place the IEF electrode strips on top of the aligned IPG gel strips at the cathodic and anodic ends.
4. When running basic IPGs (e.g. IPG 6–10 or IPG 10–12), put an extra paper strip soaked with 20 mM DTT onto the IPG gel surface near the cathodic electrode strip.
5. Apply silicone rubber frames (2×5 mm^2 inner diameter) onto the gel surface, 5 mm apart from anode or cathode, for sample application (Fig. 4.4).
6. Pipette the samples (20 μl each, preferably dissolved in lysis buffer) into the silicone frames. Protein concentration should not exceed 5–10 mg/ml. Otherwise, protein precipitation at the sample application area may occur. To avoid this, dilute the sample with lysis buffer and apply larger volumes instead (cf. Sect. 3.4.1.3).
7. Position the electrodes and press them gently down on top of the IEF electrode strips.
8. Place the lid on the electrofocusing chamber, connect the cables to the power supply and start IEF. Running conditions depend on the pH gradient and the length of the IPG gel strip used. An appropriate time schedule for orientation is given in Table 4.2 (Görg 1993). For improved sample entry, voltage is limited to 150 V (30 min) and 300 V (60 min) at the beginning. Then continue with

Table 4.2. Typical running conditions for first-dimensional IEF with IPG strips (cup loading)

Temperature	20 °C			
Current max.	0.05 mA per strip, 2 mA max. in total			
Power max.	3–5 W			
Voltage max.	Analytical		Micropreparative	
	150 V for 30 min		150 V for 1–2 h	Sample entry
	300 V for 60 min		300 V for 4–5 h	Sample entry
	1500 V for 60 min		1500 V for 12–16 h	
	3500 V See below*		3500 V See below*	To the steady state
Type of separation	Analytical	Analytical	Micropreparative	Micropreparative
Separation distance	11 cm	18 cm	11 cm	18 cm
IPG 4-7	22 000 V × h	42 000 V × h	40 000 V × h	75 000 V × h
IPG 4-8	21 000 V × h	35 000 V × h	30 000 V × h	60 000 V × h
IPG 4-9	17 000 V × h	30 000 V × h	25 000 V × h	45 000 V × h
IPG 6-10	21 000 V × h	35 000 V × h	30 000 V × h	60 000 V × h
IPG 3-10.5	11 000 V × h	18 000 V × h	15 000 V × h	30 000 V × h

*Time depends on the IPG, type of separation, and separation distance.

maximum settings of 3500 V to the steady state. Optimum focusing temperature is 20 °C (Görg et al. 1991).

9. After IEF, those IPG gel strips that are not used immediately for second dimension run and/or are kept for further reference are stored between two sheets of plastic film at -78 °C for up to several months (Fig. 4.5).

3.4.1.3
Analytical IEF and Micropreparative IEF Using Pharmacia's DryStrip Kit

The application of higher sample volumes (> 20 µl) is simplified when the sample cups of the Immobiline DryStrip Kit are used (Fig. 4.4). A sample volume of 100 µl can be applied at a time, and it is possible to apply a total of up to 200–300 µl, portion by portion, to a single IPG gel strip. Typical sample load is 60–100 µg protein/strip for silver stained 2-D patterns, whereas for micro preparative purposes up to 10 mg of protein from a cell lysate can be loaded onto a single IPG gel strip.

If very basic pH gradients (pH > 10) or narrow pH gradients (pH range < 1 unit) for micropreparative runs are used, the IPG strips ought to be protected by a layer of silicone oil. In case of broad pH gradients not exceeding pH 10 (e.g. IPG 4–7, 4–9 or 6–10), the DryStrip Kit can be used without silicone oil overlay.

Procedure

1. Place the cooling plate in the Multiphor II Electrophoresis unit. Pipette 3–4 ml of kerosene or silicone oil onto the cooling plate and position the Immobiline DryStrip tray on the cooling plate. Avoid trapping large air bubbles between the tray and the cooling plate.
2. Connect the electrode leads on the tray to the Multiphor II unit.
3. Pour about 10 ml of silicone oil into the tray.
4. Place the Immobiline strip aligner in the tray on top of the oil.

5. Transfer the rehydrated IPG gel strips (gel-side-up and acidic end towards the anode) into adjacent grooves of the aligner in the tray. Align the strips such that the anodic gel edges are lined up (Fig. 4.4).
6. Cut two IEF electrode strips or paper strips prepared from 2-mm-thick filter paper (e.g. MN 440, Macherey and Nagel, Germany) to a length corresponding to the width of all IPG gel strips lying in the tray. Soak the electrode strips with deionized water, remove excessive moisture by blotting with filter paper and place the moistened IEF electrode strips on top of the aligned strips near the cathode and anode.
7. Position the electrodes and press them gently down on top of the IEF electrode strips.
8. Put the sample cups on the sample cup bar. Place the cups high enough on the bar to avoid touching the gel surface. Put the sample cup bar in position so that there is a distance of a few millimeters between the sample cups and the anode (or cathode, in the case of cathodic sample application).
9. Move the sample cups into position, one sample cup above each IPG gel strip, and finally press down the sample cups to ensure good contact with each strip.
10. Once the sample cups are properly positioned, pour about 80 ml of silicone oil into the tray so that the IPG gel strips are completely covered. If the oil leaks into the sample cups, suck the oil out, readjust the sample cups and check for leakage again. Add approximately 150 ml of oil to completely cover the sample cups.

Note: In the case of IEF using pH gradients in the range pH 4 to 9, the oil step can be omitted.

11. Pipette the samples into the cups by underlaying. Watch again for leakage.
12. Close the lid of the Multiphor II electrophoresis chamber and start the run according to the parameters given in Table 4.2. For improved sample entry, voltage is limited to low voltages (150–300 V) for the first few hours. Then continue with maximum settings of 3500 V to the steady state. Optimum focusing temperature is 20 °C (Görg et al. 1991).
13. If sample volumes exceeding 100 µl are to be applied, pipette in 100 µl and run IEF with limited voltage until the sample has migrated out of the cup. Then apply another 100 µl and repeat the procedure until the whole sample has been applied.
14. When the IEF run is completed, remove the electrodes, sample cup bar and IEF electrode strips from the tray. Use clean forceps and remove the IPG gel strips from the tray. Those IPG gel strips which are not used immediately for second dimension run and/or are kept for further reference are stored between two sheets of plastic film at –78 °C for up to several months (Fig. 4.5).

3.4.2
IEF Following In-Gel Rehydration of Sample

Especially for micropreparative IPG-Dalt, samples can be applied by in-gel rehydration, which works for the majority of proteins (Rabilloud et al. 1994; Sanchez et al. 1997). However, it is recommended to check whether the high-molecular mass proteins, very basic proteins, and membrane proteins, respectively, have entered the gel matrix properly.

Procedure

1. Cut the dried IPG gels into 3-mm-wide strips with the help of a paper cutter (Fig. 4.3). Alternatively, use ready-made IPG dry strips.
2. Solubilize proteins with sample solubilization buffer (9 M urea, 2 % CHAPS, 0.8 % Pharmalytes 3–10, 15 mM DTT) and dilute the extract (protein concentration: $\approx$ 10 mg/ml) with sample solubilization buffer (dilution: 1+1 for micropreparative runs, and 1+19 for analytical runs, respectively) to a final volume of 400 μl for 180-mm-long IPG strips.
3. Pipette 350 μl of sample-containing rehydration solution into the grooves of the reswelling tray (Fig. 4.3). Peel off the protective cover sheets from the IPG strips and insert the IPG strips (gel-side-down) into the grooves. Avoid trapping air-bubbles. Cover the IPG strips with 1 ml of silicone oil, close the lid and let the strips rehydrate overnight.
4. After the IPG gel strips have been rehydrated, rinse them with deionized water for a second and place them, gel-side-up, on a sheet of water-saturated filter paper. Wet a second sheet of filter paper with deionized water, blot it slightly to remove excess water and put it onto the surface of the IPG gel strips. Blot them gently for a few seconds to remove excess rehydration solution in order to prevent urea crystallization on the surface of the gel during IEF.
5. Cover the flat-bed cooling block with 2–3 ml of kerosene and place the IPG gel strips, side by side and 1–2 mm apart, on it. The acidic end of the IPG gel strips must face towards the anode. If IEF is to be performed under a protective layer of silicone oil, use Pharmacia's DryStrip Kit instead (however, without sample cups) (cf. Sect. 3.4.1.3).
6. Cut two IEF electrode strips to a length corresponding to the width of all IPG gel strips (and spaces between them) lying on the cooling plate. Soak the electrode strips with deionized water and remove excessive moisture by blotting with filter paper.
7. Place the IEF electrode strips on top of the aligned IPG gel strips at the cathodic and anodic gel ends. When running basic IPGs (e.g. IPG 6–10 or 8–12), put an extra paper strip soaked with 20 mM DTT on the surface of the IPG strips next to the cathode.
8. Position the electrodes and press them gently down on top of the IEF electrode strips.
9. Place the lid on the electrofocusing chamber, connect the cables to the power supply and start IEF. Running conditions depend on the pH gradient and the length of the IPG gel strip used. An appropriate time schedule for orientation is given in Table 4.2.

For removal of salts, voltage is limited to 150, 300 and 600 V for 1 h each, followed by 3500 V to the steady state. Current is limited to 0.05 mA/IPG strip. Optimum focusing temperature is 20 °C (Görg et al. 1991).

10. After IEF, those IPG gel strips which are not used immediately for second dimension run or are kept for further reference are stored between two sheets of plastic film at –78 °C up to several months (Fig. 4.5).

3.4.3 IEF with IPGphor

IPG-IEF for 2-D electrophoresis can be simplified by the use of an integrated instrument, the *IPGphor* (Islam et al. 1998). The IPGphor includes a Peltier element for temperature control (between 18 and 25 °C) and a programmable power supply (8000 V, 1.5 mA). The central part of this instrument are so-called strip holders made from an aluminium oxide ceramic, in which IPG strip rehydration with sample solution and IEF are performed without further handling after the strip is placed in the strip holder (Fig. 4.4). The IPGphor can handle up to 12 strip holders (length 7, 11, 13 or 18 cm). The strip holder platform regulates temperature and serves as the electrical connector for the strip holders. Besides easier handling, a second advantage of the IPGphor is shorter focusing time, because IEF can be performed at rather high voltage (up to 8000 V).

The IPGphor is programmable and can store nine different programs. A delayed start is also possible, which allows the user to load the strip holders with sample dissolved in rehydration buffer in the afternoon, and then automatically starts IEF during the night so that IEF is finished the next morning.

Procedure

1. Cut the dried IPG gels into 3-mm-wide strips with the help of a paper cutter (Fig. 4.3). Alternatively, use ready-made IPG strips.
2. Solubilize proteins with sample solubilization buffer containing 9 M urea (or 7 M urea + 2 M thiourea), 2–4 % (w/v) CHAPS, 2 % (v/v) Pharmalytes 3–10, 15 mM DTT and dilute the extract (protein concentration: $\approx$ 10 mg/ml) with rehydration solution (see Sect. 3.2) (dilution: 1+1 for micropreparative runs, and 1+19 for analytical runs, respectively) to a final volume of 400 µl for 180-mm-long IPG strips.
3. Put the required number of strip holders onto the cooling plate/electrode contact area of the IPGphor.
4. Pipette 350 µl of sample-containing rehydration solution (for 180-mm-long IPG strips) into the strip holder base. For shorter IPG strips (e.g. 110 mm) in shorter strip holders use correspondingly less liquid. Peel off the protective cover sheets from the IPG strip and slowly lower the IPG strip (gel-side-down) onto the rehydration solution. Avoid trapping air-bubbles. Cover the IPG strips with 1 ml of silicone oil and apply the plastic cover. Pressure blocks on the underside of the cover ensure that the IPG strip keeps in good contact with the electrodes as the gel swells (see Fig. 4.4).

Table 4.3. IPGphor running conditions

Temperature	20 °C
Current max.	0.05 mA per IPG strip
Sample volume	350 μl (for a 180-mm-long and 3-mm-wide IPG strip)
Voltage time	analytical run 30 V for 8 h (reswelling) 60 V for 8 h (reswelling) 200 V for 1 h Gradient from 500 to 8000 V within 1 h 8000 V, to the steady state, depending on the IPG (see next lines)
IPG 4-7	8000 V for 3 h
IPG 4-9, 3-10L, 3-10 NL	8000 V for 2 h

Note: if the sample is not included in the rehydration buffer, it can be applied in a concentrated form by pipetting it into the lateral sample application wells after the IPG strip has rehydrated.

5. Program the instrument (desired rehydration time, voltage gradient, temperature) for delayed start.
6. After the IPG gel strips have been rehydrated (which requires 6 h at least), IEF starts according to the programmed parameters (Table 4.3).

Note: it is possible to apply low voltages (30–60V) during IPG strip rehydration for better entry of high M_r proteins into the polyacrylamide gel matrix (R. Westermeier, pers. comm.; Görg et al. 1999)

7. After IEF, those IPG gel strips which are not used immediately for second dimension run or are kept for further reference are stored between two sheets of plastic film at −78 °C for up to several months (Fig. 4.5).

4
Equilibration of IPG Gel Strips

The IPG gel strips are equilibrated twice, each time for 15 min in 2×10 ml equilibration buffer (Fig. 4.5). The equilibration buffer contains 6 M urea and 30 % glycerol in order to diminish electroendosmotic effects (Görg et al. 1988) which are held responsible for reduced protein transfer from the first to the second dimension. During the second equilibration step, 260 mM iodoacetamide is added to the equilibration buffer in order to remove excess DTT (responsible for the "point streaking" in silver stained patterns) (Görg et al. 1987). The equilibrated IPG gel strips are slightly rinsed and blotted to remove excess equilibration buffer and then applied onto the second dimension SDS gel. Alternatively, tributyl phosphine may be used instead of DTT and iodoacetamide (Herbert et al. 1997).

Note: Shorter equilibration times can be applied, however, at the risk that certain proteins may not migrate out of the IPG gel strip during sample entry into the SDS-gel. In this case it is advisable to check, by staining the IPG strip after removal from the SDS gel, whether all proteins have left the IPG strip.

4.1 Apparatus and Chemicals

4.1.1 Equipment

Glass tubes (200 mm long, 20 mm i.d.), Parafilm, laboratory shaker.

4.1.2 Chemicals

Sodium dodecyl sulfate (SDS) (Serva), iodoacetamide, d,l-dithiothreitol (DTT) tris(hydroxymethyl)aminomethane (Trisma base) (Sigma), Bromophenol Blue, glycerol, urea, sodium azide (Merck).

4.2 Buffer Solutions

- **Resolving gel buffer:** 1.5 M Tris-HCl, pH 8.8 and 0.4 % (w/v) SDS (Laemmli 1970). To make 100 ml, dissolve 15.15 g Trizma base and 0.4 g SDS in about 80 ml deionized water. Adjust to pH 8.8 with 4 N HCl and fill up to 100 ml with deionized water. Add 10 mg sodium azide and filter. The buffer can be stored at 4 °C for up to 2 weeks.
- **Equilibration Buffer:** 6 M urea, 30 % (w/v) glycerol and 2 % (w/v) SDS in 0.05 M Tris-HCl buffer, pH 8.8. To make 500 ml add: 180 g urea, 150 g glycerol, 10 g SDS and 16.7 ml resolving gel buffer. Dissolve in deionized water and fill up to 500 ml. The buffer can be stored at room temperature for up to 2 weeks.
- **Bromophenol Blue Solution:** 0.25 % (w/v) of Bromophenol Blue in resolving gel buffer. To make 10 ml: dissolve 25 mg Bromophenol blue in 10 ml of resolving gel buffer. Store at 4 °C.

Procedure

1. Dissolve 100 mg of DTT in 10 ml of equilibration buffer (= equilibration buffer I). Take out the focused IPG gel strips from the freezer and place them in individual test tubes (Fig. 4.5). Add 10 ml of equilibration buffer I and 50 µl of the Bromophenol Blue solution. Seal the test tubes with Parafilm, rock them 15 min on a shaker and then pour off equilibration buffer I.
2. Dissolve 400 mg of iodoacetamide in 10 ml of equilibration buffer (= equilibration buffer II). Add equilibration buffer II and 50 µl of Bromophenol Blue solution to the test tube as above and equilibrate for another 15 min on a rocker.
3. After the second equilibration, rinse the IPG gel strip with deionized water for a second and place it on a piece of filter paper to drain off excess equilibration buffer. The strip should be turned up at one edge for a few minutes to help it drain.

5
Second Dimension: SDS-PAGE

The second dimension can be run on horizontal or vertical systems (Görg et al. 1988, 1995). For horizontal setups, laboratory-made or ready-made SDS polyacrylamide gels (0.5 mm thick on GelBond PAGfilm) are placed on the cooling plate of the horizontal electrophoresis unit. Electrode wicks or buffer strips made from polyacrylamide are then applied. The equilibrated IPG strip is simply placed gel-side-down onto the surface alongside the cathodic electrode wick or polyacrylamide buffer strip without any embedding procedure. Horizontal setups are perfectly suited for the use of ready-made gels on film supports. In the vertical setup, the equilibrated IPG gel strips are placed on top of the vertical SDS gels and embedded in agarose.

5.1
Apparatus and Chemicals

5.1.1
Equipment

Multiphor II horizontal electrophoresis apparatus, power supply (1000 V), Multitemp II thermostatic circulator, ISO-Dalt vertical electrophoresis system, gradient mixer (2 × 15 ml), glass plates with a 0.5-mm-thick U-frame (200 × 260 mm^2), plain glass plates (size 200 × 260 mm^2), clamps, GelBond PAGfilm (200 × 260 mm^2), roller (Amersham Pharmacia Biotech), heating cabinet (Heraeus, Germany), laboratory shaker (rocking platform) (GFL, Germany), electrode wicks (200 × 250 mm^2, Ultra pure) (BIORAD), Milli-Q System (Millipore).

5.1.2
Ready-Made Polyacrylamide Gels

ExcelGel SDS 12–14 and buffer strips (Amersham Pharmacia Biotech).

5.1.3
Chemicals

Repel silane (Amersham Pharmacia Biotech), ammonium persulfate, acrylamide, bis-acrylamide, glycine, Serdolit MB-1 mixed bed ion exchanger resin, sodium dodecyl sulfate (SDS), tetramethylethylenediamine (TEMED) (Serva), agarose (low-melting, low electroendosmosis), tris(hydroxymethyl)aminomethane (Tris) (Sigma), Bromophenol Blue, glycerol, sodium azide (Merck).

5.2 Reagent Solutions

5.2.1 Reagent Solutions for Horizontal SDS Gels

Acrylamide/bisacrylamide solution (30 % T, 3 % C): 29.1 % (w/v) acrylamide and 0.9 % (w/v) methylenebisacrylamide in deionized water. To make 500 ml, dissolve 145.5 g acrylamide and 4.5 g methylenebisacrylamide in deionized water and fill up to 500 ml. Add 1 g Serdolit MB-1, stir for 10 min and filter. The solution can be stored for up to 2 weeks in a refrigerator.

- **Stacking Gel Buffer:** 0.5 M Tris-HCl, pH 6.8 and 0.4 % (w/v) SDS (Laemmli 1970). To make 100 ml, dissolve 6.05 g Trizma base and 0.4 g SDS in about 80 ml deionized water. Adjust to pH 6.8 with 4 N HCl and fill up to 100 ml with deionized water. Add 10 mg of sodium azide and filter. The buffer can be stored at 4 °C for 2 weeks.

Note: In most cases, stacking gel buffer can be replaced by resolving gel buffer.

- **Resolving Gel Buffer:** 1.5 M Tris-HCl, pH 8.8 and 0.4 % (w/v) SDS (Laemmli 1970). To make 250 ml, dissolve 45.5 g Trizma base and 1 g SDS in about 200 ml deionized water. Adjust to pH 8.8 with N HCl and fill up to 250 ml with deionized water. Add 25 mg of sodium azide and filter. The buffer can be stored at 4 °C for up to 2 weeks.
- **Ammonium Persulfate Solution:** 40 % (w/v) of ammonium persulfate in deionized water. To prepare 1 ml of the solution, dissolve 0.4 g of ammonium persulfate in 1 ml of deionized water. This solution should be prepared freshly just before use.
- **Electrode Buffer Stock Solution:** to make 1000 ml of a 10x solution add: 30.3 g of Trizma base, 144 g glycine, 10.0 g SDS and 100 mg sodium azide (Laemmli 1970). Dissolve in deionized water, fill up to 1000 ml and filter. Electrode buffer stock solution can be kept at room temperature for up to 1 week. Before use, mix 100 ml of the buffer with 900 ml deionized water.

5.2.2 Reagent Solutions for Vertical SDS Gels

Acrylamide/bisacrylamide solution (30.8 % T, 2.6 % C): 30 % (w/v) acrylamide and 0.8 % (w/v) methylenebisacrylamide in deionized water. To make 1000 ml, dissolve 300 g acrylamide and 8 g methylenebisacrylamide in deionized water and fill up to 1000 ml. Add 1 g Serdolit MB-1, stir for 10 min and filter. The solution can be stored for up to 2 weeks in a refrigerator.

- **Resolving gel buffer:** 1.5 M Tris-HCl, pH 8.6 and 0.4 % (w/v) SDS (Laemmli 1970). To make 500 ml, dissolve 90.85 g Trizma base and 2 g SDS in about 400 ml deionized water. Adjust to pH 8.6 with 4 N HCl and fill up to 500 ml with deionized water. Add 50 mg sodium azide and filter. The buffer can be stored at 4 °C for up to 2 weeks.

- **Ammonium Persulfate Solution:** 10 % (w/v) of ammonium persulfate in deionized water. To prepare 10 ml of the solution, dissolve 1 g ammonium persulfate in 10 ml deionized water. This solution should be prepared freshly just before use.
- **Overlay Buffer:** buffer-saturated 2-butanol. To make 30 ml, mix 20 ml resolving gel buffer (see above) with 30 ml 2-butanol, wait for a few minutes and pipette off the butanol layer.
- **Displacing Solution:** 50 % (v/v) glycerol in deionized water and 0.01 % (w/v) Bromophenol Blue. To make 500 ml, mix 300 ml glycerol (100 %) with 200 ml deionized water, add 50 mg of Bromophenol Blue and stir for a few minutes.
- **Electrode Buffer:** to make 20 l of electrode buffer, dissolve 58 g of Trizma base, 299.6 g glycine and 20 g SDS in 19.9 l deionized water.
- **Agarose Solution:** suspend 0.5 % (w/v) agarose in electrode buffer and melt it in a boiling water bath or in a microwave oven.

5.3 Gel Casting Prodecure

Horizontal SDS gels are 0.5 mm thick and cast on GelBond PAGfilm. GelBond PAGfilms are washed 6 × 10 min with deionized water prior to use to minimize spot-streaking (Görg et al. 1987). The mold, consisting of two glass plates, one covered with the GelBond PAGfilm, the other bearing the U-frame (0.5 mm thick), is loaded in a vertical position and filled from the top with the help of a gradient mixer according to the casting procedure of Görg et al. (1980, 1986) for ultrathin pore gradient gels (Fig. 4.2).

Prior to use the glass plates are thoroughly washed with a mild detergent, rinsed with deionized water and air-dried. If new glass plates are used, pipette 1–2 ml of repel silane on the glass plate that bears the U-frame and distribute it evenly with a fuzz-free filter paper (Kimwipe). Let it dry for a few minutes, rinse again with water and let it air-dry. Repeat this procedure occasionally in order to prevent the gels from sticking to the glass plates.

5.3.1 Horizontal SDS Pore Gradient Gels on Plastic Backing

- **Gel composition:** stacking gel (50 mm long): 6 % T, 3 % C, 0.1 % SDS, 125 mM Tris-HCl pH 6.8
  resolving gel (145 mm long): 9–18 or 12–15 % T, 3 % C, 0.1 % SDS, 375 mM Tris/HCl pH 8.8
- **Gel size:** 195 × 250 × 0.5 mm^3

Casting

1. Assemble the polymerization cassette consisting of two glass plates, one covered with the GelBond PAGfilm, the other bearing the U-frame (0.5 mm thick) as described in section 3.3 for casting IPG gels (Fig. 4.2). GelBond PAGfilms should be washed for 6 × 10 min with deionized water prior to use to avoid spot streaking upon silver staining.

Table 4.4. Recipes for casting SDS pore gradient gels (9–18 or 12–15% T)

	Stacking gel	Resolving gel 9–18% T		Resolving gel 12–15% T	
	6% T	Dense solution 9% T	Light solution 18% T	Dense solution 12% T	Light solution 15% T
Glycerol (100%)	3.75 g	2.5 g	–	2.5 g	–
Stacking gel buffer	2.5 ml	–	–	–	–
Resolving gel buffer	–	2.5 ml	2.5 ml	2.5 ml	2.5 m
Acrylamide/bisacrylamide	2.0 ml	3 ml	6.0 ml	4.0 ml	5.0 ml
Deionized water	2.5 ml	2.5 ml	1.5 ml	1.5 ml	2.5 ml
TEMED (100%)	5 µl	5 µl	5 µl	5 µl	5 µl
Ammonium persulfate (40%)	10 µl	10 µl	10 µl	10 µl	10 µl
Final volume	10 ml	10 ml	10 ml	10 ml	10 ml

2. Immediately before casting add 5 µl of TEMED and 10 µl of ammonium persulfate to the gel solutions (Table 4.4). For casting an SDS gel (size 250 × 195 × 0.5 mm^3) with a stacking gel length of 50 mm, pipette 6 ml of stacking gel solution into the precooled (4 °C) mold. Then cast the pore gradient on top of the stacking gel by mixing 9 ml of the dense (12% T, 25% glycerol) and 9 ml of the light (15% T, no glycerol) solution with the help of a gradient maker similarly as described in Section 3.3 for casting IPG gels. The high glycerol concentration of the stacking gel solution allows overlaying with the pore gradient mixture without an intermediate polymerization step (Fig. 4.2).

Note: pore gradient gels as well as homogeneous SDS gels (%T = constant) may be cast for the second dimension. For homogeneous gels, the stacking gel solution is simply overlayered with resoving gel solution with the help of a pipette.

3. After pouring, leave the cassette for 15 min at room temperature to allow adequate levelling of the density gradient. Then place it in a heating cabinet at 50 °C for 30 min for polymerization. The polymerized gel can be stored in a refrigerator overnight.

5.3.2 Vertical SDS Gels for Multiple Runs

Vertical SDS is usually performed in the Dalt apparatus originally described by Anderson and Anderson (1978) because this system allows a large batch of SDS slab gels (up to 20) to be run under identical conditions. A stacking gel is usually not necessary in this system.

- **Gel composition:** no stacking gel
  resolving gel: 10 %T, 12.5 %T or 15 %T homogeneous, 2.6% C, 0.1% SDS, 375 mM Tris-HCl pH 8.6
- **Gel size:** 200 × 250 × 1 mm^3

Casting

1. The polymerization cassettes (200 × 250 mm^2) are made in the shape of books consisting of two glass plates connected by a hinge strip, and two 1-mm-thick spacers in between them.
2. Stack 23 cassettes into the gel casting box with the hinge strips to the right and vertical, interspersed with plastic sheets (e.g. 0.05-mm-thick polyester sheets). The stack is firmly held in position with a 10-mm-thick polyurethane sheet on both ends.
3. Put the front plate of the casting box in place and screw on the nuts (hand tight).
4. Connect a polyethylene tube (5 mm i.d.) to a funnel held in a ring stand at a level of about 30 cm above the top of the casting box. Place the other end of the tube in the grommet in the casting box side chamber.
5. Fill the side chamber with heavy displacing solution.
6. Immediately before gel casting add TEMED and ammonium persulfate solutions to the gel solution (Table 4.5). To cast the gels, pour the gel solution (about 1400 ml) into the funnel. Avoid introduction of any air bubbles into the tube.
7. When pouring is complete, remove the tube from the side chamber grommet so that the level of the displacing solution in the side chamber falls.
8. Very carefully pipette about 1 ml of overlay buffer onto the top of each gel in order to obtain a smooth, flat gel top surface.
9. Allow the gels to polymerize for at least three hours – or, better, overnight – at room temperature.
10. Remove the front of the casting box and carefully unload the gel cassettes from the box, using a razor blade to separate the cassettes. Remove the polyester sheets which had been placed between the individual cassettes.
11. Wash each cassette with water to remove any acrylamide adhered to the outer surface and drain excess liquid off the top surface. Discard unsatisfactory gels (uneven top surfaces, etc.) and replace them with one of the extra cassettes.
12. Gels not needed at the moment can be wrapped in plastic wrap and stored in a refrigerator (4 °C) for 1–2 days.

Table 4.5. Recipes for vertical SDS gels (10, 12.5 or 15 % T homogeneous; 2.6 % C)

	10 % T	12.5 % T	15 % T
Acrylamide/bisacrylamide (30.8 % T, 2.6 % C)	455 ml	568 ml	682 ml
Gel Buffer	350 ml	350 ml	350 ml
Glycerol (100 %)	70 g	70 g	70 g
Deionized water	532 ml	419 ml	305 ml
TEMED (100 %)	66 μl	66 μl	66 μl
Ammonium persulfate (10 %)	7 ml	7 ml	7 ml
Final volume	1400 ml	1400 ml	1400 ml

5.4 Running SDS gels

5.4.1 Horizontal SDS-PAGE

5.4.1.1 SDS-PAGE on Laboratory-made Gels

1. Fill the buffer tanks of the electrophoresis unit with electrode buffer [Laemmli (1970) buffer system]. Soak two sheets of filter paper (size 250 × 200 mm^2) in electrode buffer and put them on the cooling block (15 °C). Soak the electrode wicks (size 250 × 100 mm^2) in electrode buffer. Place them at the edges of the buffer-soaked filter papers and perform a pre-run (600 V, 30 mA) for 3 h to remove impurities from the electrode wicks. Then remove the filter papers and discard them whereas the purified electrode wicks remain in the electrode buffer tanks and are used repeatedly.
2. During the equilibration step of the IPG gel strips (see Sect. 4), open the polymerization cassette, pipette a few milliliters of kerosene onto the cooling block (15 °C) of the electrophoresis unit and put the SDS gel (gel-side-up) on it. Lay the electrode wicks on the surface of the SDS gel so that they overlap the cathodic and anodic edges of the gel by about 10 mm.
3. Place the blotted IPG gel strip(s) gel-side-down onto the SDS gel surface adjacent to the cathodic wick (Fig. 4.6). No embedding of the IPG gel strip is necessary. If it is desired to co-electrophorese molecular weight (M_r) marker proteins, put a silicone rubber frame onto the SDS gel surface alongside the IPG gel strip and pipette in 5 µl of M_r marker proteins dissolved in SDS buffer.
4. Put the lid on the electrophoresis unit and start SDS-PAGE at 200 V for about 70 min with a limit of 20 mA. When the Bromophenol Blue tracking dye has completely moved out of the IPG gel strip, interrupt the run, remove the IPG gel strip and move the cathodic electrode wick forward for 4–5 mm so that it now overlaps the former sample application area (Fig. 4.6). Then continue the run at 600 V with a limit of 30 mA until the tracking dye has migrated into the anodic electrode wick. Total running time is approximately 6 h (running distance: 180 mm) (Table 4.6). The gel is then fixed in 40 % alcohol and 10 % acetic acid for at least 1 h and stained with either silver nitrate or Coomassie Blue. Alternatively, it can be removed from the plastic backing with the help of a film remover (Amersham Pharmacia Biotech) and used for blotting.

5.4.1.2 SDS-PAGE on Ready-Made ExcelGel

Ready-made SDS gels (ExcelGel SDS Gradient 12–14, size 245 × 190 × 0.5 mm^3, on plastic backing, Tris/Tricine buffer system), in combination with polyacrylamide buffer strips are used.

1. Equilibrate the IPG gel strips as described above (see Sect. 4).
2. While the strips are being equilibrated, assemble the SDS ExcelGel for the second dimension: remove the ExcelGel from its foil package. Pipette 2–3 ml of kerosene onto the cooling plate of the horizontal electrophoresis unit (15 °C).

Table 4.6. Running conditions of SDS gels

SDS pore gradient gel (12–15 % T, 3 % C, size 250 × 195 × 0.5 mm³)

Time	Voltage	Current	Power	Temperature
75 min	200 V	20 mA	30 W	15 °C
Remove the IPG gel strips and move forward the cathodic electrode wick so that it overlaps the former IG strips application area				
5 h	600 V	30 mA	30 W	15 °C

ExcelGel SDS Gradient 12–14 (12–14 % T, 3 % C, size 245 × 190 × 0.5 mm³)

Time	Voltage	Current	Power	Temperature
40 min	200 V	20 mA	50 W	15 °C
Remove the IPG gel strips and move forward the cathodic buffer strip so that it overlaps the former IPG strip application area				
160 min	800 V	40 mA	50 W	15 °C

Running conditions for 10 vertical SDS gels (13 % T homogeneous, 2.6 % C, size 250 × 200 × 1 mm³) using Anderson's (1978) Dalt tank

Time	Voltage	Current	Power	Temperature
18 h	150 V	150 mA	50 W	15 °C

Remove the protective cover from the top of the ExcelGel and place the gel on the cooling plate, cut-off edge towards the anode. Avoid trapping air bubbles between the gel and the cooling block.

3. Peel back the plastic foil of the cathodic SDS buffer strip. Wet your gloves with a few drops of deionized water and place the buffer strip on the cathodic end of the gel. Avoid trapping air bubbles between gel surface and buffer strip.
4. Repeat this procedure with the anodic buffer strip.
5. Place the equilibrated and slightly blotted IPG gel strip(s) gel-side-down on the surface of the ExcelGel, 1–2 mm apart from the cathodic buffer strip (Fig. 4.6).
6. Press gently on top of the IPG gel strip(s) with forceps to remove any trapped air bubbles.
7. Align the electrodes with the buffer strip(s) and lower the electrode holder carefully onto the buffer strip(s).
8. Start SDS-PAGE at 200 V for about 40 min with a limit of 20 mA. When the Bromophenol Blue tracking dye has moved 4–5 mm from the IPG gel strip, interrupt the run, remove the IPG gel strip and move the cathodic buffer strip forward so that it just covers the former contact area of the IPG gel strip. Readjust the electrodes and continue electrophoresis at 800 V and 40 mA for about 160 min until the Bromophenol Blue dye front has reached the anodic buffer strip (Table 4.6.)
9. Proceed with silver or Coomassie Blue staining or with blotting.

5.4.2 Vertical SDS-PAGE

First dimension IEF and equilibration step are performed as described in Sections 3 and 4, no matter whether the second dimension is run horizontally or vertically (Görg et al. 1988, 1995). After equilibration, the IPG gel strip is placed on top of the vertical SDS gel.

Procedure

1. Fill the electrophoresis chamber with electrode buffer and turn on the cooling (15 °C).
2. Support the SDS gel cassettes [gel size 200 × 250 × 1 mm^3, Laemmli buffer system (1970)] in a vertical position to facilitate the application of the first dimension IPG strips.
3. Equilibrate the IPG gel strips as described above (see Sect. 4) and immerse them in electrode buffer for a few seconds.
4. Place the IPG gel strip on top of an SDS gel and overlay it with 2 ml of hot agarose solution (75 °C). Carefully press the IPG strip with a spatula onto the surface of the SDS gel to achieve complete contact (Fig. 4.7). Allow the agarose to solidify for at least 5 min and then place the slab gel into the electrophoresis apparatus. Repeat this procedure for the remaining IPG strips.

Note: embedding is not absolutely necessary but ensures better contact between the IPG gel strip and the top of the SDS gel.

5. Insert the gel cassettes in the electrophoresis apparatus and start electrophoresis. In contrast to the procedure of horizontal SDS-PAGE it is not necessary to remove the IPG gel strips from the surface of the vertical SDS gel once the proteins have migrated out of the IPG gel strip.
6. Run the SDS-PAGE gels overnight (100–150 V maximum setting) (Table 4.6).
7. Terminate the run when the Bromophenol Blue tracking dye has migrated off the lower end of the gel.
8. Open the cassettes carefully with a spatula. Use a spatula to remove the agarose overlay from the polyacrylamide gel.
9. Peel the gel from the glass plate carefully, lifting it by the lower edge and place it in a box of stain solution or transfer buffer, respectively. Then continue with fixing, protein staining or blotting.

5.5 Fixing

After termination of the second dimension run (SDS-PAGE), fixing is necessary to immobilize the separated proteins in the gel and to remove any non-protein components which might interfere with subsequent staining. Depending on gel thickness, the gel is submersed in the fixative for 1h at least, but usually overnight, with gentle shaking.

Widely used fixatives are either 20 % (w/v) TCA or methanolic (or ethanolic) solutions of acetic acid (e.g. methanol/distilled water/acetic acid 45/45/10). A disadvantage of the latter procedure is that low molecular weight polypeptides may not be adequately fixed. When using fluorography, fixing may be carried out in 30 % isopropyl alkohol/10 % acetic acid, as methanol can interfere with detection. Several authors also recommend aqueous solutions of glutardialdehyde for covalently cross-linking proteins to the gel matrix (e.g. for diamine silver staining).

5.6 Visualization

Silver staining methods are about 10–100 times more sensitive than various Coomassie Blue staining techniques. Consequently, they are the method of choice when very low amounts of protein have to be detected on electrophoresis gels. A huge number of silver staining protocols have been published, based on the silver nitrate staining technique of Merril et al. (1981), or the silver diamine procedure of Oakely et al. (1980).

It advisable not to pre-stain gels that will be subjected to autoradiography, as it will cause quenching (this is more a problem with silver staining than with Coomassie staining). Gels are soaked in an autoradiography reagent for 30 min at room temperature if using ^{35}S or ^{14}C labelled proteins, whereas gels containing ^{32}P labeled proteins do not need to be treated that way.

Electrophoretically separated polypeptides can be visualized by "general" stains like Coomassie Blue, silver, fluorescence or autoradiography (see chapter 5 of this book), or by "specific" stains such as glycoprotein staining or immunochemical detection methods. Whereas the "general" protein stains are carried out in the electrophoresis gel directly, immunochemical detection methods are usually performed after electrophoretic transfer ("blotting") of the separated polypeptides from the electrophoresis gel onto an immobilizing membrane (Towbin et al. 1979; Kyhse-Andersen 1984).

Since the advent of highly sensitive microsequencing techniques, it is also possible to gain N-terminal or even internal amino acid sequence information of proteins separated by 2-D PAGE. Coomassie Blue stained spots excised from the electrophoresis gel or the blotting membrane with a scalpel or a razor blade can be subjected directly, or after cleavage, to Edman degradation using a gas phase amino acid sequence analyzer, amino acid composition analysis, or mass spectrometry. Usually, a single Coomassie stained spot yields sufficient protein (1–10 μg) to obtain an N-terminal amino acid sequence. To elucidate internal sequences (which requires higher protein amounts), several identical spots from different 2-D gels have to be pooled. For mass spectrometric analyses, lower amounts of protein (one silver stained spot) are sufficient (Wilm et al. 1996). For more details see Rabilloud and Charmont (Chap. 5), Lottspeich (Chap. 9) and Corthals et al. (Chap.10).

5.7 Gel Drying

Gels must be dried before autoradiography or for permanent storage. To dry Coomassie or silver stained gels, they are first soaked for 0.5 to 2 h in glycerol, sorbitol, other polyols or polyethyleneglycol. Horizontal, 0.5-mm-thin SDS gels on plastic backing are usually impregnated in 30 % glycerol for 30 min, air-dried for 2–3 h and sealed in a plastic bag. Vertical slab gels without plastic backing are soaked in 2–3 % glycerol for several hours, laid on cellophane (for scanning), or filter paper (for autoradiography), covered with cellophane and dried using a vacuum gel dryer for several hours at 50–60 °C. Problems with gel swelling or shrinking during fixing, staining and drying are usually encountered with gels not polymerized to plastic backing.

References

Anderson NG, Anderson NL (1978) Analytical techniques for cell fractions: two-dimensional analysis of serum and tissue proteins. Anal Biochem 85: 331–354

Bjellqvist B, Ek K, Righetti PG, Gianazza E, Görg A, Westermeier R, Postel W (1982) Isoelectric focusing in immobilized pH gradients: principle, methodology and some applications. J Biochem Biophys Methods 6: 317–339

Bjellqvist B, Sanchez JC, Pasquali C, Ravier F, Paquet N, Frutiger S, Hughes GJ, Hochstrasser DF (1993) Micropreparative two-dimensional electrophoresis allowing the separation of samples containing milligram amounts of proteins. Electrophoresis 14: 1375–1378

Blomberg A, Blomberg L, Fey SJ, Larsen PM, Roepstorff P, Degand P, Boutry M, Posch A, Görg A (1995) Interlaboratory reproducibility of yeast protein patterns analyzed by immobilized pH gradient two-dimensional gel electrophoresis. Electrophoresis 16: 1935–1945

Chevallet M, Santoni V, Poinas A, Rouquié D, Fuchs A, Kieffer S, Rossignol M, Lunardi J, Garin J, Rabilloud T (1998) New zwitterionic detergents improve the analysis of membrane proteins by two-dimensional electrophoresis. Electrophoresis 19, 1901–1909

Corbett J, Dunn MJ, Posch A, Görg A (1994) Positional reproducibility of protein spots in two-dimensional polyacrylamide gel electrophoresis using immobilized pH gradient isoelectric focusing in the first dimension: an interlaboratory comparison. Electrophoresis 15: 1205–1211

Corthals GL, Molloy, MP, Herbert BR, Williams KL, Gooley AA (1997) Prefractionation of protein samples prior to two-dimensional electrophoresis. Electrophoresis 18: 317–323

Dunn MJ (1993) Gel electrophoresis of proteins. BIOS Scientific Publishers, Alden Press, Oxford

Görg A (1991) Two-dimensional electrophoresis. Nature 349: 545–546

Görg A (1993) Two-dimensional electrophoresis with immobilized pH gradients: current state. Biochem Soc Trans 21: 130–132

Görg A, Weiss W (1998) High-resolution two-dimensional electrophoresis of proteins using immobilized pH gradients. In: Celis J (ed) Cell biology. A laboratory handbook. Academic Press, New York, pp 386–397

Görg A, Postel W, Westermeier R, Gianazza E, Righetti PG (1980) Gel gradient electrophoresis, isoelectric focusing, and two-dimensional techniques in horizontal ultrathin polyacrylamide layers. J Biochem Biophys Methods 3: 273–284

Görg A, Postel W, Günther S, Weser J (1986) Electrophoretic methods in horizontal systems. In: Dunn MJ (ed) Electrophoresis '86. VCH Weinheim, pp 435–449

Görg A, Postel W, Weser J, Günther S, Strahler SR, Hanash SM, Somerlot L (1987) Elimination of point streaking on silver stained two-dimensional gels by addition of iodoacetamide to the equilibration buffer. Electrophoresis 8: 122–124

Görg A, Postel W, Günther S (1988) The current state of two-dimensional electrophoresis with immobilized pH gradients. Electrophoresis 9: 531–546

Görg A, Postel W, Friedrich C, Kuick R, Strahler JR, Hanash SM (1991) Temperature-dependent spot positional variability in two-dimensional polypeptide patterns. Electrophoresis 12: 653–658

Görg A, Boguth G, Obermaier C, Posch A, Weiss W (1995) Two-dimensional polyacrylamide gel electrophoresis with immobilized pH gradients in the first dimension (IPG-Dalt): the state of the art and the controversy of vertical versus horizontal systems. Electrophoresis 16: 1079–1086

Görg A, Obermaier C, Boguth G, Csordas A, Diaz JJ, Madjar JJ (1997) Very alkaline immobilized pH gradients for two-dimensional electrophoresis of ribosomal and nuclear proteins. Electrophoresis 18: 328–337

Görg A, Boguth G, Obermaier C, Weiss W (1988) Two-dimensional electrophoresis of proteins in an immobilized pH 4–12 gradient. Electrophoresis 19: 1516–1519

Görg A, Obermaier C, Boguth G, Weiss W (1999) Recent developments in IPG-Dalt: Wide pH gradients up to pH 12, longer separation distances and simplified procedures. Electrophoresis 20: 712–717

Hanash SM, Strahler JR, Neel JV, Hailat N, Melham R, Keim D, Zhu XX, Wagner D, Gage DA, Watson, JT (1991) Highly resolving two-dimensional gels for protein sequencing. Proc Natl Acad Sci USA 88: 5709–5713

Herbert BR, Sanchez, JC, Bini L (1997) Two-dimensional electrophoresis: the state of the art and future directions In: Wilkins MR, Williams KL, Appel RD, Hochstrasser DF (eds) Proteome research: new frontiers in functional genomics. Springer, Berlin Heidelberg New York, pp 13–33

Herbert BR, Molloy MP, Gooley AA, Walsh BJ, Bryson WG, Williams KL (1998) Improved protein solubility in two-dimensional electrophoresis using tributyl phosphine as reducing agent. Electrophoresis 19: 845–851

Islam R, Ko C, Landers T (1998) A new approach to rapid immobilised pH gradient IEF for 2-D electrophoresis. Sci Tools 3: 14–15

Kyhse-Andersen J (1984) Electroblotting from multiple gels: a simple apparatus without buffer tank for rapid transfer of proteins to nitrocellulose membranes. J Biochem Biophys Methods 10: 203–209

Laemmli UK (1970) Cleavage of structural proteins during the assembly of the head of bacteriophage T4. Nature 227: 680–685

Merril CR, Goldman D, Sedman S, Ebert H (1981) Ultrasensitive stain for proteins in polyacrylamide gels shows regional variation in cerebrospinal fluid proteins. Science 211: 1437–1438

Oakely BR, Kirsch DR, Morris NR (1980) A simplified ultrasensitive silver stain for detecting proteins in polyacrylamide gels. Anal Biochem 105: 361–363

O' Farrell PH (1975) High resolution two-dimensional electrophoresis of proteins. J Biol Chem 250: 4007–4021

O'Farrell PZ, Goodman HM, O'Farrell PH (1977) High resolution two-dimensional electrophoresis of basic as well as acidic proteins. Cell 12: 1133–1142

Posch A, Weiss W, Wheeler C, Dunn MJ, Görg A (1995) Sequence analysis of wheat grain allergens separated by two-dimensional electrophoresis with immobilized pH gradients. Electrophoresis 16: 1115–1119

Rabilloud T (1996) Solubilization of proteins for electrophoretic analyses. Electrophoresis 17: 813–829

Rabilloud T, Valette C, Lawrence JJ (1994) Sample application by in-gel rehydration improves the resolution of two-dimensional electrophoresis with immobilized pH-gradients in the first dimension. Electrophoresis 15: 1552–1558

Rabilloud T, Adessi, C, Giraudel A, Lunardi J (1997) Improvement of the solubilization of proteins in two-dimensional electrophoresis with immobilized pH gradients. Electrophoresis 18: 307–316

Righetti PG (1990) Immobilized pH gradients. Theory and methodology. Elsevier, Amsterdam

Sanchez JC, Rouge V, Pisteur M, Ravier F, Tonella L, Moosmayer M, Wilkins MR, Hochstrasser D (1997) Improved and simplified in-gel sample application using reswelling of dry immobilized pH gradients. Electrophoresis 18: 324–327

Towbin H, Staehelin T, Gordon J (1979) Electrophoretic transfer of proteins from polyacrylamide gels to nitrocellulose sheets. Procedure and some applications. Proc Natl Acad Sci USA 76: 4350–4354

Weiss W, Postel W, Görg A (1992) Application of sequential extraction procedures and glycoprotein blotting for the characterization of the 2-D polypeptide patterns of barley seed proteins. Electrophoresis 13: 770–773

Weiss W, Vogelmeier C, Görg A (1993) Electrophoretic characterization of wheat grain allergens from different cultivars involved in baker's asthma. Electrophoresis 14: 805–816

Wilkins MR, Williams KL, Appel RD, Hochstrasser DF (eds) (1997) Proteome research: new frontiers in functional genomics. Springer, Berlin Heidelberg New York

Wilm M, Shevchenko A, Houthave T, Breit, S, Schweigerer L, Fotsis T, Mann M (1996) Femtomole sequencing of proteins from polyacrylamide gels by nano electrospray mass spectrometry. Nature 379: 466–469

Appendix A: General Troubleshooting

1 General Aspects

- Quality of chemicals should be at least of analytical grade (p.a.).
- Double-distilled or deionized (Millipore) water (conductivity < 2 μS) should be used.
- Urea and acrylamide/bisacrylamide solutions should be prepared freshly.
- Deionize urea prior to use.
- Do not heat urea-containing buffers > 37 °C; otherwise protein carbamylation may occur.
- Filter all solutions. Use clean and dust-free vessels.

2 Sample Preparation

- Sample extraction buffer (lysis buffer) has to be prepared freshly. Alternatively, make small portions (1 ml) and store frozen in Eppendorf vials at –78 °C. Lysis buffer thawed once should not be refrozen.
- Add protease inhibitors during cell lysis if necessary.

Note: several protease inhibitors are inactivated by DTT and/or mercaptoethanol.

- To remove insoluble material, the protein extract should be spun for 1 h at 40000 *g*.

3 Gel Casting

- Ammonium persulfate solution should be prepared freshly. A 40 % solution of ammonium persulfate may be used for 2–3 days if stored in a refrigerator, whereas less concentrated solutions should be prepared the day you use them.
- TEMED should be stored under nitrogen and replaced every 6 months.

3.1 Observed Problems

3.1.1 Gel Did Not Polymerize Properly

Probable reasons	*Remedies*
• TEMED or ammonium persulfate too old	Replace TEMED and persulfate
• SDS gel: Tris-buffer has not been titrated with HCl	Titre stacking or resolving gel buffer with HCl (pH 6.8 or 8.8)

4 First dimension (IEF)

4.1 Observed Problems

4.1.1 Zero or Low Voltage; Voltage Readings Rapidly Change

Probable reasons	*Remedies*
• No or bad contact between electrodes (or electrode buffer strips) and the IEF gel	Check contact
• Lid not properly connected with power supply	Check connection
• Malfunction of electrodes, lid, power supply or electrodes	Check accordingly

4.1.2 Current Does Not Drop During Initial Stage of IEF

Probable reasons	*Remedies*
• Wrong electrode solutions	Preferably use deionized water
• High salt concentration in the sample	Desalt or dilute sample; replace electrode strips after 1–2 h of IEF

4.1.3 Protein Precipitation Near Sample Application Zone

Probable reasons	*Remedies*
• Sample too concentrated	Dilute sample with Lysis buffer
• Initial field strength too high	Start with low field strength (10 V/cm)
• Proteins poorly soluble	Add high amounts of urea (> 8 M) and/or proper detergent (> 1 %)

5 Second Dimension (SDS-PAGE)

5.1 Observed Problems

5.1.1 Horizontal Streaks on SDS Gel

Probable reasons	*Remedies*
• Focusing time too short (especially for high M_r proteins) or too long (proteins are not stable for an unlimited period of time)	Perform time-course to find out optimum focusing time
• Concentration of detergent too low; or inappropriate detergent used	Check concentration of detergent; test different detergents
• Urea concentration too low	Urea concentration in IEF gel: > 9 M
• Insufficient amount of DTT in the sample	Add 1 % DTT to the sample buffer
• Different oxidation forms of a single protein	Add sufficient amount of DTT
• Wrong sample application area	Check whether anodic or cathodic sample application gives better results. Alternatively apply sample by in-gel rehydration
• Artifacts due to endogenous proteolytic activity in the sample	Inactivate proteases by TCA-acetone treatment, boiling with SDS and/or adding protease inhibitors
• Interference of atmospheric carbon dioxide	Perform IEF under a layer of silicone oil flushed with argon
• M_r 68 and/or 55-kDa streaks: possible contamination due to keratin and/or albumin, or caused by mercaptoethanol	Use clean glassware only; filter all buffers (membrane filter). Keep the lab dust-free. Use DTT instead of mercaptoethanol
• Protein extract contains insoluble material which slowly redissolves during IEF	Thoroughly centrifuge the extract (40 000 *g*; 1 h)

5.1.2 Vertical streaks on the SDS gel

Probable reasons	*Remedies*
• Proteins insufficiently loaded with SDS	SDS-concentration in equilibration buffer > 1 %; equilibrate 2 × 15 min
• Glycoproteins	Use borate buffer instead of Tris buffer in SDS gel; use steep pore gradient; deglycosylate proteins
• Partial re-oxidation of free SH groups leads to disulfide bonded aggregates	Add sufficient amount of DTT to equilibration buffer, alkylate proteins; use tributylphosphine instead of DTT and iodoacetamide
• Carbamylation trains	Never heat urea containing solutions > 37 °C; deionize urea prior to use
• Endogenous proteolytic enzymes have not been inactivated during sample preparation	Inactivate proteases by TCA-acetone treatment, boiling with SDS and/or adding protease inhibitors
• "Point steaking" caused by dust particles or excess DTT	Filter all buffers (membrane filter); add iodoacetamide to the second equilibration step to remove excess DTT

5.1.3 Patterns Partially Distorted

Probable reasons	*Remedies*
• NP-40 or Triton X-100 concentration in the IEF gel too high	If possible, reduce amount of NP-40/Triton or volume of IEF gel (diameter for tube gels, width for IPG strips); use CHAPS instead of Triton/NP-40

5.1.4 No, or Only Few, Proteins Visible on SDS Gel

Probable reasons	*Remedies*
• Inappropriate sample extraction procedure (low protein concentration)	Perform protein assay (or SDS-PAGE) to estimate the protein concentration of the sample
• Wrong electrode/IEF system orientation	Check proper connection of the electrodes with the power supply
• Second dimension: Poor protein transfer from IEF gel onto SDS gel	Use detergents other than NP-40, Triton or CHAPS, or use tributylphosphine (TBP) for improved protein solubilization

Probable reasons	Remedies
• Erroneous silver staining protocol	Check protocol
• Formaldehyde oxidized	Use fresh formaldehyde
• Improper pH of developing solution	Check pH of developer
• Insufficient amount of buffer solutions	SDS gel has to be completely covered with buffer solutions during the silver staining procedure

5.1.5 Low or High M_r proteins Missing on SDS Gel

Probable reasons	Remedies
• Low M_r proteins not adequately fixed after SDS-PAGE	Use 20 % TCA or glutardialdehyde as fixative instead of 40 % alcohol and 10 % acetic acid
• High M_r proteins missing due to proteolytic degradation	Inactivate endogenous proteases in the sample

5.1.6 Diffuse Background Smear

Probable reasons	Remedies
• Endogenous proteases in the sample not inactivated	Inactivate proteases during sample preparation procedure
• Insufficient washing steps during silver staining procedure	Perform sufficient number of washing steps
• Complex formed between carrier ampholytes and SDS and/or other detergents	Fix the gel > 3 h or overnight and wash it intensely to remove SDS-carrier ampholyte complexes
• Poor quality of chemicals	Use analytical grade (or better)
• Poor water quality	Conductivity < 2 μS
• IEF apparatus may have been contaminated with proteins	Clean thoroughly after use

5.1.7 Negatively Stained Spots

Probable reasons	Remedies
• Inappropriate silver staining procedure	Change silver staining method
• Protein concentration too high	Reduce amount of protein to be loaded onto the gel or pre-stain with Coomassie Brilliant Blue

Appendix B: IPG and Horizontal SDS-PAGE Troubleshooting

1 Gel Casting

- The glass plate which bears the U-shaped frame should be treated with Repel-Silane to avoid sticking of the gel to the glass plate after polymerization.
- Glycerol (37.5 %) is incorporated into the stacking gel of horizontal SDS gels in order to diminish electroendosmotic effects.
- If SDS gels are cast onto GelBond PAGfilm, GelBondPAGfilm should be washed 6 × 10 min prior to use to avoid "spot streaking" upon silver staining.
- For proper polymerization, acidic as well as basic Immobiline starter solutions should be titrated to pH 7 with NaOH and HCl, respectively, prior to IPG gel casting.
- After polymerization, IPG gels have to be washed thoroughly (6 × 10 min) with deionized water to remove buffer ions and any unpolymerized material.
- Washed IPG gels are impregnated with glycerol (2 %) for 30 min and dried overnight at room temperature in a dust-free cabinet with the help of a fan.
- The surface of the dried IPG gels has to be covered with a sheet of plastic film prior to storage at –20 °C.

1.1 Observed Problems

1.1.1 IPG Gel Did Not Polymerize Properly

Probable reasons	*Remedies*
• IPG gels: Immobiline starter solutions have not been titrated to pH 7	Titrate to pH 7 with NaOH and HCl, respectively

1.1.2
Gel Is Released From the Plastic Support When Gel Casting Cassette Is Opened

Probable reasons	*Remedies*
• Gel has been polymerized onto the hydrophobic side of the GelBond PAGfilm	Gel has to be polymerized onto the hydrophilic side of the gel support
• Wrong support matrix (e.g. for agarose gels)	Use support matrices designed for polyacrylamide gels exclusively
• GelBondPAGfilm too old	Do not use GelBondPAGfilms which are older than 12 months

1.1.3
When Gel Casting Cassette Is Opened, Gel Sticks to Glass Plate

Probable reasons	*Remedies*
• Glass plate is not hydrophobic	Treat glass plate with RepelSilane prior to the assembly of the gel casting cassette

2
Rehydration of IPG Strips

Prior to IEF, IPG dry gels have to be cut into individual IPG strips with the help of a paper cutter. During cutting, the surface of the IPG strips has to be protected by a sheet of plastic film to avoid damage to the gel surface.

- IPG strips have to be rehydrated to their original thickness of 0.5 mm.
- IPG gel reswelling time depends on the composition of the rehydration buffer. If the rehydration solution contains high concentrations of urea (> 8 M) and detergents (> 1 %), rehydration should be performed for 6 h at least or, better, overnight.
- Rinse and blot the rehydrated IPG gel strips to remove excess rehydration solution (not to be done with the IPGphor); otherwise urea crystallization on the surface of the IPG gel strips might occur and lead to disturbed IEF patterns.
- If IPG strips are rehydrated in the IPG reswelling tray or in IPGphor strip holders, avoid trapping air bubbles between the IPG strip and the bottom of the tray or the strip holder.
- Distribute the sample solution evenly beneath the IPG gel strip. Cover the IPG strips with a layer of silicone oil during reswelling to prevent evaporation of the rehydration solution.

3 First Dimension (IEF-IPG)

3.1 Application of Rehydrated IPG Strips onto Cooling Plate of the Electrophoresis Apparatus

- Use kerosene exclusively to facilitate contact between IPG strips and cooling block.
- Use distilled water as electrode solution preferably.
- Make sure that the orientation of the IPG gel strips on the cooling block of the IEF chamber is correct (acidic end facing towards anode).

3.2 Sample Application

- Sample may be applied by in-gel rehydration or by cup loading.

3.2.1 Sample Application by Cup Loading

- When sample is applied in cups, do not apply less than 20 μl of sample solution.
- Samples may be applied near anode or cathode. In the case of unknown samples it should be checked which sample application area provides better results.
- Sample solution should not be too concentrated (max. 10 mg protein/ml) to avoid protein precipitation at the sample application point. If you are in doubt, better dilute the sample with lysis buffer and apply a larger volume instead.
- Sample solution should not contain too high concentrations of salt. Either desalt or dilute with lysis buffer and apply a larger volume instead. Apply low voltage for slow sample entry.

3.2.2 Sample Application by In-Gel Rehydration

- When using the reswelling tray for in-gel rehydration, the sample volume has to be limited to the size of the IPG strip so that no superfluous sample solution is left in the tray. For a 180-mm-long and 3-mm-wide IPG strip, the correct sample volume is about 350 μl. When the reswelling tray is used for sample application, it should also be checked whether high molecular mass, alkaline and/or membrane proteins have entered the IPG gel matrix properly.

3.3 Isoelectric Focusing

- Never pre-focus IPG strips; otherwise poor sample entry occurs due to the very low conductivity of the gels.
- Electrode strips should be humid, but not too wet. Remove superfluous liquid by blotting with filter paper.
- Keep temperature of the cooling block at 20 °C.
- For better sample entry, start IEF with a low voltage gradient (150 V for 30 min, followed by 300 V for 60 min). For micropreparative runs (cup loading) with a high sample volume (100 μl) start IEF at low voltage for several hours (200 V for 5–6 h), followed by 1500 V overnight, before voltage is raised to 3500 V.
- IPGphor: for improved sample entry, apply low voltage (30–60 V) during rehydration. Then raise voltage gradually and continue with maximum 8000 V up to the steady state (see Table 4.3).
- Focusing time depends on gel length, pH gradient and gel additives (carrier ampholytes etc.). Focusing time is shorter when separation distance is shorter, or when wide-range pH-gradients are used, or when carrier ampholytes are added to the reswelling solution.
- When running very basic and/or narrow-range IPGs, cover the IPG strips with a layer of degassed silicone oil flushed with argon.
- After completion of IEF, IPG strips should be stored frozen at –78 °C (unless immediately used for the second dimension).

3.4 Observed Problems

3.4.1 IPG Gel Strips "Burn" Near the Electrodes (Or Electrode Strips)

Probable reasons	*Remedies*
• Gel strips have dried out at the anodic or cathodic ends due to electroendosmotic flow	Add 10 % glycerol to the reswelling buffer
• Wrong electrode solutions	Use deionized water only

3.4.2 Current Does Not Drop During Initial Stage of IEF

Probable reasons	*Remedies*
• Wrong orientation of IPG strips (acidic end facing towards cathode)	Check orientation
• Wrong electrode solutions	Preferably use deionized water
• High salt concentration in the sample	Desalt or dilute sample; replace electrode strips after 1–2 h of IEF

3.4.3 Water Condensating on Gel Surface or on Lower Side Of Glass Plate Which Carries Electrodes

Probable reasons	*Remedies*
• High salt concentration in the sample	Desalt or dilute sample; add carrier ampholytes; reduce initial current
• Humidity too high	Cover IPG strips with a layer of silicone oil. If IEF is performed in the Multiphor apparatus, seal the holes in the lid with adhesive tape
• Current or power too high	Maximum 0.05 mA/IPG strip; maximum 5 W (in total)

3.4.4 Water Exudation Near Sample Application Area

Probable reasons	*Remedies*
• High salt concentration in the sample	Desalt or dilute sample; add carrier ampholytes; limit voltage (100 V) during sample entry; prolong sample entry time

3.4.5 Formation of Urea Crystals on IPG Gel Surface

Probable reasons	*Remedies*
• Temperature of cooling plate too low	Temperature optimum: 20 °C
• Very low humidity resulting in evaporation of water from the gel	Add glycerol or sorbitol to the reswelling buffer; seal the Multiphor apparatus; put wet filter paper into the IEF chamber
• Excess reswelling buffer sticking to the IPG strips	Rinse IPG strips with deionized water for a second after rehydration and blot them with moistened filter paper

3.4.6 Protein Precipitation Near Sample Application Zone

Probable reasons	*Remedies*
• Very low initial conductivity	Never pre-focus IPG gels

4 IPG Strip Equilibration and Second Dimension (SDS-PAGE)

- Equilibration time should be sufficiently long (2 × 10 min at least).
- Equilibration buffer contains Tris-HCl buffer (pH 8.8), SDS (1 %), high amounts of urea (6 M) and glycerol (30 %) for improved protein solubility and to suppress electroend-osmotic effects. In the first equilibration step, DTT (1 %) is added to the equilibration buffer for proper unfolding of proteins, and iodoacetamide (4 %) during the second step to remove excess DTT held responsible for "point streaking" during silver staining.
- For very hydrophobic and/or disulphide containing proteins, tributylphosphine may be advantageous compared to DTT and iodoacetamide.
- Horizontal SDS-PAGE: high amounts of glycerol (37.5 %) are incorporated into the stacking gel to suppress electroendosmotic effects.
- Horizontal SDS-PAGE: stacking gel length should at least exceed 25 mm.
- Protein transfer from the first dimension (IPG-Strip) to the second (SDS-gel) should be performed rather slowly (field strength: < 10 V/cm) in order to avoid streaking and to minimize loss of high M_r proteins.
- Horizontal SDS-PAGE: remove IPG gel strips from the surface of the SDS gel as soon as the Bromophenol Blue dye front has migrated 4–5 mm off the IPG gel strip. Then move the cathodic electrode wick (or buffer strip) forward so that it overlaps the area the IPG gel strip once covered.

4.1 Observed Problems

4.1.1 Horizontal Streaks on SDS Gel

Probable reasons	*Remedies*
• IPG strip has not been reswollen to its original thickness	Reswell IPG strip to a thickness of 0.5 mm prior to IEF
• Urea concentration too low	Urea concentration in IPG reswelling solution: > 8 M
• Insufficient amount of DTT in the sample or in the IPG strip rehydration solution	Add 1 % DTT to the sample buffer and 0.25 % DTT to the rehydration solution
• Different oxidation forms of a single protein	Add sufficient amount of DTT; perform IEF under a protective layer of degassed silicone oil (flushed with argon or nitrogen)
• Depletion of DTT in IPGs exceeding pH 10 due to migration of DTT towards the anode	Add an "extra paper strip" soaked in 20 mM DTT near the cathode

Probable reasons	*Remedies*
• Precipitation at the IPG strip application area	Start IEF at low field strength (< 10 V/cm)
• IPG strip rehydration time too short	Rehydrate > 6h (or overnight)

4.1.2
Vertical Streaks on SDS Gel

Probable reasons	*Remedies*
• Horizontal SDS-PAGE: stacking gel length too short	Effective stacking gel length > 25 mm
• GelBond PAGfilm had not been washed and causes "spot streaking"	Wash GelBond PAGfilm prior to SDS gel casting

4.1.3
Patterns Partially distorted

Probable reasons	*Remedies*
• IPG strip has not been removed from the surface of the horizontal SDS gel after protein transfer from the IPG strip onto the SDS gel	Remove IPG strip after protein transfer from the IPG strip onto the SDS gel
• Electrode wick (or buffer strip) has not been moved forward to cover the former IPG strip application area	Cover the IPG strip application area with electrode wick after the IPG strip has been removed from the surface of the horizontal SDS gel

4.1.4
Uneven Migration of Bromophenol Blue Front

Probable reasons	*Remedies*
• Horizontal SDS-PAGE: large air bubbles trapped between GelBond PAGfilm and cooling plate; air bubble trapped in cooling plate	Check for and remove air bubbles
• Horizontal SDS-PAGE: improper contact between electrode wicks (or buffer strips) and surface of the SDS gel	Check electrode contact

Probable reasons	*Remedies*
• SDS pore gradient gel: gel casting device had not been levelled horizontally during gel casting and polymerization	Level gel casting device

4.1.5
No, or Only Few, Proteins Visible on SDS Gel

Probable reasons	*Remedies*
• Insufficient sample entry into IPG strip	Start IEF with low field strength
• Acidic end of IPG strip facing towards the cathode; anode connected with the cathodic outlet of the power supply	Make sure that the orientation of the IPG gel strips on the cooling block of the IEF chamber is correct. Check proper connection of the electrodes with the power supply
• Second dimension: Poor protein transfer from IPG strip onto SDS gel	Perform protein transfer at low field strength (< 10 V/cm); use detergents other than NP-40, Triton or CHAPS, or use tributylphosphine (TBP) for improved protein solubilization
• Second dimension: IPG strip applied with GelBond side onto the SDS gel	The surface of the IPG strip must be in contact with the surface of SDS gel
• Large air bubbles between IPG strip and surface of SDS gel	Remove air bubbles by pressing on upside the IPG strip with forceps

4.1.6
Low or High M_r Proteins Missing on SDS Gel

Probable reasons	*Remedies*
• Poor transfer of high M_r proteins from IPG strip onto SDS gel	Perform protein transfer at low field strength (< 10 V/cm)

CHAPTER 5

Detection of Proteins on Two-Dimensional Electrophoresis Gels

T. Rabilloud[1] and S. Charmont[1]

1 Introduction

In 2-D electrophoresis, the second dimension is most commonly the SDS separation. Consequently, detection on 2-D gels uses the methods developed for SDS gels. However, some issues which are often not so important in 1-D electrophoresis must be carefully taken care of in proteome-based approaches. Apart from the detection threshold, which is always a very important parameter, these issues include the dynamic range of detection, the linearity, the reproducibility between different staining experiments and the inter-protein variability. In addition, the interfacing of the staining methods with subsequent protein analysis methods (microsequencing, amino acid analysis, mass spectrometry) is more and more often an important parameter.
Detection methods of proteins on 2-D gels can be divided into five types:

1. Detection by organic dyes
2. Detection by differential precipitation of salts
3. Detection by metal ion reduction
4. Detection by fluorescence
5. Detection of radioactive isotopes

It must also be kept in mind that the detection issue and the solubility issue are strongly interconnected in proteome approach. If very sensitive and general detection and identification methods are available, very minute amounts of proteins can be analyzed and solubility problems are minimized. Conversely, any progress with protein solubility allows us to load larger amounts of proteins on the gels, and thus to detect and analyse less and less abundant proteins.

2 Detection by Organic Dyes

Various dyes, generally arising from the textile industry, can be used to stain proteins on 2-D gels. To be able to stain proteins differentially from the background (polyacrylamide gel), the dye molecules must have a higher binding to proteins.

[1] CEA- Laboratoire de Bioénergétique Cellulaire et Pathologique, EA 2019, DBMS/BECP CEA-Grenoble, 17 rue des martyrs, F-38054 GRENOBLE CEDEX 9, France

This binding can be achieved by covalent attachment (reactive dyes) on the reactive groups of the proteins (amine or thiol groups) (e.g. in Bosshard and Datyner 1977). As this covalent attachment generally changes the charge of the molecule and thus its pI, the grafting process must take place either between the IEF and SDS dimensions or after the SDS electrophoresis. In both cases, amine- or thio-containing compounds (e.g. Tris, carrier ampholytes) must be removed prior to dye coupling by selective fixation of proteins in the support gel. This cumbersome process, coupled with generally low coupling yields and thus poor sensitivity (detection threshold 5–10 μg), has made these methods rather unattractive. The situation is completely different when dye binding to the proteins is made by adsorption (non-covalent interactions). In this case, the staining process takes place after electrophoresis, generally after a short fixation to remove compounds interfering with dye binding (especially SDS). In most methods, the staining process is a regressive staining, which means that the gel is saturated with the dye solution and then destained by an appropriate process which takes profit from the higher affinity of the dye for the proteins to destain the acrylamide gel first. Numerous dyes have been described for staining of proteins, but the most popular ones belong to the triphenylmethane family. Among this family, the Brilliant Blue G and R dyes (respectively C.I. 42655 and 42660) also known as Coomassie Blue dyes are the most popular due to their high extinction coefficient and high affinity for proteins, resulting in good detection limits (ca. 1 μg) (Fazekas de Saint Groth et al. 1963). Moreover, these dyes are anionic dyes, so that they are most efficiently used in acidic media, which are also the most efficient to fix proteins in gels and remove interfering compounds.

A typical staining protocol with these dyes is indicated below:

Protocol 1

1. After electrophoresis, fix the gels at least 1 h in a solution containing 300 ml ethanol and 50 ml acetic acid per liter (30 % ethanol, 5 % acetic acid).
2. Stain for 1h in the same solution containing 0.2 % Brilliant Blue G or R.
3. Recover the stain for further use, and rinse briefly with water.
4. Destain in 30 % ethanol, 5 % acetic acid.

Note: when sequencing is planned after staining, the above protocol must be slightly altered. The acetic acid concentration in the fixing and staining baths must be reduced to 0.5 %. Destaining is carried out in 30 % ethanol only.

A number of variegated protocols have been described in the literature over the years, varying in the concentration of the acid and in the nature and concentration of the alcohol. In addition to its good sensitivity as a dye, Brilliant Blue has gained popularity because it stains proteins rather uniformly, with the notable exception of proteins with extreme pIs (often not analysed by 2-D gels) and of glycoproteins.

The main limitation of this staining process is linked to its sensitivity. As long as high quantities of proteins were needed for identification, staining with dyes was sensitive enough. It is still widely used in this respect, as the sensitivity matches the requirements for peptide microsequencing. However, the detection limit is not adequate for staining of analytical 2-D gels. In addition, the regressive staining process (overstaining followed by destaining) makes the batch-to-batch reproducibility difficult to achieve. Three different approaches have been suggested to overcome this drawback. The first one is to stain proteins during their migration in the SDS dimension (Schõgger et al. 1988). The gels are ready to use immediately after the migration, the proteins being of a deeper blue tint than the background. As the proteins are not fixed by this method, they can be easily eluted. However, eluted proteins still bind dye molecules, so that they cannot be used as such in mass spectrometry of whole proteins. The main limitation of this method is a heavily stained background, so that quantitation is difficult. Moreover, the detection limit is not as good as with protocol 1.

The second way to achieve a progressive staining is to use a dilute solution of the dye and to stain for a long time, so that dye molecules will adsorb to the proteins, while the background is kept clear by the low overall concentration of the dye (30 mg/l) (Chen et al. 1993). This approach is very efficient, but a large relative volume of staining solution compared to gel volume must be used, to ensure that sufficient dye molecules will be present to saturate the proteins present in the gel.

A very elegant extension of this approach was devised by Neuhoff's group (Neuhoff et al. 1985, 1988). In this approach, advantage is taken of the fact that Brilliant Blue G (C.I. 42655 or Coomassie Blue G-250) can form microprecipitates in acidic media containing ammonium sulfate, making a colloidal staining possible. This is however restricted to Brilliant Blue G and does not apply to Brilliant Blue R. Consequently, there is a very low concentration of free dye, so that background staining is minimal and can most often be removed by a simple water wash. However, the microprecipitates act as a reservoir of dye molecules, so that enough dye is present to occupy all the binding sites on all the proteins, provided that staining is prolonged to the steady state. In the former version of the staining process (Neuhoff et al. 1985), steady state was obtained after several days of staining. In the improved version, the steady-state is reached between 24 and 48 h, which is long but still practicable. Apart from the very good reproducibility afforded by this type of staining, an additional benefit in sensitivity is due to the fact that staining goes to its end-point in a solvent containing the minimal amount of alcohols, so that more binding sites for the dye are available on the proteins. In fact, the presence of high concentration of alcohols, required for classical destaining, decreases the strength of the hydrophobic interactions involved in dye-protein binding. Conversely, the presence of ammonium sulfate in the colloidal staining solution increases the strength of hydrophobic interactions, resulting in better sensitivity. The increase in sensitivity can easily be seen by comparing Fig. 5.1A, 5.1B, and a typical colloidal staining protocol is given in protocol 2

Protocol 2

1. After electrophoresis, fix the gels 3 × 30 min in 30 % ethanol, 2 % (v/v) phosphoric acid (Note 1)
2. Rinse 3x 20 min in 2 % phosphoric acid.
3. Equilibrate for 30 min in a solution containing 2 % phosphoric acid, 18 % (v/v) ethanol and 15 % (w/v) ammonium sulfate (Note 2).

4. Add to the gels and solution 1 % (v/v) of a solution containing 20 g of Brilliant Blue G per liter (Note 3). Let the stain proceed for 24 to 72 h.
5. If needed, destain the background with water. Avoid alcohol-containing solutions.

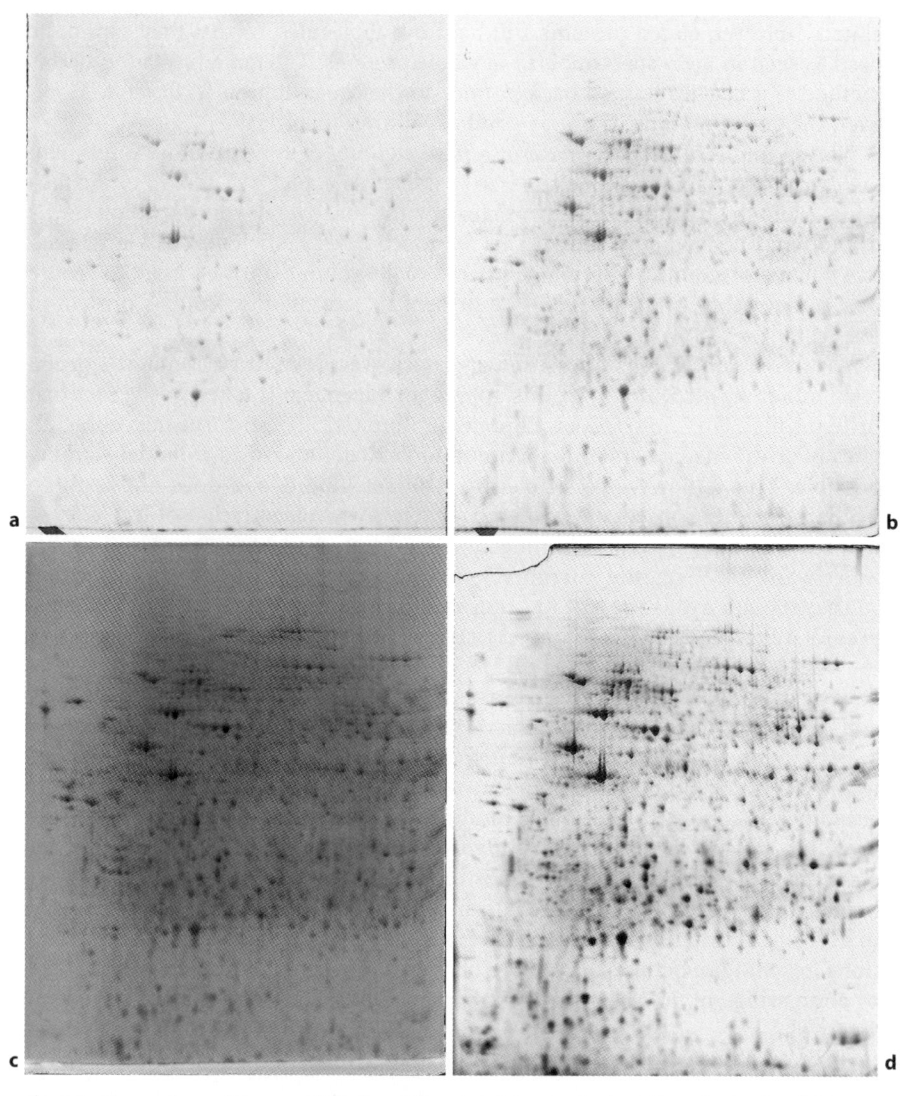

Fig. 5.1. Comparison of the sensitivity for different staining methods. 400μg of mitochondrial proteins were loaded and separated by 2-D electrophoresis. The resulting gels were stained with: Brilliant Blue G in acid-alcohol medium (*A*), colloidal Brilliant Blue G (*B*), imidazole zinc (*C*), and silver as described in protocol 8 (*D*)

Note 1: The concentrated phosphoric acid used is 85 % phosphoric acid. Percentages are expressed in volumes. For example, the fixing solution contains 20 ml of 85 % phosphoric acid per liter.

Note 2: The solution is prepared as follows. For 1 l of solution, place 500 ml of water in a flask with magnetic stirring. Add 20 ml of 85 % phosphoric acid, then 150 g of ammonium sulfate. Let dissolve, transfer into a graduated cylinder and adjust to 800 ml with water. Add 20 ml additional water, retransfer into the flask with stirring and add 180 ml ethanol while stirring.

Note 3: The Brilliant Blue G solution is prepared by dissolving 2 g of pure Brilliant Blue (e.g. Serva Blue G) in 100 ml of *hot* water with stirring. Dissolution is complete after 30 min. Let the solution cool, then add 0.2 g/l sodium azide as a preservative. Store at room temperature.

On the whole, staining with dyes is compatible with most subsequent protein analysis methods, provided that alterations to the protocols given here are made to maximize the yield in the subsequent protein microanalysis step (see Chapter 9 on microsequencing). We have, however, experienced difficulties when trying to perform internal peptide sequencing from colloidal blue-stained spots, while analysis by mass spectrometry (peptide mass fingerprinting) did not give any problem. The main limitation, even for colloidal staining, lies in rather limited sensitivity.

3 Detection by Differential Precipitation of Salts

The basis of the use of differential precipitation of salts as a staining method lies in the fact that protein-bound salts (e.g. dodecyl sulfate or heavy cations) are less reactive than the free salt in the gel. Consequently, precipitation of an insoluble salt is slower on the sites occupied by proteins than in the gel background, provided that proteins have an affinity for at least one of the components of the insoluble salt. This differential precipitation speed can then be used to devise negative staining methods where the proteins stay translucent while the background becomes opaque through salt precipitation. In such staining protocols, the limiting factor is the affinity of the proteins for one of the ions. The higher and the more general the affinity, the better the stain. A general affinity is required to avoid stains highly variable from one protein to another. A high affinity will give rise to a high sensitivity stain, as a minute amount of proteins will be able to delay salt precipitation sufficiently to give a signal.

The first insoluble salt used in negative staining was potassium dodecyl sulfate (Nelles and Bamburg 1976). The staining protocol was quite simple and only involved dipping the SDS-containing gel in a concentrated KCl solution for a minute or so, and then rinsing with water to stop precipitation and thus avoid masking of the proteins by slow potassium dodecyl sulfate precipitation.

Owing to the rather low affinity of proteins for potassium, the stain relies only on the affinity for dodecyl sulfate. The sensitivity is therefore rather low, much inferior to staining with Brilliant Blue, for example. In addition, removal of dodecyl sulfate and potassium from the proteins is not very easy. This brings to attention the fact that the proteins zones are translucent but not devoid of the components involved in the stain, which are in fact bound to the proteins. Quite often, these components must be removed for subsequent analysis of the proteins (blotting, microsequencing, mass spectrometry).

A major advance in sensitivity and versatility of negative stains was achieved by using heavy divalent cations (copper, zinc) for making a precipitate with dodecyl sulfate (Lee et al. 1987 ; Dzandu et al. 1988; Adams and Weaver 1990). Sensitivities as good as those achieved with colloidal Brilliant Blue were easily obtained within min, as shown in protocol 3. Moreover, destaining was easily achieved by the use of chelators such as EDTA.

Protocol 3 (Lee et al. 1987)

1. After electrophoresis, rinse gels in water for 2–3 min.
2. Dip the gel in 0.3 M cupric chloride, and incubate for 5 min.
3. Rinse in several changes of water.

A further increase in sensitivity was achieved by a modification of a zinc stain, in which the precipitated salt is no longer zinc dodecyl sulfate, but a complex salt of zinc and imidazole (Ortiz et al. 1992; Fernandez-Patron et al. 1995). The stain protocol is still simple, but optimal results depend on careful timing of the steps (Ferreras et al. 1993). A variegated protocol intended to minimize these difficulties is shown in protocol 4.

Protocol 4

1. After electrophoresis, transfer the gel(s) for 5 min in 1% sodium carbonate (Note 1).
2. Replace the solution by 0.2 M imidazole containing 0.1% SDS. Incubate for 15–20 min.
3. Rinse the gel for 10 s in water (Note 2).
4. Transfer to 0.2 M zinc acetate and agitate for 40–70 s (Note 3). The gel should become white and opaque at this stage.
5. Rinse thoroughly with water (2×2 min + 2×5 min + at least 3×15 min).

Note 1: Gels can be grouped (up to four per dish) for the first two steps (carbonate and imidazole). The change of solution is made by pressing a plastic sheet on the gel pile with one hand, inclining the whole assembly to pour off the liquid while keeping the gels in the vessel. The following solution is poured on the gels with the plastic sheet in place to avoid any disturbance of the gels. The sheet is then removed, rinsed and kept in a vessel with water.

Note 2: The best staining setup comprises the vessel containing the gels in imidazole-SDS, one dish with water, one with the zinc solution (250–300 ml per gel) and a last dish with water for collecting stained gels. Gels are handled with powder-free nitrile gloves. One gel at a time is transferred first in water, then in the zinc solution, then in the collecting dish. The water for rinsing and the zinc solution are changed for each gel.

Note 3: Too short a time in the zinc solution decreases sensitivity, but so does too long a time, as the acidity of the zinc salt dissolves the precipitate (Ferreras et al. 1993). To enhance the robustness of the method, we recommend the use of zinc acetate in place of the original zinc sulfate. Zinc acetate is more basic than zinc sulfate and gives more flexibility in the timing.

Detection sensitivities approach those reached with silver staining (cf Fig. 5.1C, 5.1D), with the great advantage that no fixing compound except zinc is present. Proteins stay precipitated as colorless zinc salts in the gel, which can be kept for several days at 4 °C without changes, provided that they have been thoroughly rinsed with water prior to storage. Zinc can be removed with any of the solutions described above when needed for further processing (e.g. in Fernandez-Patron et al. 1995).

Owing to its simplicity, speed and versatility, zinc-imidazole staining is a highly valuable tool for proteomics. Its drawbacks lie mainly:

1. In some difficulties with the recognition of the translucent zones by eye. This leads to difficulties in spot excision for protein processing.
2. In the fact that the stain is poorly linear if at all, so that it cannot be used as a general tool for looking for quantitative variations on 2-D gels.
3. Poor detection is often experienced for glycoproteins (e.g. fetuin in Courchesne et al. 1997) and low molecular weight proteins.

4 Detection by Metal Ion Reduction (Silver Staining)

Although it has been described before zinc staining, silver staining shares common principles with staining by differential precipitation. Here again, the driving force for staining is the affinity of proteins for the cation, here silver. However, many substances, among which SDS, chloride and amino acids show high affinities for silver, so that they must be removed prior to contacting proteins with silver ion. This means that a fixation step is required for silver staining. Consequently, subsequent elution of whole proteins from the gel after silver staining is hardly possible.

In the simplest form of silver staining, the fixed gel is impregnated with silver nitrate, rinsed briefly and immersed in a dilute photographic developer, resulting in a negative silver stain of poor sensitivity (Merril and Goldman 1984). The trick for increasing the sensitivity is to take advantage of the affinity of proteins for silver ions and of the strong autocatalytic character of silver reduction. This is achieved by the use of a very weak developer (dilute solutions of formaldehyde at a precise pH) and by the use of various compounds (sensitizers) between fixation and silver impregnation. Sensitizers bind to proteins and react with or bind to silver. By this fact, silver reduction is primed where the sensitizer is bound (i.e. at the protein level) and a positive image is obtained due to autocatalysis. This process has been reviewed elsewhere (Rabilloud 1990; Rabilloud et al. 1994).

Owing to various fixation and sensitization processes, more than 80 different silver staining methods have been described in the literature. They can be divided into two families defined by the silvering agent, which can be either a solution of silver nitrate, or a silver-ammonia complex. In the former case (silver nitrate), the developer is formaldehyde in an alkaline carbonate solution. In the latter case (silver–ammonia), the developer is formaldehyde in a weakly acidic solution, generally containing citric acid.

Although numerous silver staining kits are available on the market, none of them is as sensitive as home-made protocols can be. The reason for this is quite simple. All high-performance silver staining methods use at least one solution of limited shelf life (1 week for thiosulfate, a month or so for ammonia, a few days for silver-ammonia complex). This make these protocols unmarketable as kits for simple production and storage reasons. It is therefore much more efficient to have a good laboratory silver staining protocol and to stick to it.

Four different silver staining methods are given in protocols 5 to 8. The criteria for selecting a method among these four are as follows:

1. Protocols using fixation with an aldehyde (6 and 7) are more sensitive and the sensitivity is more uniform from one protein to another. They are also slightly more reproducible. However, aldehyde fixation precludes any further analysis of the stained protein, while analysis by peptide mass fingerprinting can be performed on silver-stained gels without aldehyde fixation (protocols 5 and 8).
2. Protocols using very short steps (1 min or less, e.g. protocol 5) are less reproducible than protocols using long steps only. They are, however, much faster.
3. Protocols using ammoniacal silver have a slightly higher sensitivity and give darker hues than those based on silver nitrate. They are also much more sensitive for basic proteins, but less for highly acidic proteins. They are, however, more prone to negative staining phenomena (hollow or „doughnut“ spots) (especially protocol 8). In addition, adequate results are obtained only when thiosulfate is included in the gel matrix at the polymerization step (Hochstrasser and Merril 1988), so that ready-made gels cannot be used. Further, some gel systems (e.g. Tricine-based systems) give very high backgrounds with ammoniacal silver. Last but not least, the performances of ammoniacal silver methods strongly depend on the ammonia-silver ratio (Eschenbruch and Burk 1982).

   A precise and reproducible ammonia-silver ratio is rather difficult to achieve over a long time, as ammonia escapes from ammonium hydroxide solutions upon storage either in the lab or on the shelves of the supplier. A good solution is to use a titrated ammonium hydroxide solution (Aldrich) and to work very close to the precipitation limit of silver hydroxide. When the ammoniacal silver cannot be prepared with the defined volumes any longer, the ammonia stock solution must be changed for a new one (see Note 9 below).
4. For gels supported on a plastic backing, best results are obtained with protocols 5 and 7.

Protocol 5:

Fast Silver Nitrate Staining (see Note 1 first) This protocol is based on the protocol of Blum et al. (1987), with modifications (Rabilloud 1992).

1. Fix the gels ($\geq 3\times 30$ min) in 5 % acetic acid/30 % ethanol (v/v).
2. Rinse in water for 4×10 min.
3. To sensitize, soak gels for 1 min (1 gel at a time) in 0.8 mM sodium thiosulfate (See Notes 2 and 3).

4. Rinse 2× 1 min in water (see Note 3) .
5. Impregnate for 30–60 min in 12 mM silver nitrate (0.2g/l). The gels may become yellowish at this stage.
6. Rinse in water for 5–15 s (see Note 4).
7. Develop image (10–20 min) in 3 % potassium carbonate containing 250 μl formalin and 125 μl 10 % sodium thiosulfate per liter (see Note 5).
8. Stop development (30–60 min) in a solution containing 40 g of Tris and 20 ml of acetic acid per liter.
9. Rinse with water (several changes) prior to drying or densitometry.

Protocol 6: Long Silver Nitrate Staining (see Note 1 first)

1. Fix the gels in 5 % acetic acid/30 % ethanol (v/v) (3 × 30 min).
2. Sensitize overnight in 0.5 M potassium acetate containing 250 ml ethanol, 3 g potassium tetrathionate and 5 g glutaraldehyde per liter (see Note 6).
3. Rinse in water (6 × 20 min).
4. Impregnate for 1–2 h with 12 mM silver nitrate containing 0.5 ml formalin per liter.
5. Rinse with water for 5–10 s (see Note 4).
6. Develop image (10–20 min) in 3 % potassium carbonate containing 250 μl formalin and 125 μl 10 % sodium thiosulfate per liter (see Note 5).
7. Stop development (30–60 min) in a solution containing 40g of Tris and 20 ml of acetic acid per liter.
8. Rinse with water (several changes) prior to drying or densitometry.

Protocol 7:

Ammoniacal Silver Staining with Aldehyde Fixation (see Note 1 first). This protocol is based on the original protocol of Eschenbruch and Bürk (1982) and Mold et al. (1983), with modifications (Rabilloud 1992). For optimal results, alterations must be made at the level of gel casting. Thiosulfate is added at the gel polymerization step (Hochstrasser and Merril 1988). Practically, the initiating system is composed of 1 μl of TEMED, 7 μl of 10 % sodium thiosulfate solution and 8 μl of 10 % ammonium persulfate solution per milliliter of gel mix. This ensures correct gel formation and gives minimal background upon staining.

1. Immediately after electrophoresis, place the gels in water, and let rinse for 5–10 min.
2. Soak gels in 20 % ethanol containing 10 % (v/v) formalin for 1 h.
3. Rinse 2 × 15 min in water.
4. Sensitize overnight in 0.05 % naphtalene disulfonate (see Note 7).
5. Rinse 6 × 20 min in water.
6. Impregnate for 30–60 min in the ammoniacal silver solution. (see Notes 8 and 9).
7. Rinse 3 × 5 min in water.
8. Develop image (5–10 min) in 350μM citric acid containing 1 ml formalin per liter.
9. Stop development in 2 % acetic acid (v/v)/0.5 % ethanolamine (v/v). Leave in this solution for 30–60 min.
10. Rinse with water (several changes) prior to drying or densitometry.

Protocol 8:
Ammoniacal Silver staining Without Aldehyde Fixation (see Note 1 first). This protocol is derived from the above-described protocol. Thiosulfate is also added to the gel during polymerization. After electrophoresis, silver staining processes as follows:

1. Fix in 5 % acetic acid/30 % ethanol and 0.05 % naphtalene disulfonate for 1 h.
2. Fix overnight in the same solution.
3. Rinse 7 × 10 min in water.
4. Impregnate for 30–60 min in the ammoniacal silver solution (see Note 8). The solution is prepared as described in protocol 7.
5. Rinse 3 × 5 min in water.
6. Develop image (5–10 min) in 350 μM citric acid containing 1 ml formalin per liter.
7. Stop development in 2 % acetic acid (v/v)/0.5 % ethanolamine (v/v). Leave in this solution for 30–60 min.
8. Rinse with water (several changes) prior to drying or densitometry.

Note 1: General practice: Batches of gels (up to five gels per box) can be stained. For a batch of three to five medium-sized gels (e.g. 160 × 200 × 1.5 mm), 1 l of the required solution is used, which corresponds to a solution/gel volume ratio of 5 or more; 500 ml of solution is used for one or two gels. Batch processing can be used for every step longer than 5 min, except for image development, where one gel per box is required. For steps shorter than 5 min, the gels should be dipped individually in the corresponding solution.

For changing solutions, the best way is to use a plastic sheet. The sheet is pressed on the pile of gels with the aid of a gloved hand. Inclining the entire setup allows the emptying of the box while keeping the gels in it. The next solution is poured with the plastic sheet in place, which prevents the solution flow from breaking the gels. The plastic sheet is removed after the solution change and kept in a separate box filled with water until the next solution change. This water is changed after each complete round of silver staining. The above statements are not true when gels supported by a plastic film are stained. In this case, only one gel per dish is required. A setup for multiple staining of supported gels has been described elsewhere (Granier and De Vienne 1985)

When gels must be handled individually, they are manipulated with gloved hands. The use of powder-free, nitrile gloves is strongly recommended, as powdered latex gloves are often the cause of pressure marks. Except for development or short steps, where occasional hand agitation of the staining vessel is convenient, constant agitation is required for all the steps. A reciprocal ("ping-pong") shaker is used at 30–40 strokes per minute.

Dishes used for silver staining can be made of glass or plastic. It is very important to avoid scratches in the inner surface of the dishes, as scratches promote silver reduction and thus artefacts. Cleaning is best achieved by wiping with a tissue soaked with ethanol. If this is not sufficient, use instantly prepared Farmer's reducer (50 mM ammonia, 0.3 % potassium ferricyanide, 0.6 % sodim thiosulfate). Let the yellow-green solution dissolve any trace of silver, discard, rinse thoroughly with water (until the yellow color is no longer visible), then rinse with 95 % ethanol and wipe.

Formalin stands for 37 % formaldehyde. It is stable for months at room temperature. However, solutions containing a thick layer of polymerized formaldehyde must not be used. Never put formalin in the fridge, as this promotes polymerization. 95 % ethanol can be use instead of absolute ethanol. Do not use denatured alcohol. It is possible to purchase 1M silver nitrate ready-made. The solution is cheaper than solid silver nitrate on a silver weight basis. It is stable for months in the fridge.

Last, but not least, the quality of water is critical. Best results are obtained with water treated with ion exchange resins (resistivity higher than 15 mega ohms/cm). Distilled water gives more erratic results.

Note 2: 0.8 mM sodium thiosulfate corresponds to 2 ml/l of 10 % sodium thiosulfate (pentahydrate). The 10 % thiosulfate solution is made fresh every week and stored at room temperature.

Note 3: The optimal setup for sensitization is the following: prepare four staining boxes containing respectively the sensitizing thiosulfate solution, water (two boxes), and the silver nitrate solution. Put the vessel containing the rinsed gels on one side of this series of boxes. Take one gel out of the vessel and dip it in the sensitizing and rinsing solutions (1 min in each solution). Then transfer to silver nitrate. Repeat this process for all the gels of the batch. A new gel can be sensitized while the former one is in the first rinse solution, provided that the 1-minute time is kept (use a bench chronometer). When several batches of gels are stained on the same day, it is necessary to prepare several batches of silver solution. However, the sensitizing and rinsing solutions can be kept for at least three batches, and probably more.

Note 4: This very short step is intended to remove the liquid film of silver solution carried over with the gel.

Note 5: When the gel is dipped in the developer, a brown microprecipitate of silver carbonate should form. This precipitate must be redissolved to prevent deposition and background formation. This is simply achieved by **immediate** agitation of the box. Do not expect the appearance of the major spots before 3 min of development. The spot intensity reaches a plateau after 15–20 min of development; then background appears. Stop development at the very beginning of background development. This ensures maximal and reproducible sensitivity.

Note 6: The sensitization solution is prepared as follows: dissolve potassium tetrathionate (3 g) in 500 ml water. After complete dissolution, add 100 ml of 5 M potassium acetate, 250 ml ethanol, and 10 ml of 50 % glutaraldehyde. Fill up to 1 liter with water. Use a good grade of glutaraldehyde (avoid yellowish solutions), but a microscopy grade is not needed. Keep the glutaraldehyde stock solution at 4 °C.

Note 7: 2,7 naphtalene disulfonate, disodium salt (Acros), is the standard. However, 1,5 naphtalene disulfonic acid or its sodium salt (both available from Aldrich) can also be used with similar results.

Note 8: The solution or the gels may become yellowish or slightly brown during this step. Dark brown colors or heavy precipitates are indicative of a severe staining problem (mistake in a solution, insufficient rinses, poor quality of the formalin used for fixation). Very poor results are to be expected in this case.

Note 9: The ammoniacal silver solution is prepared as follows: for ca. 500 ml of staining solution, 475 ml of water are placed in a flask with strong magnetic stirring. First, 7 ml of 1 N sodium hydroxide is added, followed first by 7.5 ml of 5 N ammonium hydroxide (Aldrich) and then by 12 ml of 1 N silver nitrate. A transient brown precipitate forms during silver nitrate addition. It should disappear in a few seconds after the end of silver addition. Persistence of a brown precipitate or color indicates exhaustion of the stock ammonium hydroxide solution. Attempts to correct the problem by adding more ammonium hydroxide solution generally lead to poorer sensitivity.

The ammonia-silver ratio is a critical parameter for good sensitivity (Eschenbruch and Bürk 1982). The above proportions give a ratio of 3.1, which is one of the lowest practicable ratios. This ensures highest sensitivity and good reproducibility control of the ammonia concentration through silver hydroxide precipitation.

Flasks used for preparation of silver-ammonia complexes and silver-ammonia solutions must not be left to dry out, as explosive silver azide may form. Flasks must be rinsed at once with distilled water, while used silver solutions should be put in a dedicated waste vessel containing either sodium chloride or a reducer (e.g. ascorbic acid) to precipitate silver.

The sensitivity of these protocols is consistently in the low nanogram range. Silver staining is ca. 100 times more sensitive than classical Brilliant Blue staining, 10 times more sensitive than colloidal Brilliant Blue staining and ca. 2 times more sensitive than zinc staining, with a much better contrast (see Fig. 5.1). With the protocols described above, the protein to protein variation is minimal (except for protocol 8) and glycoproteins or low molecular weight proteins are correctly detected. The linear range covers more than 1 order of magnitude, with a slope generally smaller than 1 (see Fig. 5.2). This means that variations detected with silver staining are always minimized. However, there is a strong toe to the staining curve, so that rare proteins are generally overdetected compared to more abundant ones. This must be kept in mind while assessing the purity of a protein.

Owing to protein fixation prior to staining and to the fact that development uses formaldehyde (a very efficient protein crosslinker), interfacing of silver-stained gels with identification methods is generally difficult. Fixation with aldehydes precludes any further analysis. When it is avoided, good results are often obtained with methods where the protein is digested or hydrolyzed (e.g. peptide mass fingerprinting). This success can be attributed, at least in part, to the fact that silver staining is rather a surface phenomenon (see Heukeshoven and Dernick 1985). This means that the inner part of a silver-stained gel is generally transparent and that the proteins present here has not been damaged by the staining process. Protocols 5 and 8 have been successfully used with peptide mass fingerprinting. However, proteins stained with protocol 5 may fail to give any result with this method on some occasions, especially if staining is prolonged too much. This is tentatively attributed to crosslinking of proteins by formaldehyde

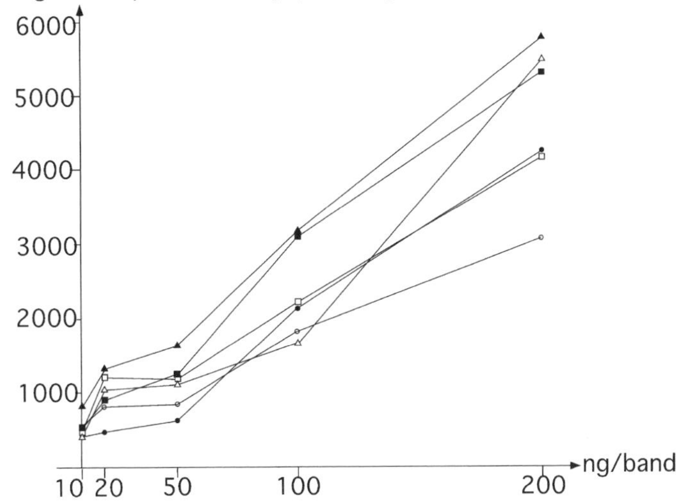

Fig. 5.2. Linearity of silver staining. Serial dilutions of standard proteins were separated by SDS PAGE and stained with silver (protocols 6 and 7). The resulting signals were quantified by densitometry (signal = optical density × area) and plotted against the amount of protein to give the dose-response curves presented here. □ Carbonic anhydrase stained with ammoniacal silver; ■ carbonic anhydrase stained with silver nitrate; △ ovalbumin stained with ammoniacal silver; ▲ ovalbumin stained with silver nitrate; ○ soybean trypsin inhibitor stained with ammoniacal silver; ● soybean trypsin inhibitor stained with silver nitrate

during development. This problem has not been experienced with staining by protocol 8, probably because development takes place in an acidic medium with ammonia remaining from the silver ammonia solution. Both factors contribute to decreasing the activity of formaldehyde toward proteins.

In some cases, blotting of proteins stained with silver has been described (Wise and Lin 1991), thereby showing the possibility of analyzing whole proteins after silver staining. However, it must be kept in mind that analysis of silver-stained proteins often looks rather disappointing. In most cases, this is not due to a flaw linked to the staining process per se, but rather to the fact that a nice-looking silver-stained spot contains only a few nanograms or even sub-nanogram quantities of proteins. This is often too low to give any positive result in protein identification methods, except maybe with mass spectrometry. The only solution would be to pool silver-stained spots, but this possibility does not seem to have been successfully experienced yet.

5
Detection by Fluorescence

In principle, fluorescent staining is closely related to staining with dyes. The aim is to be able to bind a molecule to proteins, covalently or not, and to make it fluoresce. This is, however, much more difficult than binding a dye molecule, as fluorescence is a delicate phenomenon. Light must be absorbed by the fluorophore without destroying it (destruction gives rise to fading), and the fluorophore must be in a suitable chemical environment so that the quantum yield (i.e. the ratio of photons re-emitted by fluorescence over absorbed photons) is maximized. This is one of the major parameters in fluorescence, as some fluorophores have a high quantum yield in water (e.g. fluorescein), while some others (e.g. pyrene derivatives) have a high quantum yield only in apolar solvents.

Last but not least, detection by fluorescence implies a rather delicate optical hardware. This is due to the fact that the emitted light is usually very low in intensity, especially when compared to excitation light. The illumination source must be as monochromatic as possible in order to minimize background illumination at the emission wavelength. This will be of course simpler if the shift between the excitation and emission wavelengths is large. The emitted light must then be collected, as it is isotropically emitted, filtered to eliminate any residual excitation light and then quantitated.

The process is rather complicated, but the results may be astonishingly good. In the field of DNA sequencing, the use of fluorescent probes (fluorescein derivatives) allows the detection of a few attomoles of probes (Smith et al. 1986). In addition, detection by fluorescence can be linear over several orders of magnitude (up to five). Last but not least, the use of a set of different probes combined to pre-electrophoresis labelling opens up the possibility of multiplexing, i.e. analysis of several samples mixed on the same gel (Unlü et al. 1997).

These enormous advantages have prompted continuous research on protein detection with fluorophores, centered on two different pathways. The first way is to covalently couple a fluorophore to the proteins. In the case of 2-D electrophoresis, pre-electrophoresis labelling is complicated by the fact that the pI should not be modified. Pre labelling has therefore been carried out either with neutral, thiol-reactive probes (Urwin and Jackson 1993) or with amino-reactive probes bearing a single positive charge in order to replace the protein charge lost by coupling the probe (Unlü et al. 1997). Other probes must be grafted either between IEF and SDS PAGE (Urwin and Jackson 1991) or after electrophoresis. In the latter case, molecules which are non fluorescent but become fluorescent after coupling are generally chosen, such as fluorescamine (Jackowski and Liew, 1980), or 2-methoxy-2,4-diphenyl furanone (MDPF) (Jackson et al., 1988). This is due to the fact that any fluorescent molecule remaining in the gel will induce a very high background.

The second method of fluorescent detection is to use non-covalent, environment-sensitive probes. Generally, these probes are very weakly fluorescent in water but highly fluorescent in apolar media, such as detergent. Advantage is then taken of the binding of detergent to proteins to build a fluorescence-promoting environment at the protein sites in the gel. One of the best-known probes of this type is ANS (anilino naphtalene sulfonate). Its application to fluorescent detection of proteins in gels was described some years ago, with a sensitivity close to the one of dye detection (Daban and Aragay 1984). The use of more sensitive probes, such as BisANS (Horowitz and Bowman 1987), and more recently Sypro dyes from Molecular Probes (Steinberg et al. 1996a, b), or Nile Red (Alba et al.,1996) has considerably improved the sensitivity of this approach.

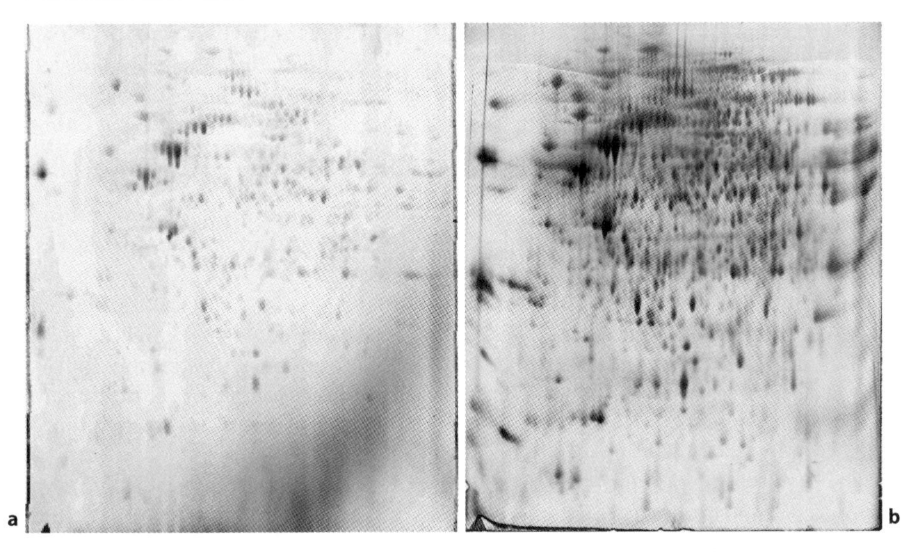

Fig. 5.3. Comparison of detection by fluorescence with silver staining. 100 g of total cell extract is loaded on the 2-D gel. The second dimension is run in 0.05% SDS as recommended for fluorescent detection. After staining with Sypro orange according to manufacturer's instructions, the gel is quantified with a Fluorimager (Molecular Dynamics). The resulting image is shown in *A*. The gel is then fixed with ethanol and acetic acid and then silver-stained (silver nitrate, protocol 6). The resulting image is shown in *B*

Both methods have their advantages and drawbacks. The non-covalent method has by definition a limited contrast. Ideally detergent should be present on the proteins, but no apolar environment (i.e. no micelles) should be present in the gel. This is almost impossible to achieve, as any treatment disrupting the micelles (and thus reducing the background) will also decrease the amount of detergent bound to the proteins (and thus the signal). The commonly suggested lowering of SDS concentration in the gel to avoid the presence of micelles (e.g. in Alba et al. 1996) is acceptable for 1D electrophoresis but often leads to vertical streaking when applied to 2-D electrophoresis (e.g. in Fig. 5.3). However, as labelling is non covalent and made after electrophoresis, there are neither migration artifacts linked to coupling of a probe nor problems linked to modification of the proteins in subsequent analyses.

These artifacts are the plague of methods involving covalent coupling. Even with probes used for pre-labelling, problems will be encountered when basic proteins are analyzed. In this case, the real pKa's of thiol and amino group must be respected to have access to the true pI, and this is not the case with the probes used to date for pre-labelling. Even for neutral and acidic proteins, artifacts in the apparent molecular weight are quite common (Unlü et al. 1997). Consequently, the position of the fluorescent spot will be slightly offset to the position of the bulk of the protein, leading to problems when the spot must be excised for subsequent analysis. However, the very high sensitivity and contrast shown by this type of detection in the DNA sequencing field make this approach worth a trial.

Another major problem linked to the use of 2-D gels is the hardware problem. In DNA sequencing, the optical apparatus is fixed and the labelled bands go through it. This cannot be the case in 2-D electrophoresis. If the excitation light is a point (e.g. laser excitation), there must be a scanning device moving either the gel or the optical device (or both) so that all the gel will be scanned. This is the case with laser fluorescence scanners (e.g. Molecular Dynamics Fluorimagers). Such scanners are best used with probes having an excitation maximum close to an available laser wavelength (fluorescein and rhodamine derivatives, Sypro orange).

In another setup, the illumination is made in the gel plane and the fluorescent light collected by a camera, usually a CCD (charge-coupled device) camera (Jackson et al. 1988; Urwin and Jackson, 1991). In this case, molecules excited by UV are preferred, as UV light can be very efficiently filtered just before the camera optics. Such probes are generally less sensitive and/or more prone to photobleaching than visible light-excited probes.

Owing to these difficulties in chemistry and hardware, the current situation of fluorescent detection of proteins in 2-D gels is much less glorious than it should be. Although some papers claim that fluorescent detection is at least as sensitive as silver staining (Urwin and Jackson 1991; Steinberg et al. 1996a; Unlü et al. 1997), the reference silver-staining methods are generally kits or old methods of poor sensitivity. A comparison of fluorescent detection with silver staining is shown in Fig. 5.3. Fluorescent detection appears to be much less sensitive than zinc or silver staining, and rather as sensitive as colloidal Brilliant Blue staining.

These poor performances can be explained by hardware problems linked to the constraints brought by 2-D gels, but are also explained by chemical reasons. The limited contrast of non-covalent environment-sensitive probes certainly limits sensitivity, but other problems arise for covalent probes. For example, thiol-reactive probes will only detect cysteine-containing proteins. Some proteins (e.g. carbonic anhydrase) do not contain cysteine and will thus be undetected by these probes. Other problems arise when amine-reactive probes are coupled to proteins between IEF and SDS PAGE. Carrier ampholytes containing numerous reactive amino groups are present in the IEF gel, even in IPG gels where ampholytes are used to smooth conductivity. These carrier ampholytes must be first removed by fixation (Urwin and Jackson 1991). This leads to protein losses, as one cannot be sure that proteins will be resolubilized at the SDS PAGE step, especially in IPG gels which are prone to protein binding problems (Adessi et al. 1997).

Last but not least, the quenching problem also decreases the performances of fluorescent detection in 2-D electrophoresis. In DNA sequencing, there is a single fluorescent probe per DNA molecule. This cannot be guaranteed in protein labelling, as every protein molecule has several probe-reactive sites, especially for amino-reactive probes. Consequently, a high coupling reaction scheme will lead to most proteins bearing several fluorophores. They unfortunately interact one with another to decrease the efficiency of fluorescence (quenching). In a low coupling reaction scheme, the fluorescence is decreased by the fact that most protein molecules do not bear any fluorophore at all. In addition, the interaction between the fluorophore and the protein residues or the SDS bound to the protein is still unknown and may also give rise to quenching.

6 Detection of Radioactive Isotopes

For a long time, prior to the use of highly sensitive silver staining methods, detection of proteins labelled with radioisotope was the only type of sensitive detection available. Apart from the safety problems associated with the use of radioactivity, this method had several other problems.

First, labelling is best performed by giving radioactive metabolites to the sample source. This means that in vitro labelling is rather simple (e.g. O'Farrell 1975), while labelling in vivo is much less convenient and efficient, and of course impossible for human tissues. To overcome this limitation, in vitro labelling after protein extraction has been proposed (Rabilloud et al. 1986a; Rabilloud and Therre 1986; Boxberg 1988), but it never gained widespread use because of the concurrence of silver staining.

A second limit to detection of radioisotopes arose from difficulties in detecting some isotopes emitting low-energy radiations, e.g. tritium. This precluded the use of tritium labelling, which was by far one of the most versatile and convenient labels available. This limitation was overcome by the introduction of fluorography (Bonner and Laskey 1974). In this method, a scintillator is precipitated in the gel prior to drying. It converts the low energy radiations emitted by tritium, unable to go through the dry polyacrylamide layer, into visible light which is then detected with an autoradiographic film. Best results are obtained with 2,5 diphenyloxazole (PPO) dissolved either in DMSO (Bonner and Laskey 1974) or in acetic acid (Skinner and Griswold 1983).

Another important problem in radioactivity detection is linearity. With the classical film detection, linearity was obtained with preflashing (Laskey and Mills 1975). This procedure was however difficult to standardize, and linearity was obtained over a limited range. A major change in sensitivity and linearity was afforded by film-less detection methods, depending on completely different principles. One of these methods is based on phosphor storage (Johnston et al. 1990). In this technique, the beta radiations induce an energy change in a europium salt. This accumulated energy is converted into visible light upon laser excitation, the emitted light being quantified with a photomultiplier. Compared to detection with an autoradiographic film, sensitivity is increased 20- to 100-fold. However, the major improvement lies in linearity, with a linear dynamic range of four orders of magnitude.

The other new detection method is based on amplification detectors similar to those used in high energy physics. Here again, linear performances are obtained over a wide range, with a 100-fold increase in sensitivity compared to film detection (Charpak et al. 1989). Moreover, dual isotope detection can be easily carried out, provided that the two radiation energies are different enough (e.g. ^{3}H vs ^{14}C or ^{35}S, ^{14}C or ^{35}S vs ^{32}P). Furthermore, sensitive detection of tritium can be easily carried out (Tribollet et al. 1991), provided that the radiation is not absorbed by the separation medium. The use of blots or ultrathin gels is therefore highly recommended. The main drawback of this detection method is that the gel must be in the detector during all the data acquisition time, thereby precluding any parallel detection such as those afforded by films or multiple phosphor storage plates.

Compared to film detection, these new methods suffer from a lack of resolution. Resolutions are in the range 200–400 μM, while a good resolution for the analysis of a 2-D gel is rather in the range 50–100 μm.

7 Conclusions and Future Prospects

It can be easily deduced from the above text that numerous, widely different, detection methods are available for the visualization of proteins separated on 2-D gels. The choice of the detection method will therefore depend mainly on the constraints dictated by the experiment. Radioactive detection probably offers the best signal to noise ratio and the best ultimate sensitivity. With its latest developments, it is ideally suited for dual labelling techniques, allowing duplex analysis of samples on the same gel or determination of amino acid ratios (see Chapter 8 by Labarre and Perrot). However, limitations in scope (difficulties for tissues), time needed for detection (usually several days) and, more importantly, safety regulations increasing in strength have confined this method to specialized niches. In addition to the determination of amino acid ratio, the identification of post-translational modifications with radioactive, dedicated precursors is a field of choice.

The limitations inherent in the use of radioactivity and the availability of reliable and highly sensitive staining techniques have dramatically increased the scope of direct staining techniques. For proteomics studies, where quantitative data must be produced, the staining techniques of choice are silver staining and colloidal Brilliant Blue staining. These techniques are of great interest, as they are compatible with subsequent analysis of the proteins of interest, for example with mass spectrometry. In difficult cases where fixation precludes any further analysis [e.g. mass spectrometry of whole proteins (Cohen and Chait 1997)], dedicated techniques such as heavy metal precipitation are used. Due to the expensive associated hardware and to rather disappointing performances (sensitivity and spatial resolution) compared to its possibilities, fluorescence has not gained widespread use in 2-D gel staining yet.

Owing to the ever increasing sensitivity of identification methods, detection methods have to keep the pace. However, most staining methods have come to a plateau close to the theoretical maximum. For dye staining, this maximum is reached when a highly absorbing dye is bound at all the available sites on a protein. Colloidal Brilliant Blue staining achieves conditions very close to this maximum. A maximum also seems to have been reached for silver staining, where no progress in sensitivity has been made over the last 5 years. This indicates that all bound silver ion reducible under contrasting conditions is reduced with the current protocols. Further increase in the sensitivity of silver staining will result from techniques based on different principles to those used nowadays, thereby requiring heavy methodological developments.

Finally, the method with the highest progression potential is fluorescence. However, difficult problems remain to be solved. These include, for example, the extent of labelling. Extensive labelling of reactive groups induces quenching and loss of solubility (Unlü et al. 1997), while limited labelling is difficult to achieve reproducibly. In addition, quenching may occur from the very chemical nature of proteins, thereby severely limiting sensitivity. Other problems to be solved in fluorescent detection from 2-D gels will come from hardware problems in connection with the high spatial resolution required for correct 2-D gel analysis. As the current performances are quite remote from the experimental maxima achieved for example in DNA detection, we are quite confident that the future should be bright.

References

Adams LD, Weaver KM (1990) Detection and recovery of proteins from gels following zinc chloride staining. Appl Theor Electrophoresis 1: 279–282

Adessi C, Miege C, Albrieux C, Rabilloud T (1997) Two-dimensional electrophoresis of membrane proteins: a current challenge for immobilized pH gradients. Electrophoresis 18: 127–135

Alba FJ, Bermudez A, Bartolome S, Daban JR (1996) Detection of five nanograms of protein by two-minute Nile red staining of unfixed SDS gels. Biotechniques 21: 625–626

Blum H, Beier H, Gross HJ (1987) Improved silver staining of plant proteins, RNA and DNA in polyacrylamide gels. Electrophoresis 8: 93–99

Bonner WM, Laskey RA(1974) A film detection method for tritium-labelled proteins and nucleic acids in polyacrylamide gels. Eur J Biochem 46: 83–88

Bosshard HF, Datyner A (1977) The use of a new reactive dye for quantitation of prestained proteins on polyacrylamide gels. Anal Biochem 82: 327–333

Boxberg YV (1988) Protein analysis on two-dimensional polyacrylamide gels in the femtogram range: use of a new sulfur-labeling reagent. Anal Biochem 169: 372–375.

Charpak G, Dominik W, Zaganidis N (1989) Optical imaging of the spatial distribution of beta-particles emerging from surfaces. Proc Natl Acad Sci USA 86: 1741–1745

Chen H, Cheng H, Bjerknes M (1993) One step Coomassie Brilliant Blue R250 staining of proteins in polyacrylamide gels. Anal Biochem 212: 295–296

Cohen SL, Chait BT (1997) Mass spectrometry of whole proteins eluted from sodium dodecyl sulfate-polyacrylamide gel electrophoresis gels. Anal Biochem 247: 257–267

Courchesne PL, Luethy R, Patterson SD (1997) Comparison of in-gel and on-membrane digestion methods at low to sub-pmol level for subsequent peptide and fragment-ion mass analysis using matrix-assisted laser-desorption/ionization mass spectrometry. Electrophoresis 18:369–381

Daban JR, Aragay AM (1984) Rapid fluorescent staining of histones in sodium dodecyl sulfate-polyacrylamide gels. Anal Biochem 138: 223–228

Dzandu JK, Johnson JF, Wise GE (1988) Sodium dodecyl sulfate-gel electrophoresis: staining of polypeptides using heavy metal salts. Anal Biochem 174: 157–167

Eschenbruch M, Bürk RR (1982) Experimentally improved reliability of ultrasensitive silver staining of protein in polyacrylamide gels. Anal Biochem 125: 96–99

Fazekas de St Groth S, Webster RG, Datyner A (1963) Two new staining procedures for quantitative estimation of proteins on electrophoresis strips. Biochim Biophys Acta 71: 377–391

Fernandez-Patron C, Calero M, Collazo PR, Garcia JR, Madrazo J, Musacchio A, Soriano F, Estrada R, Frank R, Castellanos-Serra LR, Mendez E (1995) Protein reverse staining: high-efficiency micro-analysis of unmodified proteins detected on electrophoresis gels. Anal Biochem 224: 203–211

Ferreras M, Gavilanes JG, Garcia-Segura JM (1993) A permanent Zn^{2+} reverse staining method for the detection and quantification of proteins in polyacrylamide gels. Anal Biochem 213: 206–212

Granier F, De Vienne D (1986) Silver staining of proteins: standardized procedure for two-dimensional gels bound to polyester sheets. Anal Biochem 155: 45–50

Hardy E, Sosa AE, Pupo E, Casalvilla R, Fernandez-Patron C (1996) Zinc-imidazole positive: a new method for DNA detection after electrophoresis on agarose gels not interfering with DNA biological integrity. Electrophoresis 17: 26–29

Heukeshoven J, Dernick R (1985) Simplified method for silver staining of proteins in polyacrylamide and the mechanism of silver staining. Electrophoresis 6: 103–112

Hochstrasser DF, Merril CR (1988) 'Catalysts' for polyacrylamide gel polymerization and detection of proteins by silver staining. Appl Theor Electrophoresis 1: 35–40

Horowitz PM, Bowman S (1987) Ion-enhanced fluorescence staining of sodium dodecyl sulfate-polyacrylamide gels using bis (8-p-anilino-1-naphthalenesulfonate). Anal Biochem 165: 430–434.

Jackowski G, Liew CC (1980) Fluorescamine staining of nonhistone chromatin proteins as revealed by two-dimensional polyacrylamide gel electrophoresis. Anal Biochem 102: 321–325

Jackson P, Urwin V, Mackay CD (1988) Rapid imaging, using a cooled charge-coupled-device, of fluorescent two-dimensional polyacrylamide gels produced by labelling proteins in the first dimensional isoelectric focusing gel with the fluorophore 2-methoxy-2,4-diphenyl-3(2H)furanone. Electrophoresis 9: 330–339

Johnston RF, Pickett SC, Barker DL (1990) Autoradiography using storage phosphor technology. Electrophoresis 11: 355–360
Laskey RA, Mills AD (1975) Quantitative film detection of 3H and 14C in polyacrylamide gels by fluorography. Eur J Biochem 56: 335–341
Lee C, Levin A, Branton D (1987) Copper staining: a five-minute protein stain for sodium dodecyl sulfate-polyacrylamide gels. Anal Biochem 166: 308–312
Merril CR, Goldman D (1984) Detection of polypeptides in two-dimensional gels using silver staining. In: Celis JE, Bravo R (eds) Two-dimensional gel electrophoresis of proteins, Academic Press, London, pp93–109
Mold DE, Weingart J, Assaraf J, Lubahn DB, Kelner DN, Shaw BR, McCarty KS Sr (1983) Silver staining of histones in Triton-acid-urea gels. Anal Biochem 135:44–47
Nelles LP, Bamburg JR (1976) Rapid visualization of protein–dodecyl sulfate complexes in polyacrylamide gels. Anal Biochem 73: 522–531
Neuhoff V, Stamm R, Eibl H (1985) Clear background and highly sensitive protein staining with Coomassie Blue dyes in polyacrylamide gels: a systematic analysis. Electrophoresis 6: 427–448
Neuhoff V, Arold N, Taube D, Ehrhardt W (1988) Improved staining of proteins in polyacrylamide gels including isoelectric focusing gels with clear background at nanogram sensitivity using Coomassie Brilliant Blue G-250 and R-250. Electrophoresis 9: 255–262
O'Farrell PH (1975) High resolution two-dimensional electrophoresis of proteins. J Biol Chem 250:4007–4021
Ortiz ML, Calero M, Fernandez-Patron C, Castellanos L, Mendez E (1992) Imidazole-SDS-Zn reverse staining of proteins in gels containing or not SDS and microsequence of individual unmodified electroblotted proteins. FEBS Lett 296: 300–304
Rabilloud T (1990) Mechanisms of protein silver staining in polyacrylamide gels: a 10-year synthesis. Electrophoresis 11: 785–794
Rabilloud T (1992) A comparison between low background silver diammine and silver nitrate protein stains. Electrophoresis 13: 429–439
Rabilloud T, Therre H. (1986) Suitability of sulfur labelling reagent (Amersham) as an in vitro protein radiolabelling reagent for two-dimensional electrophoresis. Electrophoresis 7: 49–51
Rabilloud T, Hubert M, Tarroux P (1986) Procedures for two-dimensional electrophoretic analysis of nuclear proteins. J. Chromatogr 351: 77–89
Rabilloud T, Vuillard L, Gilly C, Lawrence JJ (1994) Silver-staining of proteins in polyacrylamide gels: a general overview. Cell Mol Biol 40: 57–75
Schägger H, Aquila H, Von Jagow G (1988) Coomassie Blue-sodium dodecyl sulfate-polyacrylamide gel electrophoresis for direct visualization of polypeptides during electrophoresis. Anal Biochem 173: 201–205
Skinner MK, Griswold MD (1983) Fluorographic detection of radioactivity in polyacrylamide gels with 2,5-diphenyloxazole in acetic acid and its comparison with existing procedures. Biochem J 209: 281–284
Smith LM, Sanders JZ, Kaiser RJ, Hughes P, Dodd C, Connell CR, Heiner C, Kent SB, Hood LE (1986) Fluorescence detection in automated DNA sequence analysis. Nature 321: 674–679
Steinberg TH, Jones LJ, Haugland RP, Singer V (1996a) SYPRO orange and SYPRO red protein gel stains: one-step fluorescent staining of denaturing gels for detection of nanogram levels of protein. Anal Biochem 239: 223–237
Steinberg TH, Haugland RP, Singer V (1996b) Applications of SYPRO orange and SYPRO red protein gel stains. Anal Biochem 239: 238–245
Tribollet E, Dreifuss JJ, Charpak G, Dominik W, Zaganidis N (1991) Localization and quantitation of tritiated compounds in tissue sections with a gaseous detector of beta particles: comparison with film autoradiography. Proc Natl Acad Sci U A 88: 1466–1468
Unlü M, Morgan EM, Minden JS (1997) Difference gel electrophoresis: a single gel method for detecting changes in protein extracts. Electrophoresis 18: 2071–2077
Urwin VE, Jackson P (1991) A multiple high-resolution mini two-dimensional polyacrylamide gel electrophoresis system: imaging two-dimensional gels using a cooled charge-coupled device after staining with silver or labelling with fluorophore. Anal Biochem 195: 30–37
Urwin VE, Jackson P (1993) Two-dimensional polyacrylamide gel electrophoresis of proteins labelled with the fluorophore monobromobimane prior to first-dimensional isoelectric focusing: imaging of the fluorescent protein spot patterns using a cooled charge-coupled device. Anal Biochem 209: 57–62
Wise GE, Lin F (1991) Transfer of silver-stained proteins from polyacrylamide gels to polyvinylidene difluoride membranes. J Biochem Biophys Methods 22:223–231

CHAPTER 6

Protein Blotting and Immunoblotting

L. Bini[1], S. Liberatori[1], B. Magi[1], B. Marzocchi[1],
R. Raggiaschi[1] and V. Pallini[1]

1 Introduction

Immunoblotting was first devised by Towbin et al. (1979) to exploit the specificity of the reaction between antibodies and proteins transferred from SDS-polyacrylamide electrophoretic gels onto nitrocellulose. Immunoreactive bands were evidenced by labelled "second antibody" or protein A.

When combined with 2-D electrophoretic separation, immunoblotting can conveniently be used to identify proteins by using antibodies of known specificity. As different protein antigens can be distinguished by their position in 2-D gels, blotting from a single gel with a mixture of several antibodies can be exploited. However, the potentialities of 2-D immunoblotting in post-genomic biological research are fully understood if progress in related areas of technological research is also taken into account. To begin with, the introduction of immobiline-based electrofocusing has dramatically increased the reproducibility and the resolving power of 2-D PAGE, which now plays the role of core technique in proteomics. Updated 2-D PAGE also allows us to measure both M_r and pI of denatured proteins, providing experimental values comparable to those calculated from gene sequences (Bjellqvist et al. 1993). On the other hand, genome projects have promoted the development of tools for sequence analysis, including "cyber proteomic tools" (http://www.expasy.ch/) to predict post-translation modifications which, in turn, affect protein electrophoretic parameters. Techniques for antibody production have also been improved so as to include immunisation with recombinant proteins and synthetic oligopeptides, and, as an alternative to animal immunisation, selection from CDR (complimentarity determining region)-expressing phage libraries. As to "second antibody" labelling, the sensitivity of chemiluminescent procedures is several-fold higher than that of radioactive labelling (Leong et al. 1986).

In the area of proteome studies (Kahn 1995), 2-D immunoblotting is profitably exploited to monitor protein heterogeneity. Protein isoforms, although differing by pI and/or M_r, tend to be cross-reactive, especially when probed with polyclonal antibodies. In case of heterogeneity due to post-translational modifications, primary gene products can be tentatively distinguished from modified iso-

[1] Dipartimento Biologia Molecolare, Università degli Studi di Siena, Via Fiorentina, 1, 53100 Siena, Italy. Fax + 39 0577 234 903, e-mail: bini@unisi.it.

forms by comparing experimentally determined electrophoretic parameters to values predicted from coding sequences (Bini et al. 1996a; Magi et al. 1998). The extent of the discrepancy between calculated and experimental pI values can be related with the extent of modification, i.e. with the number of phosphoryl, sialic acid groups, etc. per protein molecule. Protein heterogeneity can also derive from alternative splicing of transcripts. In this case, too, the gene sequence can be scanned for splice sites, and pI and M_r values of products can be predicted and compared with experimental coordinates of immunostained spots. Both post-transcriptional and post-translational variations can sum up to generate tens of electrophoretically different polypeptides deriving from a single coding region, as it happens for tau factor and low molecular weight heat shock proteins (Janke et al. 1996; Scheler et al. 1997). Protein polymorphism deriving from a limited number of genes can be impressive. Thousands of isoforms of brain neurexin are generated from three genes by alternative promoters and alternative splicing (Missler and Suedhof 1998).

It appears that transcriptional and post-transcriptional mechanisms, as well as post-translational modifications, generate defined subsets of cross-reactive electrophoretic spots which can be distinguished by immunoblotting with adequate antibodies. Similar cross-reactive spot subsets defined by more or less extensive sequence homology conceivably derive from protein families. On the other hand, spot subsets can be generated by proteins whose similarity is limited to single epitopes selectively recognised by monoclonal antibodies, such as antibodies to phosphotyrosine, phosphoserine, penicilloyl groups (Magi et al. 1995; Marzocchi et al. 1995). Separation of whole electrophoretic patterns into subsets of immunologically related spots may be a profitable strategy for proteome research, as the high sensitivity of immunodetection coupled with chemiluminescent or radioactive labelling can extend the analysis to low copy number proteins (Sanchez et al. 1997). Dissection of humoral immune response and large-scale identification of immunogenic proteins are also possible with 2-D immunoblotting (e.g. Posch et al. 1997).

In the following pages, procedures for 2-D immunoblotting routinely used in our laboratory are presented. The main complication encountered in the transition from mono-dimensional to two-dimensional immunoblotting, i.e. the way to match immunoreactivity patterns to silver-stained patterns, is described in particular detail.

2 Protein Transfer onto Membranes

The first step of immunoblotting consists of transferring the separated proteins from the gel to an inert membrane, generally made from nitrocellulose or polyvinylidene difluoride (PVDF). This technique is also profitably used in many other protein analysis techniques, such as microsequencing or mass spectrometry. This is due to the fact that these membranes provide a very open environment for microchemistry on the proteins (see the corresponding chapters in this book). This open environment is also very efficient for immunodetection, as the antibodies can easily access their epitopes on a protein immobilised on the surface of a membrane, while penetration of antibodies is rather poor in the restrictive gel media used for electrophoresis.

Ideally, the blotting process should transfer all the proteins from the gel to the membrane, irrespective of their physico-chemical characteristics. This is, however, not easily achieved by current blotting techniques. This is not due to the driving force used in protein blotting, which is almost always performed with an electric field (electroblotting) and not by passive elution or fluid motion, common practices for nucleic acid blotting. The problems rather arise from incompatibilities between the conditions required for efficient elution of the proteins from the gel (high pH, presence of detergents) and those required for efficient immobilisation of the proteins on the membranes (low pH, absence of detergent, presence of organic cosolvents). Moreover, low molecular weight proteins tend to elute easily from gels but also to go through membranes without being retained, while high molecular weight proteins are difficult to elute but are easily immobilised on membranes. It is consequently very difficult to achieve correct and even transfer of the whole spectrum of proteins with widely different molecular masses onto a membrane under a simple set of transfer conditions.

Electrotransfer from gels to membranes can be carried out with a "wet" method, in a tank where the gel is completely submerged in a single transfer buffer (see Fig. 6.1). This method is recommended when antigens are present in small quantities and/or their molecular weight is high (Okamura et al. 1995). Alternatively, "semidry" methods can be used. In this method, the liquid reser-

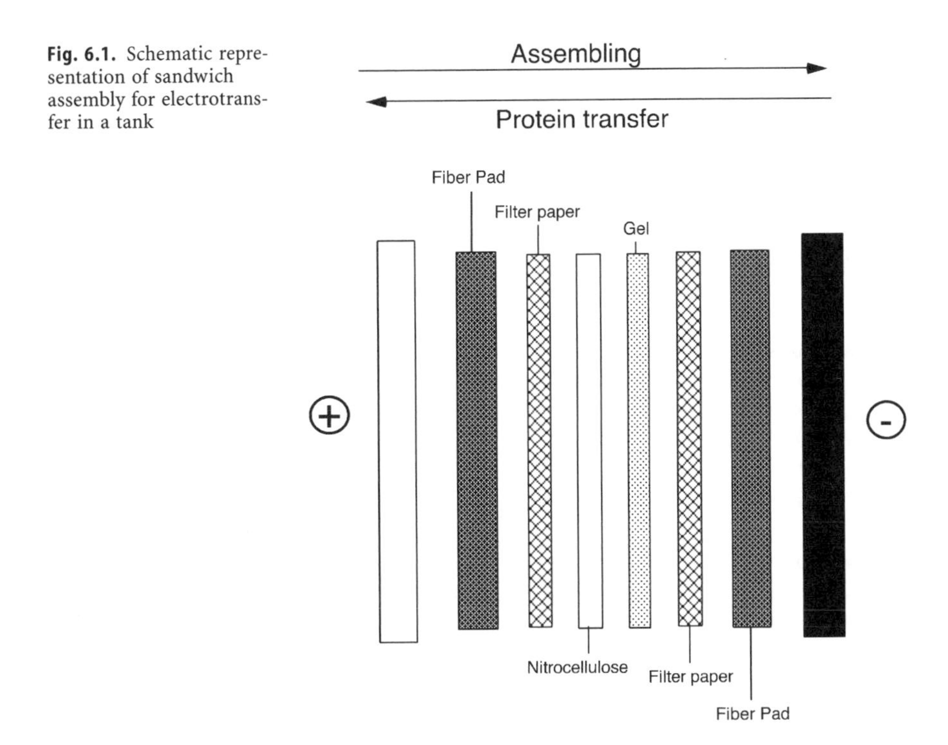

Fig. 6.1. Schematic representation of sandwich assembly for electrotransfer in a tank

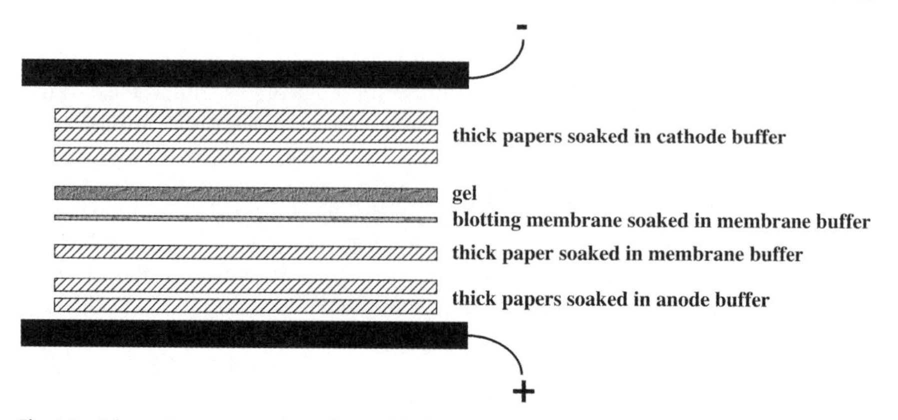

Fig. 6.2. Schematic representation of assembly for semidry electrotransfer

voirs are only paper layers wetted with the appropriate buffers and transfer occurs in a horizontal setup (see Fig. 6.2). Semidry blotting is faster and can be performed with smaller volumes of buffer as membrane and filter paper only are to be wet. In addition, different buffers can be used on the gel side and on the membrane side. It is therefore possible to use discontinuous buffers and/or an "elution promoting" buffer on the gel side and a "retention-promoting" buffer on the membrane side. An example of a sophisticated but highly efficient buffer system for semidry blotting has been described by Laurière (1993).

Protocol 1. Tank Transfer onto Nitrocellulose Membranes

Prepare transfer buffer before the end of the electrophoretic run and cool to 4 °C.

- Transfer Buffer: 25 mM Tris, 192 mM glycine, 20 % (v/v) methanol, pH 8.3. Dissolve Tris and glycine in half the final volume of water. Add the required amount of methanol and complete to final volume.

In order to avoid membrane contamination during all steps, clean gloves must be used throughout the procedure.

1. At the end of the electrophoretic run, wash the gel for 3 min in distilled water.
2. Equilibrate in transfer buffer for 10–15 min in a clean glass box. The nitrocellulose membrane must be soaked in transfer buffer for the same time.
3. After the equilibration step, wet a filter paper sheet (Whatman 17Chr) and put it on a glass plate.
4. Place the nitrocellulose carefully on the filter paper. In all steps of sandwich assembly air bubbles must be avoided to achieve a correct transfer in the whole membrane.
5. Lay down the gel on top of the nitrocellulose membrane.
6. Cut the membrane and filter paper exactly to the gel dimension with a scalpel. In order to recognise the transferred side of nitrocellulose, cut a small piece of the lower right corner of the membrane.
7. Cover the gel with another filter paper of the same dimension, pre-wetted in transfer buffer.

8. Wet two fiber pads in transfer buffer.
9. Place a fiber pad, on the anode (white) side of the gel holder (BioRad).
10. Place the sandwich, consisting of filter paper, nitrocellulose, gel and filter paper on the fiber pad and cover with the second pre-wetted fiber pad.
11. Close the gel holder firmly to ensure a tight contact between membrane and gel. It is important to have a good contact between the sandwich layers to avoid distortion of protein spots during transfer. In this way the final sandwich consists of: white side of gel holder (anode), fiber pad, filter paper, membrane, gel, filter paper, fiber pad and black side (cathode) of gel holder (Fig. 6.1).
12. Fill the transfer tank with buffer and place a stir bar inside to mix the buffer during electro-transfer (uniform temperature and conductivity).
13. Introduce the gel holder in the transfer cell with the sandwich oriented as follows: ANODE (red electrode)/white side of gel holder; CATHODE (black electrode)/black side of gel holder.
14. Run at 100 V constant voltage for 1 h at 4 °C by means of a thermostatic circulator. It is important to refrigerate the transfer tank because high voltage is applied. If two sandwiches are used on the same tank, to ensure the same transfer conditions in both membranes, the position of the two sandwiches in the tank is inverted after 30 min.
15. At the end of the run, open the gel holder and disassemble the sandwich; remove the nitrocellulose membrane and stain the transferred gel with a suitable method. The latter is important when two-dimensional electrotransfer is carried out in order to evaluate the transfer efficiency and to assign exactly the immunoreactive spots to the corresponding silver-stained polypeptides (see Sect. 5 of this chapter).
    The membrane can be processed immediately for the next step or can be air-dried and stored at −20 °C, within parafilm sheets for extended periods (Bernstein et al. 1985).

Protocol 2. Tank Transfer onto PVDF Membrane

- Transfer Buffer: 10 mM CAPS (3[cyclohexamino]-1-propanesulfonic acid), 10% (v/v) methanol, pH 11.

The procedure first described by Matsudaira (1987) is performed. The electrotransfer is quite similar to that described for nitrocellulose membrane (see protocol 1) except for the composition of the buffer. In addition, PVDF membranes must be washed in methanol for 1 min and then equilibrated for 10–15 min in transfer buffer. The run is carried out at 100 V constant voltage for 2 h, and if two sandwiches are running, their position is inverted after 1 h.

Protocol 3. Semidry Transfer

This protocol uses the discontinuous buffer described by Kyhse-Andersen (1984), with some modifications introduced by Svoboda et al. (1985).

- Anode Buffer: 0.3 M Tris, 30% (v/v) ethanol.
- Membrane Buffer: 0.03 M Tris, 30% (v/v) ethanol.
- Cathode Buffer: 0.03 M Tris, 0.04 M 6-aminocaproic acid, 0.05% (w/v) SDS.

1. Cut six extra thick blotting papers (BioRad) and a blotting membrane to the gel dimension.
2. Equilibrate the membrane in membrane buffer (after wetting in pure methanol or ethanol for 5 min for PVDF membranes).
3. Wet two paper sheets in anode buffer and lay down on the anode on the transfer device. Avoid any air bubble by rolling a disposable pipette on the papers.
4. Wet one paper sheet in membrane buffer and lay down on top of the previous sheets. Eliminate air bubbles.
5. Lay down the transfer membrane (cut at one corner) on top of the papers.
6. Lay down the gel, aligning with the transfer membrane.
7. Lay down the remaining three paper sheets, wetted in cathode buffer, to complete the setup.
8. Place the cathode, close the apparatus and transfer at 0.8 mA/cm^2 of gel for 1 h (voltage should be less than 60 V).
9. Dismantle the apparatus, remove the gel and stain appropriately. Stain the membrane, and then process.

3 Membrane Staining

Several staining procedures, not interfering with antigen-antibody reaction, can be chosen for chemical staining of proteins pattern onto nitrocellulose or other membranes before immunodetection. This step is crucial when 2-D gel is used, as it provides "landmarks" for the assignment of immunoreactive spots to the complex silver-stained pattern. Permanent staining, e.g. with Amido Black (Clegg 1982), Coomassie Blue (Burnette 1981) or India ink (Hancock and Tsang 1983), can be used if immunoreactivity pattern is collected from enhanced chemiluminescent (ECL)-impressed films. PVDF membrane can be stained with Coomassie Blue without using organic solvent (Houen et al. 1997), after preblocking the membrane with Tween 20. This method allows a very sensitive staining of the total protein pattern in the membrane with a low background and without the need for the destaining step.

Reversible staining procedures (e.g. metal chelates or dyes like Ponceau S or Fast Green) can also be employed (Reinhart and Malamud 1982; Lin and Kasamatsu 1983; Salinovich and Montelaro 1986; Root and Reisler 1989; Patton et al. 1994), but major spots must be circled with a waterproof pen to maintain some "landmarks" on the membrane or the membrane must be scanned or photographed before immunodetection. The latter staining procedures can also be used when immunodetection is carried out by colorimetric methods.

The use of low-sensitivity staining methods usually requires high sample loading on gels. The most sensitive detection procedure consists of radioactive protein labelling (e.g. Celis et al. 1995; Birkelund et al. 1997). In this case, detection by phosphor-imaging produces images of global protein pattern on transblotted membranes, while ECL-impressed films detect immunoreactive spots. The two images have the same dimensions, so that general alignment and immunoreactive spots recognition on total protein pattern can be easily achieved.

However, the requirement of radioactive proteins is a strong limitation. Perfect alignment of immunoreactive spots to total protein pattern can be achieved, at least on PVDF membranes, by the conjunction of colloidal gold staining for total protein detection and ECL for immunoreactivity displaying on the same membrane (Chevallet et al. 1997). This procedure produces multiple ECL-impressed films: low exposure allows the detection of immunoreactive spots; strong exposure produces a background pattern. The final result is a single image where immunoreactivity appears as dark black spots and general protein pattern appears as light grey spots. Colloidal gold or India ink staining can be also applied after immunodetection (Moeremans et al. 1985; Daneels et al. 1986; Glenney 1986), provided that membrane saturation is achieved by protein-free solutions (e.g. Tween 20) (Batteiger et al. 1982).

Zeindl-Eberhart et al. (1997) have also proposed an easy method for immunodetection of specific antigens and their localisation on 2-D gel. The gel is transferred onto PVDF membrane, immunostained with specific antibodies using Fast Red or 5-bromo-4-chloro-3-indolyl phosphate/nitroblue tetrazolium as detection systems and then counterstained with Coomassie Brillant Blue. The membrane appears with immunostained spots coloured in red or black and the total protein pattern in blue. With this method, the blocking protein (milk or serum) used to saturate non-specific sites of membrane before immunodetection does not create a disturbing background. This seems to be due to weak binding to the surface of the membrane, so that the blocking protein is removed during the Coomassie Blue staining process.

Transfer of proteins from polyacrylamide gels after Coomassie Blue or silver staining has also been reported (Wise and Lin 1991; Thompson and Larson 1992; Ranganathan and De 1995). The transferred proteins remain stained during immunostaining, providing many "landmarks" for spot location. Staining of membranes with Coomassie Blue or silver (for nitrocellulose) is described in the chapters devoted to microsequencing and mass spectrometry (Chaps 10 and 11), respectively. A protocol for reversible staining with Ponceau S is described here.

Protocol 4. Reversible Ponceau S Staining

Reagent Solutions

- Trichloroacetic Acid (TCA) Solution: 50 % (w/v) TCA. Dissolve 100 g of TCA in 100 ml of distilled water. This solution can be conserved at 4 °C for months.
- Ponceau S Solution: 0.2 % (w/v) Ponceau S in 3 % (v/v) TCA. To prepare 500 ml of this solution dissolve 0.1 g of Ponceau S in water and add 30 ml of trichloroacetic acid (TCA) solution. Fill up with distilled water to 500 ml. This solution can be reused several times.

Procedure

1. Place the nitrocellulose membrane on a clean and flat glass box and stain for 3 min with the Ponceau S solution.
2. To decrease background colour, use two changes of distilled water.
3. Circle the red spots obtained with a waterproof pen, because the Ponceau S staining disappears completely during immunodetection. This operation is important to have some landmarks to match immunoreactive spots with corresponding silver-stained polyacrylamide gel pattern.

4 Immunodetection

Staining with appropriate antibodies allows the identification of the membrane-bound proteins. This procedure is very useful to assign polypeptides in two-dimensional electrophoresis. Before any incubation with antibodies, all unoccupied binding sites of the membrane must be blocked with a quencher, such as an inert protein (e.g. non-fat dry milk, bovine serum albumin, gelatin). However, special conditions and reagents are required to perform immunoblotting with particular immunoglobulins, for example anti-phospho-tyrosine antibodies (Kamps 1991; Birkelund et al. 1997). Quenching with non-ionic detergent, such as Tween 20, without inert protein has also been introduced with the advantage that the membrane can be stained for total protein pattern after immunodetection (Batteiger et al. 1982; Mohammad and Esen 1989). On the other hand, the latter can cause loss of proteins as reported by Flanagan and Yost (1984) and Hoffman and Jump (1986).

A two-stage immunodetection method is generally used in order to increase the sensitivity. Briefly, the primary antibody raised against a specific protein is bound to the antigen on the membrane. Subsequent detection of bound antibody is produced with a specific probe binding to immunoglobulin such as staphylococcal protein A (Towbin et al. 1979; Burnette 1981), IgG binding streptococcal protein G (Harper et al. 1990) or specific antibodies directed against the immunoglobulin of interest (Landini et al. 1986; Porath et al. 1987; Harper et al. 1988).

The latter is the most generally used second probe and it can be marked by radioisotopic or chromogenic substances. ^{125}I is almost universally used as radioisotopic labelling for the secondary antibody. On the other hand, the most commonly used chromogenic system consists of labelling the second antibody with an enzyme and uses precipitation of a coloured reaction product to locate bound antibody. A typical example is the use of horseradish peroxidase-conjugated antibodies, combined with a substrate such as 4-chloro-1-naphthol which gives purple bands or spots (Bers and Garfin 1985). A novel variation on the use of chromogenic labelling is the use of a luminescent reaction, allowing the detection of immunoreactive spots on an X-ray film. This allows re-exposure of membrane at different time, as in radiolabelled probe, with the time advantage of a chromogenic system (Leong at al. 1986; Vachereau 1989). Enhanced luminescent reaction with a higher sensitivity is now commercially available; this method provides detection of less than 1 pg of antigen, which is at least ten times more sensitive than other non-radioactive or radioactive detection systems (Bini et al. 1996b). This procedure permits reprobing of membranes with a variety of antibodies and the complete removal of primary and secondary antibodies (stripping) from membrane without antigen damage.

Protocol 5. Immunodetection with Chemiluminescent Method and/or Chromogenic Substrate

Reagent Solution

- Phosphate Buffered Saline (PBS) Solution: 0.15 M NaCl, 10 mM NaH_2PO_4, adjusted to pH 7.4 with NaOH. This solution can be stored for 2 weeks at 4 °C.

- 10 % Triton X-100 Stock Solution: weigh 10 g of Triton X-100 and complete to 100 ml with distilled water. Heat at 50 °C to achieve complete dissolution. This solution could be stored for at least 1 month refrigerated at 4 °C.
- Blocking Solution: 3 % (w/v) non-fat dry milk, 0.1 % Triton X-100 in PBS. To prepare 600 ml, put 18 g of non-fat dry milk in 500 ml of PBS. To avoid foam, add 6 ml of 10 % Triton X-100 stock solution when the milk is well dissolved, and complete to the right volume. This solution should be made freshly, just before use and kept for 1 day at 4 °C.
- Primary Antibody Solution (primary antibody in blocking solution): mix the right quantity of primary antibody with 50 ml of blocking solution. Prepare it just before use.

Note:
(1) the optimal dilution of antibody should be determined for each antibody by monodimensional immunoblotting or dot blot analysis.
(2) After incubation with the membrane and with high-titer of antibodies, this solution can be stored at –20 °C and used again at least ten times (Krajewski et al. 1996).

- Secondary Antibody Solution (secondary antibody in blocking solution): dilute, to the right concentration, a secondary antibody against primary immunoglobulin used, conjugated with horseradish peroxidase (HRPase), in 50 ml of blocking solution.

Note:
(1) the optimal dilution of antibody should be determined by monodimensional immunoblotting or dot blot analysis.
(2) After incubation with the membrane and with high-titer of antibodies, this solution can be stored at –20 °C and used again at least ten times (Krajewski et al. 1996).
(3) If ECL detection system is used, the dilution of the second antibody should be optimised to give the highest signal with minimal background.
(4) On a membrane, test the secondary antibody alone to discriminate cross-reactive spots which may be present.

- Washing Solution: 0.5 % Triton X-100 in PBS. Mix 2.5 ml of 10 % Triton X-100 stock solution with 47.5 ml PBS.
- Tris Stock Solution: 0.5 M Tris adjusted with HCl to pH 6.8.
- Enhanced Chemiluminescent (ECL) Solution: just before use, mix 7.5 ml detection reagent 1 with 7.5 ml detection reagent 2 (Amersham Pharmacia Biotech).
- Stripping Buffer: 100 mM 2-mercaptoethanol, 2 % (w/v) SDS, 62.5 mM Tris-HCl pH 6.8. Prepare the solution just before use.
- Washing Solution for Stripping: 0.1 % Triton X-100 in PBS. To prepare 150 ml of solution, mix 1.5 ml of 10 % Triton X-100 stock solution with 148.5 ml PBS.
- 0.3 % 4-Chloro-1-Naphthol Stock Solution: 0.3 % (w/v) 4-chloro-1-naphthol in methanol.
- 4-Chloro-1-Naphthol Developing Solution: add 2 ml of Tris stock solution to 18 ml water, add 5 ml of 0.3 % 4-chloro-1-naphthol stock solution, and then add 7 µl of 30 % H_2O_2 (v/v).

Procedure for Immunoblotting

- Incubation with Antibodies. All steps are carried out at room temperature, with gentle agitation on a rocking agitator. During immunodetection, sufficient solution should be used to adequately cover the membrane. For an 18 × 20 cm membrane, take 50 ml of each solution, using a flat glass box. To decrease the volume of solutions to be used, rotating glass cylinders can also be used. The use of sealed plastic envelope can result in high background staining, especially when ECL method is used for immunodetection.

1. The membrane is washed 3 × 10 min in blocking solution, in order to block non-specific binding sites. During the last washing step the primary antibody is diluted in blocking solution.
2. The primary antibody solution is put on the membrane for overnight incubation.
3. The membrane is washed 3 × 10 min in blocking solution. During the last washing step, the HRPase-labelled second antibody is diluted in blocking solution.
4. The membrane must be incubated with the secondary antibody solution for at least 2 h.
5. The membrane is washed 3 × 10 min in blocking solution and 1 × 30 min in washing solution.
6. Washing is continued in 50 mM Tris pH 6.8 (2 × 30 min).

After this step, ECL or chromogenic substrate-based detection methods can be applied.

- *Detection of Immunoreactive Spots.* The ECL method is one of the most sensitive detection systems. To detect the immunoreactive spot(s) with ECL kit, it is necessary to work in a dark room and to wear gloves to prevent hand contact with film and reagents. If possible powder-free gloves must be used as the powder can inhibit the ECL detection reagents, leading to white patches on the film.

1. Detection reagent 1 and detection reagent 2 of the ECL kit are mixed in equal volumes; the membrane is immersed in this solution for 1 min, ensuring that all the surface is covered.
2. The excess of detection reagent is drained off by holding the membrane vertically and touching the edge of the membrane against tissue paper. The membrane is carefully placed, protein side up, on a clean glass or on a film cassette and covered with a layer of Saran wrap, avoiding air pockets. The delay between incubating the membrane in the detection reagent and exposing it to the film should be minimised.
3. The right low corner from an autoradiographic film (X-ray film) is cut away to define its orientation. Then the autoradiography film is superimposed and aligned to the nitrocellulose membrane beginning from the upper left corners. As nitrocellulose membrane and X-ray film may have different dimensions, superimposing at the upper left corner will allow a more accurate matching of images.

4. Time of film exposure can vary from 5 s to 1 h. It is convenient to begin with short exposure, develop the film, and on the basis of spot appearance estimate the next exposure of the film.
5. The film is then developed and fixed with the suitable reagents.

In case chemiluminescent reaction is too strong, or background is too high, it is possible to switch to detection with a chromogenic substrate, such as 4-chloro-1-naphthol (Hawkes et al. 1982), according to the following protocol:
1. After ECL detection the membrane is briefly washed with Tris solution and then soaked in 4-chloro-1-naphthol developing solution until the colour appears. The reaction is stopped by washing in distilled water.
2. If ECL detection has not been performed, the membrane can be directly soaked in developing solution after the last step of the procedure described in this chapter (Prot. 5, Incubation with Antibodies).
3. The membrane is air-dried and photographed or scanned as soon as possible, because the colour fades with time.

- *Stripping.* Primary and secondary antibodies may be stripped from membranes several times. Membrane should be stored wet wrapped in Saran wrap in a refrigerator after each immunodetection. For long time storage it is suitable to keep it at –20 °C. The procedure for the complete removal of primary and secondary antibodies is outlined below:

1. The membrane is submerged in stripping buffer at 50 °C (or 70 °C if more stringent conditions are required) for 30 min, with occasional shaking.
2. The nitrocellulose is washed 2 × 10 min in large volumes of washing solution for stripping at room temperature.
3. At the end of the washing steps the membrane is submitted to immunodetection as described previously (Prot. 5, Incubation with Antibodies), starting with the blocking solution step.

5 Assignment of Two-Dimensional Immunoreactive Spots by Matching

One of the major problems is the identification of the immunostained spots on corresponding chemically stained gels. For one-dimensional gel it is possible to use two lanes of a gel with the same sample, one for immunodetection after electrotransfer and the other stained with Coomassie Blue or silver. The resulting pattern is aligned and the immunodetected bands are easily found.

Two-dimensional electrophoresis can separate several hundred spots on the same gel, each corresponding to a different polypeptide. For this reason it is very difficult to assess the immunostained with the corresponding silver-stained pattern. We, therefore, suggest running two gels of the same sample, transferring and immunostaining the first gel and staining the second with a general protein staining method (usually silver staining). Then the gel and the membrane can be matched by computer or by simple eye inspection in order to assess the correct spots (Bini et al. 1996b). For accurate comparison of the membrane and the gel, we suggest loading the gel that will be transferred with twice the amount of sample (about 100 μg) loaded on the silver-stained gel and then staining the gel after electrotransfer. Most spots will still be present on the latter and can be used to make the first matching with the membrane. In this way we can assess correctly the immunostained spot because we use the same gel from which the membrane was obtained. Then, the gel stained after transfer can be easily compared with the silver-stained gel of the same sample by computer matching or by simple eye matching. These procedures are described below.

5.1 Computer-Aided Matching

We perform computer-aided matching using the software Melanie II (Appel et al. 1997, Appel et al. 1997a); it permits us to match the digitised images, using as landmarks the spots stained with Ponceau S. The following procedure may be applied to perform a correct matching:

1. With an appropriate densitometer, such as the computing densitometer 300 S from Molecular Dynamics (4000 × 5000 pixels, 12 bits/pixel), the ECL-developed film, the Ponceau S-stained nitrocellulose membrane, the gel silver-stained after the transfer and the silver-stained gel of the same sample are scanned. After the scanning process the nitrocellulose membrane needs to be rotated left to right, with an appropriate program, in order to have the four images with the same orientation (cut lower corner on the right). In fact, the nitrocellulose membrane has the spots only in one face and the scanning process generates an image with the cut lower corner placed on the left.
2. With the stacking option of the program the film and nitrocellulose membrane images are put together, the upper left corners and the two corresponding borders are aligned, and the cut lower right corner is placed in the same orientation for both.
3. With an appropriate software tool, the Ponceau S spots chosen as landmarks are added "manually" to the image of the ECL film.
4. The ECL film with the landmarks reported is superimposed with the gel, silver stained after transfer, and the spots corresponding to landmarks on the film are marked on the gel image.
5. The size of the silver nitrate image is adjusted to the smaller one of film by means of adequate software. Actually the gels are larger than the film due to the silver staining procedure. The two images superimposed have the landmarks on the same place. The automatic match program can be run and the silver-stained spots paired, with the immunoreactive ones highlighted. The immunoreactive spots are marked on the image, silver-stained after transfer, and the matching process is performed between this image and the fourth image corresponding to the silver-stained gel with the same sample and with the correct sample load. In this way the immunoreactive spots can be easily recognised on the chemically stained gel.

5.2 Matching by Simple Eye Inspection

Matching can also be done by simple eye inspection directly on nitrocellulose and ECL film when all the spots visible on the gel are detected by chemical staining of the membrane. Otherwise, matching by computer is mandatory, in order to identify immunoreactive spots in silver-stained patterns. The scope of manual matching is greatly increased either when the general pattern can also be visualised on a film (Celis et al. 1995; Chevallet et al. 1997), and therefore superimposable on the ECL film or when a colorimetric assay is performed on the membrane on which the general pattern is visible (Daneels et al. 1986; Glenney 1986; Zeindl-Eberhart et al. 1997). In the latter case, however, the benefit of the exquisite sensitivity of ECL is lost.

In the other cases, the following manual procedure is suggested:

1. The exposed film and the nitrocellulose membrane are matched, aligning the upper left corner and the two corresponding borders and placing the cut lower right corner in the same orientation for both. With a waterproof pen, the other two borders of the nitrocellulose are marked on the film and the chemically stained spots present on nitrocellulose are transferred on the ECL film in order to use them as landmarks for the next matching with the silver nitrate-stained gel.
2. The size of the gel increases after silver staining; therefore equalisation to the size of nitrocellulose membrane and film must be performed. This can be done by photographic or photocopy procedures. With a transilluminator the landmarks on the film are superimposed with the corresponding spots on the silver nitrate-stained gel and the immunoreactive spots are identified.

Note that, due to the higher sensitivity of immunodetection, some immunoreactive spots can not be detected in the silver stained gel.

References

Appel RD, Palagi PM, Walther D, Vargas JR, Sanchez JC, Ravier F, Pasquali C, Hochstrasser DF (1997) Melanie II-a third-generation software package for analysis of two-dimensional electrophoresis images: I. Features and user interface. Electrophoresis 18: 2724–2734

Appel RD, Vargas JR, Palagi PM, Walther D, Hochstrasser DF (1997a) Melanie II–a third-generation software package for analysis of two-dimensional electrophoresis images: II. Algorithms. Electrophoresis 18: 2735–2748

Batteiger B, Newhall WJV, Jones RB (1982) The use of Tween 20 as a blocking agent in the immunological detection of proteins transferred to nitrocellulose membrane. J Immunol Methods 55: 297–307

Bernstein DI, Garraty E, Lovett MA, Bryson YJ (1985) Comparison of Western blot analysis to micro neutralization for the detection of type-specific antibodies to *Herpes simplex* virus antibodies. J Med Virol 15: 223–230

Bers G, Garfin G (1985) Protein and nucleic acid blotting and immunobiochemical detection. Biotechniques 3: 276–288

Bini L, Magi B, Marzocchi B, Cellesi C, Berti B, Raggiaschi R, Rossolini A, Pallini V (1996a) Two-dimensional electrophoretic patterns of acute-phase human serum proteins in the course of bacterial and viral diseases. Electrophoresis 17: 612–616

Bini L, Sanchez-Campillo M, Santucci A, Magi B, Marzocchi B, Comanducci M, Christiansen G, Birkelund S, Cevenini R, Vretou E, Ratti G, Pallini V (1996b) Mapping of *Chlamydia trachomatis* proteins by immobiline-polyacrylamide two-dimensional electrophoresis: spot identification by N-terminal sequencing and immunoblotting. Electrophoresis 17: 185–190

Birkelund S, Bini L, Pallini V, Sanchez-Campillo M, Liberatori S, Clausen JD, Stergaard S, Holm A, Christiansen G (1997) Characterization of *Chlamydia trachomatis* L2 induced tyrosine phosphorylated HeLa cell proteins by two-dimensional gel electrophoresis. Electrophoresis 18: 563–567
Bjellqvist B, Hughes GJ, Pasquali C, Paquet N, Ravier F, Sanchez JC, Frutiger S, Hochstrasser DF (1993) The focusing positions of polypeptides in immobilized pH gradients can be predicted from their amino acid sequences. Electrophoresis 14: 1023–1031
Burnette WN (1981) "Western blotting": electrophoretic transfer of proteins from sodium dodecylsulfate-polyacrylamide gels to unmodified nitrocellulose and radiographic detection with antibody and radioiodinated protein A. Anal Biochem 112: 195–203
Celis JE, Rasmussen HH, Gromov P, Olsen E, Madsen P, Leffers H, Honore B, Dejgaard K, Vorum H, Kristensen DB, Ostergaard M, Haunso A, Jensen NA, Celis A, Basse B, Lauridsen JB, Ratz GP, Andersen AH, Walbum E, Kjaergaard I, Andersen I, Puype M, Van Damme J, and Vandekerckove J (1995) The human keratinocyte two-dimensional gel protein database (update 1995): mapping components of signal transduction pathways. Electrophoresis 16: 2177–2240
Chevallet M, Procaccio V, Rabilloud T (1997) A nonradioactive double detection method for the assignment of spots in two-dimensional blots. Anal Biochem 251: 69–72
Clegg JCS (1982) Glycoprotein detection in nitrocellulose transfers of electrophoretically separated protein mixtures using concanavalin A and peroxidase: application to arenavirus and flavivirus proteins. Anal Biochem 127: 389–394
Daneels G, Moeremans M, De Raeymaeker M, De Mey J (1986) Sequential immunostaining (gold/silver) and complete protein staining (Aurodye) on Western blots. J Immunol Methods 89: 89–91
Flanagan SD, Yost B (1984) Calmodulin-binding proteins: visualization by ^{125}I-Calmodulin overlay on blots quenched with Tween 20 or bovine serum albumin and polyethylene oxide. Anal Biochem 140: 510–519
Glenney J (1986) Antibody probing of Western blots which have been stained with India ink. Anal Biochem 156: 315–319
Hancock K, Tsang VCW (1983) India ink staining of proteins on nitrocellulose paper. Anal Biochem 133: 157–162
Harper DR, Kangro HO, Heath RB (1988) Serological responses in varicella and zoster asseyed by immunoblotting. J Med Virol 25: 387–398
Harper DR, Ming-Liu K, Kangro HO (1990) Protein blotting: ten years on. J Virol Methods 30: 25–40
Hawkes R, Niday E, Gordon, J (1982) A dot-immunobinding assay for monoclonal and other antibodies. Anal Biochem119: 142–147
Hoffman WL, Jump AA (1986) Tween 20 removes antibodies and other proteins from nitrocellulose. J Immunol Methods 94: 191–197
Houen G, Bruun L, Barkholt V (1997) Combined immunostaining and Coomassie Brilliant Blue staining of polyvinylidene difluoride membranes without organic solvent. Electrophoresis 18: 701–705
Janke C, Holzer M, Klose J, Arendt T (1996) Distribution of isoforms of the microtubule-associated protein tau in grey and white matter areas of human brain: a two dimensional gel electrophoretic analysis. FEBS Lett 379: 222–226
Kahn P (1995) From genome to proteome: looking at a cell's proteins. Science 270: 369–370
Kamps MP (1991) Generation of anti-phosphotyrosine antibodies for immunoblotting. Methods Enzymol 20: 101–110
Krajewski S, Zapata JM, Reed JC (1996) Detection of multiple antigens on Western blots. Anal Biochem 236: 221–228
Kyhse-Andersen J (1984) Electroblotting of multiple gels: a simple apparatus without buffer tank for rapid transfer of proteins from polyacrylamide to nitrocellulose. J Biochem Biophys Methods 10: 203–209
Landini MP, Mirolo G, Coppolecchia P, Re MC, La Placa M (1986) Serum antibodies to individual cytomegalovirus structural polypeptides in renal transplant recipients during viral infection. Microbiol Immunol 30: 683–695
Laurière M (1993) A semidry electroblotting system efficiently transfers both high- and low-molecular weight proteins separated by SDS-PAGE. Anal Biochem 212: 206–211
Leong MML, Milstein C, Pannel R (1986) Luminescent detection method for immunodot, Western and Southern blots. J Histochem Cytochem. 34: 1645–1650
Lin W, Kasamatsu H (1983) On the electrotransfer of polypeptides from gels to nitrocellulose membrane. Anal Biochem 128: 302–311
Magi B, Marzocchi B, Bini L, Cellesi C, Rossolini A, Pallini V (1995) Two-dimensional electrophoresis of human serum proteins modified by ampicillin during therapeutic treatment. Electrophoresis 16: 1190–1192
Magi B, Bini L, Liberatori S, Marzocchi B, Raggiaschi R, Arcuri F, Tripodi SA, Cintorino M, Tosi P, Pallini V (1998) Charge heterogeneity of macrophage migration inhibitory factor (MIF) in human liver and breast tissue. Electrophoresis 19: 2010–2013

Marzocchi B, Magi B, Bini L, Cellesi C, Rossolini A, Massidda O, Pallini V (1995) Two-dimensional gel electrophoresis and immunoblotting of human serum albumin modified by reaction with penicillins. Electrophoresis 16: 851–853
Matsudaira P (1987) Sequence from picomole quantities of proteins electroblotted onto polyvinylidene difluoride membranes. J Biol Chem 262: 10035–10038
Missler M, Südhof TC (1998) Neurexin: three genes and 1001 products. TIG 14: 20–26
Moeremans M, Daneels G, De Mey J (1985) Sensitive colloidal metal (gold or silver) staining of protein blots on nitrocellulose membranes. Anal Biochem 145: 315–321
Mohammad K, Esen A (1989) A blocking agent and a blocking step are not needed in ELISA, immunostaining dot-blots and Western blots. J Immunol Methods 117: 141–145
Okamura H, Sigal CT, Alland L, Resh MD (1995) Rapid high-resolution Western blotting. Methods Enzymol 254: 535–550
Patton WF, Lam L, Su Q, Lui M, Erdjument-Bromage H, Tempst P (1994) Metal chelates as reversible stains for detection of electroblotted proteins: application to protein microsequencing and immunoblotting. Anal Biochem 220: 324–335
Porath A, Hanuka N, Keynan A, Sarov I (1987) Virus-specific IgG, IgM, and IgA antibodies in cytomegalovirus mononucleosis patients as determined by immunoblotting technique. J Med Virol 22: 223–230
Posch A, Chen Z, Dunn MJ, Wheeler CH, Petersen A, Leubner-Metzger G, Baur X (1997) Latex allergen database. Electrophoresis 18: 2803–2810
Ranganathan V, De PK (1995) Western blot of proteins from Coomassie-stained polyacrylamide gels. Anal Biochem 234: 102–104
Reinhart MP, Malamud D (1982) Protein transfer from isoelectric focusing gels: the native blot. Anal Biochem 123: 229–235
Root DD, Reisler E (1989) Copper iodide staining of protein blots on nitrocellulose membranes. Anal Biochem 181: 250–253
Salinovich O, Montelaro RC (1986) Reversible staining and peptide mapping of proteins transferred to nitrocellulose after separation by sodium dodecyl sulfate-polyacrylamide gel electrophoresis. Anal Biochem 156: 341–347
Sanchez JC, Wirth P, Jaccoud S, Appel RD, Sarto C, Wilkins MR, Hochstrasser DF (1997) Simultaneous analysis of cyclin and oncogene expression using multiple monoclonal antibody immunoblots. Electrophoresis 18: 638–641
Scheler C, Müller EC, Sthal J, Müller-Werdan U, Salnikow J, Jungblut P (1997) Identification and characterization of heat shock protein 27 protein species in human myocardial two-dimensional electrophoresis patterns. Electrophoresis 18: 2823–2831
Svoboda M, Meuris S, Robyn C, Christophe J(1985) Rapid electrotransfer of proteins from polyacrylamide gel to nitrocellulose membrane using surface-conductive glass as anode. Anal Biochem 151: 16–23
Thompson D, Larson G (1992) Western blots using stained protein gels. Biotechniques 12: 656–658
Towbin H, Staehlin T, Gordon J (1979) Electrophoretic transfer of proteins from polyacrylamide gels to nitrocellulose sheets: procedure and some applications. Proc Natl Acad Sci 76: 4350–4354
Vachereau A (1989) Luminescent immunodetection of Western-blotted proteins from Coomassie-stained polyacrylamide gel. Anal Biochem 179: 206–208
Wilkins MR, Sanchez JC, Williams KL, Hochstrasser DF (1996) Current challenges and future applications for protein maps and post-translational vector maps in proteome projects. Electrophoresis 17: 830–838
Wise GE, Lin F (1991) Transfer of silver-stained proteins from polyacrylamide gels to poly vinylidene-difluoride membranes. J Biochem Biophys Methods 22: 223–231
Zeindl-Eberhart E, Jungblut PR, Rabes HM (1997) A new method to assign immunodetected spots in the complex two-dimensional electrophoresis pattern. Electrophoresis 18: 799–801

CHAPTER 7

Identification of Proteins by Amino Acid Composition After Acid Hydrolysis*

M. I. Tyler[1] and M. R. Wilkins[2]

1 Introduction

In a proteome project, hundreds or even thousands of spots are produced on a single two-dimensional (2-D) gel or blot, each spot being a different protein or an isoform. Identification of these proteins from the few picomoles present in the spot presents a challenge to protein chemists. One technique for identification available in many laboratories and which does not incur great cost is amino acid composition. Latter et al. (1984) first described identification of 2-D separated proteins with their amino acid compositions using radiolabelling of amino acids. They determined the composition by radiolabelling proteins with one radioactive amino acid at a time, calculating the ratios of amino acids by quantitative optical density measurements from different gels, and comparing these ratios against databases. This approach is described in Chapter 8 (this vol.) by Labarre and Perot. An alternative approach is to use acid hydrolysis of proteins followed by quantitation of the resulting free amino acids by chromatography, and the matching of data against databases (Eckerskorn et al. 1988; Jungblut et al. 1992; Shaw, 1993; Hobohm et al. 1994; Galat et al. 1995; Wilkins et al. 1996a). The matching procedure is aided by computer programs, some that are now accessible via the internet, which use amino acid composition together with pI and molecular weight estimates to produce lists of likely protein identifications. Some programs also allow for the inclusion of an N-terminal sequence tag and species.

There are several steps involved in identifying a protein which has been separated by 2-D gel electrophoresis. Firstly the protein spots are electroblotted from the gel onto a polyvinylidene difluoride (PVDF) membrane. The blotting of samples to PVDF must produce samples which contain a minimum of salts, detergents and amino acid contamination such as glycine. Hydrolysis of the protein into amino acids is then carried out with the protein bound to the PVDF. After hydrolysis, the amino acids are extracted from the membrane so that they can be derivatized and injected onto a high performance liquid chromatography column

* This chapter is dedicated to the memory of Yik Fung.

[1] Australian Proteome Analysis Facility (APAF), School of Biological Sciences, Macquarie University, Sydney, New South Wales, 2109 Australia. E-mail: mtyler@rna.bio.mq.edu.au.
[2] Macquarie University Centre for Analytical Biotechnology (MUCAB), School of Biological Sciences, Macquarie University, Sydney, New South Wales 2109, Australia. E-mail: mwilkins@proteome.org.au.

for separation, identification and quantitation. Generally, a minimum of 250 ng of protein in a spot is required for the procedure. This is equivalent to 25 pmol of a 10-kDa protein or 4 pmol of a 60-kDa protein. However, some 5–10 μg of protein is desirable to get consistent, high confidence protein identification.

This chapter will give step-by-step methods for the identification of proteins by their amino acid compositions. Whilst this can, of course, be done in many ways, we have found the methods here to be particularly well suited to provide the high sensitivity, high accuracy data that are desirable when undertaking identification of proteins purified by 2-D gel electrophoresis. The procedures can be undertaken on a small scale, or automated to provide high sample throughput (e.g. Ou et al. 1996). This chapter also provides details of how proteins can be first subjected to an Edman degradation "sequence tag" procedure, following which the same sample can be used for amino acid analysis (Wilkins et al. 1996b). This method is useful for increasing confidence in protein identifications assigned by amino acid composition alone.

2 Blotting of Proteins from Gels to PVDF Membranes

Proteins must be pure before they can be submitted for identification by amino acid analysis. In proteome projects, the most popular means of protein purification is by 2-D gel electrophoresis. Here we will not describe methods for 2-D gel electrophoresis, as they are available elsewhere (Bjellqvist et al. 1993; Sanchez et al. 1997a). However, as all proteins must be blotted to PVDF membranes before amino acid analysis, and as the blotting procedure can be a source of salt, detergent and amino acid contamination, Protocol 1 shows a method for electroblotting (modified from Sanchez et al. 1997b). Note that the transfer buffer is 3-[cyclohexamino]-1-propanesulfonic acid (CAPS), which contains no amino acids. Electroblotting to PVDF prior to amino acid analysis should not be done with the Towbin buffer system, as it contains high concentrations of the amino acid glycine, which becomes a contaminant during subsequent amino acid analysis.

Protocol 1. *Blotting of Proteins from 2-D Gels to PVDF with CAPS Buffer System*

1. After two-dimensional electrophoresis, soak gels in deionized water for 3 min.
2. Equilibrate gels in a solution containing 10 mM CAPS pH 11 for 30 min. During the same 30 min, wet PVDF membranes in methanol for 1 min and equilibrate them in a solution containing 10 mM CAPS pH 11 and methanol (10 % v/v).
3. Carry out the blotting in either:
   - a wet transfer tank with a solution containing 10 mM CAPS pH 11 and methanol (10 % v/v) at constant 90 V for 3 h at 15°C; or
   - a semi-dry apparatus with a solution containing 10 mM CAPS pH 11 and methanol (20 % v/v cathodic side; 5 % v/v anodic side) at 1 mA/cm^2 constant current for 3 h at 15°C or as described by the manufacturer.
4. After blotting, wash membranes thoroughly in MilliQ water. Membranes can be stained with Amido black, Ponceau S or Coomassie Blue, all of which are suitable for subsequent amino acid analysis.

Note: to reduce protein contamination, gloves must be worn and all filter papers should be washed three times for 3 min in transfer buffer. These two steps are important in order to avoid any protein or amino acid contamination.

3 Hydrolysis of PVDF-Bound Proteins

When a protein is hydrolysed in 6 M hydrochloric acid, the amide bonds joining adjacent amino acids are broken, resulting in a mixture of 17 individual amino acids (Ala, Arg, Asp, Cys, Glu, Gly, His, Ile, Leu, Lys, Met, Phe, Pro, Ser, Thr, Tyr, Val). During hydrolysis, asparagine is converted to aspartic acid and glutamine is converted to glutamic acid because their side chain amide groups are hydrolysed by the HCl. The final quantity of aspartic acid will thus have come from both from asparagine and aspartic acid, and the final amount of glutamic acid have come both from glutamine and glutamic acid. Also during acid hydrolysis, tryptophan is destroyed. Other hydrolysis conditions (e.g. Yamada et al. 1991) are needed if it is to be determined. Cysteine is generally ignored in amino acid analysis because it is partially destroyed in hydrolysis, needing different hydrolysis conditions (Hoogerheide and Campbell 1992) or chemical modification prior to hydrolysis (e.g. Yan et al. 1998) to make it easily detected. To prevent tyrosine from modification by chloride ions, a crystal of phenol is often added to hydrolysis vessels as a scavenger.

Good hydrolysis is a critical step in obtaining reliable data for the identification of proteins. Our current method, modified from Yan et al. (1996) is shown in Protocol 2 and Fig. 7.1. Traditionally, proteins have been hydrolyzed in the vapour phase for 24 h at 110 °C. More recently, the hydrolysis time has been shortened to 1 h at a temperature of 155 °C with equally good results (Yan et al. 1996). These hydrolysis conditions can be used for protein samples bound to PVDF and also for proteins in solution, when the samples are first dried down into an hydrolysis tube. The design and robustness of the hydrolysis vessel is very important. A vessel that can be made by glass blowers has been described (Meyer et al. 1991), and is illustrated in Fig. 7.2. Waters Corporation (Milford, Massachusetts) also supply a protein hydrolysis work station and vessels which perform reliably.

Protocol 2. *Vapour Phase Hydrolysis of PVDF-Bound Proteins*

1. Cut spots from the PVDF membrane with a scalpel or cold punch. Cleanliness is of utmost importance. Include a known calibration protein on PVDF as a control, preferably from the same blot.
2. Place each spot into a glass hydrolysis tube, preferably a 200-μl clear glass autosampler vial, which has been labelled using a diamond pen or glass engraving tool.
3. Place 300 μl of 6 M constant boiling HCl (BDH Aristar Grade or Pierce 1-ml Sequanal Ampoules) and a small crystal of analytical reagent grade phenol (about 0.1 % w/v) into the bottom of the hydrolysis vessel.
4. Place the autosampler vials into the hydrolysis vessel using stainless steel tweezers. Remember that one vial should contain a calibration protein. To

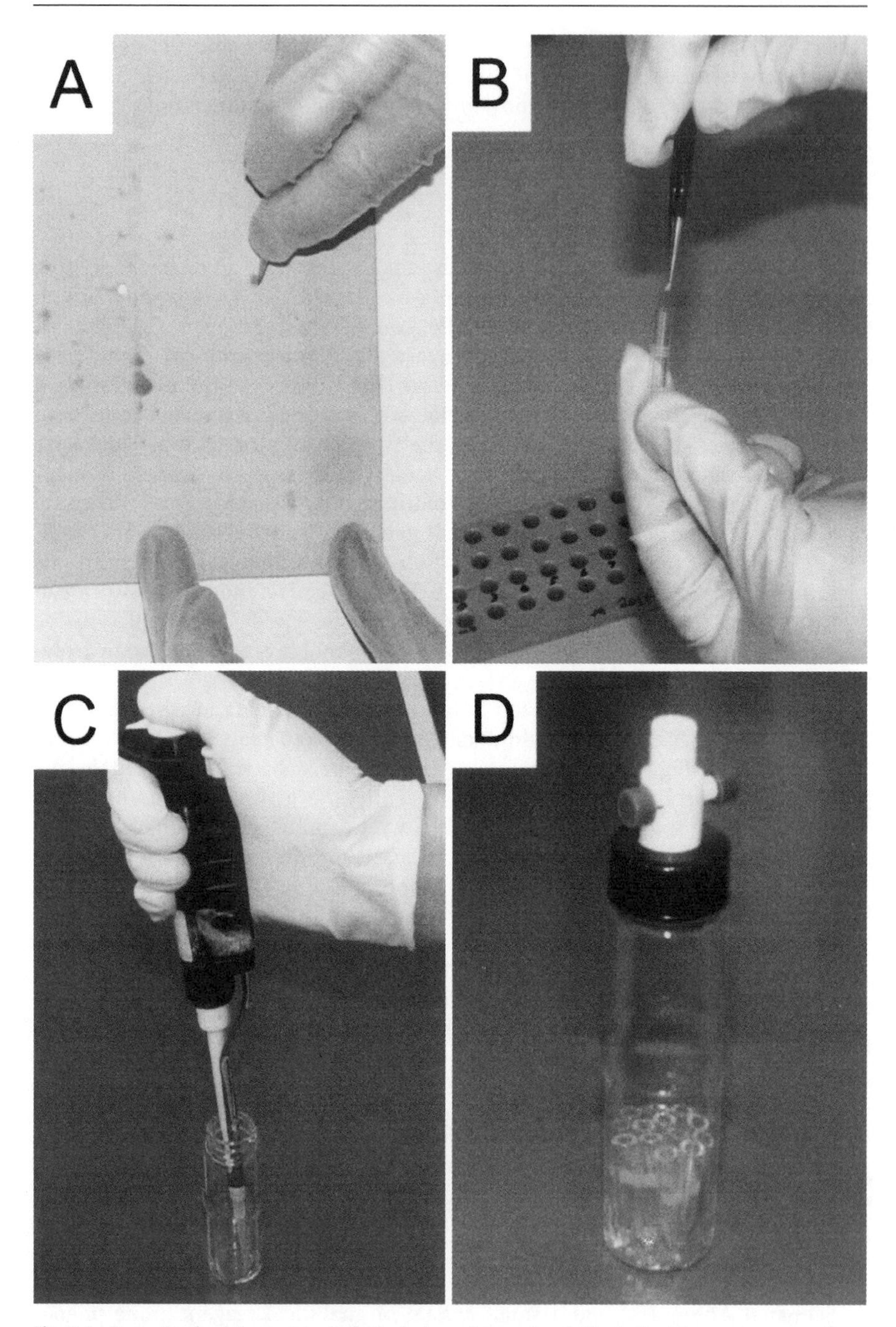

Fig. 7.1. Steps involved in protein hydrolysis (see also protocol 2). A. Protein spots are cut from PVDF membranes using a new scalpel blade. B. Spots are placed into autosampler vials using stainless steel tweezers. C. Acid is added to the bottom of the hydrolysis vessel. D. The vials are then added to the vessel and it is then assembled and sealed

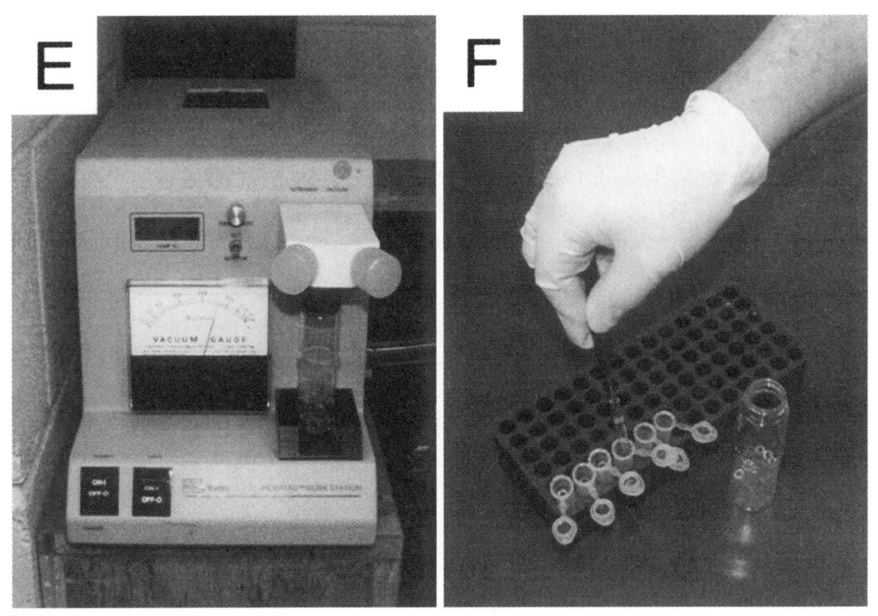

Fig. 7.1. E. The Waters Picotag hydrolysis station. Here the vessel is evacuated and flushed with argon, this is repeated twice, and evacuated once more before being subjected to the heating step. F. After hydrolysis and opening of the vessel in a fume hood, vials are removed, placed into Eppendorf tubes, and dried in a vacuum centrifuge to remove residual HCl

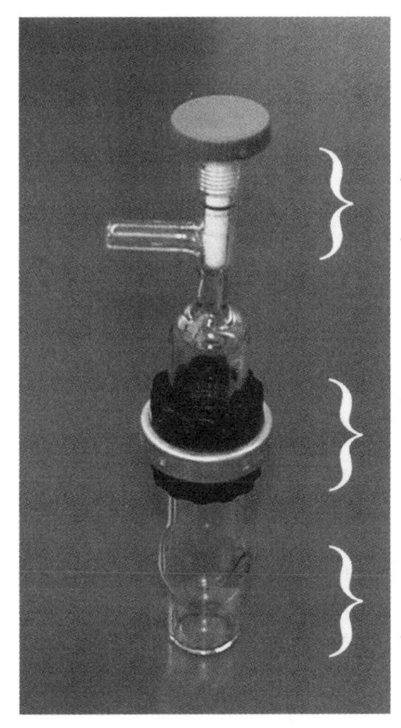

Fig. 7.2. An hydrolysis vessel that can be made by glass blowers. The hydrolysis vessel consists of a belly-bottomed flask, a threaded Corning SVL-30 fitting, and an acid and heat resistant vacuum tap. A large Kal-Rez o-ring, which is situated within the SVL-30 fitting, provides a vacuum-tight seal between the belly bottom and the top tap. Materials other than Kal-Rez were found not to be resistant to the acid and heat generated during hydrolysis conditions. (Modified from Meyer et al. 1991)

avoid getting liquid acid into the vials, they must remain upright during the hydrolysis. Empty vials should be included in the vessel if necessary for support.

5. Assemble the vessel. Evacuate for 30 s and flush with high purity nitrogen or argon. Repeat the evacuation and flush steps twice more, then evacuate and seal the vessel. Place in the 155°C oven for 1 h.
6. Remove the hydrolysis vessel from the oven, place immediately in the fume cupboard, and open the vacuum tap to release HCl vapour. If no vapour appears the vessel has leaked and the hydrolysis is likely to not have been successful.
7. Remove the vials with stainless steel tweezers and place them in 1.5-ml Eppendorf tubes. Leaving the Eppendorf tube lids open, dry under vacuum for 10 min in a Savant Speedvac to remove residual HCl.
8. Store at -20°C if unable to continue with the analysis immediately.

Note: gloves should be worn when cutting spots and handling sample vials. One should avoid breathing on samples and be aware that skin and hair can easily be a source of contamination. Safety glasses and heavy gloves must be worn when handling concentrated acid and during steps 5 and 6.

4 Extraction of Amino Acids from PVDF Membranes

After hydrolysis, the amino acids must be completely extracted from the PVDF membrane for derivatization, chromatography and quantitation. The extraction conditions must be capable of extracting all the acidic, basic, hydrophobic and hydrophilic amino acids or results will not give accurate proportions of amino acids for identification. For the extraction of amino acids (see Protocol 3), the PVDF-bound samples are kept in the same vials as used in the hydrolysis procedure, thus minimizing sample handling and loss. The protocol below is a single step extraction, and is essentially as described by Yan et al. (1996).

Protocol 3. *Extraction of Amino Acids from PVDF Membranes*

1. Prepare a fresh extraction solution of 60 % v/v acetonitrile in 0.01 % v/v TFA. Use MilliQ grade water or equivalent.
2. Add 150μl of the extraction solution to each autosampler vial containing a PVDF spot.
3. Place glass hydrolysis tubes into 1.5 ml Eppendorf tubes, close the lids and place upright in a foam rack floating in an ultrasonic water bath. Sonicate for 10 min.
4. Remove PVDF membranes from each vial with a stainless steel hypodermic needle. Discard the PVDF. Rinse the needle well with extraction solution between vials. It is essential to use a stainless steel needle as needles plated with other metals can lead to metal contamination, which inhibits the derivatization of several amino acids.
5. Place the vials carrying extraction solution in a vacuum centrifuge and evaporate completely to dryness.

6. Resuspend amino acids in a buffer appropriate to your amino acid analysis system. If using the Fmoc derivatization method (protocol 4) add 10 μl 250 mM sodium borate buffer pH 8.8 to each vial, and mix carefully by pipetting.

5 Derivatization and Chromatography

There are many methods for the derivatization and chromatography of amino acids which can give satisfactory results with hydrolysates from PVDF membranes. Derivatization chemically modifies the amino acids' amino groups with a reagent which makes the resulting amino acid derivative detectable via fluorescence or UV absorbance. There are precolumn and postcolumn derivatization chemistries. Ninhydrin, for example, is used as a postcolumn derivatization method in which the amino acids are derivatized after they elute from an ion-exchange column but before detection. By contrast, in the precolumn derivatization methods the amino acids are derivatized before they are chromatographed. The common derivatization methods are summarised in Table 7.1.

We have had experience with three methods that couple precolumn derivatization chemistry with reversed phase chromatography: Fmoc (9-fluorenylmethyl chloroformate), PTC (phenylthiocarbamyl) and ACQ (6-aminoquinolyl-N-hydroxysuccinimidyl carbamate). All three methods have been used successfully to generate data for protein identification. Table 7.2 shows a comparison of analysis results and the AACompIdent identification score we obtained for 1 μg (15 pmol) of membrane-bound bovine serum albumin (BSA) analysed by each of these methods. The data are also presented graphically in Fig. 7.3 where it can be seen that there is very little variation for individual amino acids by these three methods. The Fmoc method proved to be the most sensitive in our hands and was able to obtain AACompIdent scores of < 10 (see Sect. 7) with as little as 2 pmol of BSA. Note that Table 7.2 shows how amino acid composition data are presented for protein identification by matching against protein databases.

The Association of Biomolecular Resource Facilities (ABRF) undertook studies in 1995 (Mahrenholz et al. 1996) and 1997 (Mahrenholz et al. 1997) to investigate the accuracy of various amino acid analysis methods for the analysis of proteins bound to PVDF membranes. Fmoc, ACQ, PTC, OPA or ninhydrin methods were used by participants. Some laboratories performed well with any of

Table 7.1. Comparison of different derivatization chemistries for amino acid analysis

Derivatization	Ninhydrin	OPA	OPA	PITC	Fmoc	ACQ
type	Postcolumn	Postcolumn	Precolumn	Precolumn	Precolumn	Precolumn
Detection mode	Col.	Fluo.	Fluo.	UV	Fluo.	Fluo.
Sensitivity	pmol	fmol	fmol	pmol	fmol	fmol
Derivative stability	n/a	n/a	Poor	Good	Excellent	Excellent
Reaction kinetics	Slow	Rapid	Rapid	Moderate	Rapid	Rapid
Chromatography	I.E.	I.E.	R.P.	R.P.	R.P.	R.P.

OPA, orthophthalaldehyde; PITC, phenylisothiocyanate; Fmoc, 9-fluorenylmethyl chloroformate; ACQ, 6-aminoquinolyl-N-hydroxysuccinimidyl carbamate; Col., colorimetry; Fluo., fluorescence; UV, ultra-violet absorbance; n/a not available; I.E., ion exchange; R.P., reversed phase.

these methods but there was an overall average mol% error of about 20 % in both trials, which was two-fold higher than that found in a previous study of non-membrane bound proteins. However, it was found that even with an average mol % error of 15 %, it was possible to achieve correct identification of proteins. Most laboratories had difficulty with background amino acids, particularly Gly, Ser and Glu, when analysing a 1-μg sample. This highlights the need for great care to be taken with every step of analysis of PVDF-bound protein.

Table 7.2. Amino acid analysis of 1 ug PVDF-bound BSA using different chemistries. The AACompIdent score is discussed in Section 8 of this chapter

Amino acid	Mol% BSA expected	Mol% Fmoc	Mol% PITC	Mol% AQC
Asx	9.9	9.4	9.5	10.1
Glx	14.4	13.3	16.6	14.5
Ser	5.1	5.5	5.5	4.7
His	3.1	2.3	3.4	2.7
Gly	2.9	4.1	4.9	3.8
Thr	6.1	6.1	5.4	5.9
Ala	8.6	8.9	8.1	8.9
Pro	5.1	5.7	3.8	5.4
Tyr	3.6	3.8	4.0	3.7
Arg	4.2	4.7	4.4	3.9
Val	6.6	6.7	5.5	6.5
Met	0.7	0.6	0.9	0.7
Ile	2.6	2.8	2.1	2.6
Leu	11.2	10.9	10.3	11.6
Phe	4.9	5.2	4.6	5.1
Lys	10.8	9.9	11.0	10.0
AA CompIdent score	5	14	2	

Fig. 7.3. Amino acid analysis of 1μg (15 pmol) of PVDF-bound bovine serum albumin using three different derivatization chemistries. See Table 7.2 for a numerical presentation of this data

5.1
Derivatization of Amino Acids with Fmoc

Here we will present methods for the derivatization of protein hydrolysates with Fmoc. We will not describe protocols for other methods, as these are adequately documented by equipment suppliers. Fmoc was first used for precolumn derivatization by Einarsson et al. (1983). The method was later simplified to avoid the necessity for removal of the Fmoc reagent prior to chromatography (Haynes et al. 1991a,b). The Fmoc chemistry has been found to be particularly suitable for proteome studies because of its sensitivity, and because the method has proven amenable to automation and thus high throughput (Ou et al. 1996; Yan et al. 1996). A method for the manual derivatization of amino acids using Fmoc (following Haynes et al. 1991a) is shown in protocol 4, and the automation of this for the GBC Aminomate System (Dandenong, Victoria, Australia) is described in Table 7.3 (Ou et al. 1996). These methods can be adapted to other instrumentation if needed.

Protocol 4. *Manual Derivatization of Amino Acids with Fmoc*
1. Make reagents according to Table 7.4.
2. Ensure all reagents and samples are at room temperature.
3. To the hydrolysate dissolved in 10 μl 250 mM sodium borate buffer pH 8.8 (step 6, protocol 3) add 10 μl Fmoc reagent and mix thoroughly. Allow to stand 90 s.
4. Add 10 μl cleavage reagent and mix thoroughly. Allow to stand for 3.5 min.
5. Add 10μl of quenching reagent and mix thoroughly.
6. The derivatized amino acids are now ready for injection into the HPLC system

Table 7.3. Sequenze for automated derivatization of amino acids, as described in Out et al. (1996). In automated processing of a batch of samples, the first instruction set used is steps 1–34. This derivatizes and injects the first sample, and then derivatizes the second sample during the chromatography of the first. The second instruction set used is steps 15–34, which injects the second sample and derivatizes the third sample during the chromatography of the second. Steps 15–34 are used for all further samples with the exception of the last, which only requires steps 15–16

Autosampler event	
1. Wash needle	18. Needle out of line
2. Get Fmoc	19. Wait 23 min
3. Put Fmoc to sample vial *n*	20. Wash needle
4. Mix solution	21. Get Fmoc
5. Wait 60 s	22. Put Fmoc to sample vial *n*+1
6. Wash needle	23. Mix solution
7. Get cleavage reagent	24. Wait 60 s
8. Put cleavage reagent to sample vial *n*	25. Wash needle
9. Mix solution	26. Get cleavage reagent
10. Wait 200 s	27. Put cleavage reagent to sample vial *n*+1
11. Wash needle	28. Mix solution
12. Get quenching reagent	29. Wait 200 s
13. Put quenching reagent to sample vial *n*	30. Wash needle
14. Mix solution	31. Get quenching reagent
15. Wash needle	32. Put quenching reagent to sample vial *n*+1
16. Inject sample (needle in line)	33. Mix solution
17. Wait 45 s	34. Wait for gradient finish

Table 7.4. Stock solutions for derivatization of amino acids with Fmoc

Solution	To be made
Fmoc (4 mg/ml in acetonitrile)	Daily
Cleavage reagent: 340 μl of 0.85 M sodium hydroxide, 150 μl of 0.5 M hydroxylamine hydrochloride, and 10 μl of 2-(methylthio) ethanol	Daily
NaOH stock (0.85 M)	Weekly
Hydroxylamine hydrochloride stpck (0.5 M)	Weekly
Quenching Reagent: 20 % v/v glacial acetic acid in acetonitrile	Weekly

Table 7.5. Ternary gradient for separation of Fmoc amino acids, using 5.5 mM ammonium dihydrogen orthophosphate (Ou et al. 1996)

Time (min)	% A	% B	% C
0	18	66	16
1	18	66	16
31	11	43	46
31.05	0	0	100
34	0	0	100
34.05	20	64	16
35.05	18	66	16

Solution A, 30 mM ammonium dihydrogen orthophosphate pH 6.5 in 15 % (v/v) methanol.
Solution B, 15 % (v/v) methanol.
Solution C, 90 % (v/v) acetonitrile.

5.2 Chromatography and Detection

The HPLC separation of Fmoc amino acids is achieved using a 4.6 × 150 mm inner diameter column of 5-μm particle size of ODS-Hypersil (Keystone Scientific or SGE), with a flow rate of 1 ml/min (Ou et al. 1996). The gradients and mobile phases are described in Table 7.5. Detection of Fmoc amino acids is achieved via fluorescence, with excitation of 270 nm and detection of 316 nm. A typical chromatogram of 250-pmol standard amino acids is shown in Fig. 7.4, as is the separation of a protein hydrolysate.

6 Amino Acid Analysis Troubleshooting Guide

Amino acid analysis of PVDF-bound proteins is a challenging procedure, and it may be difficult to know how to solve a problem if one does arise. Below we present a troubleshooting guide to assist in the diagnosis of some of the common problems in amino acid analysis of PVDF-bound proteins.

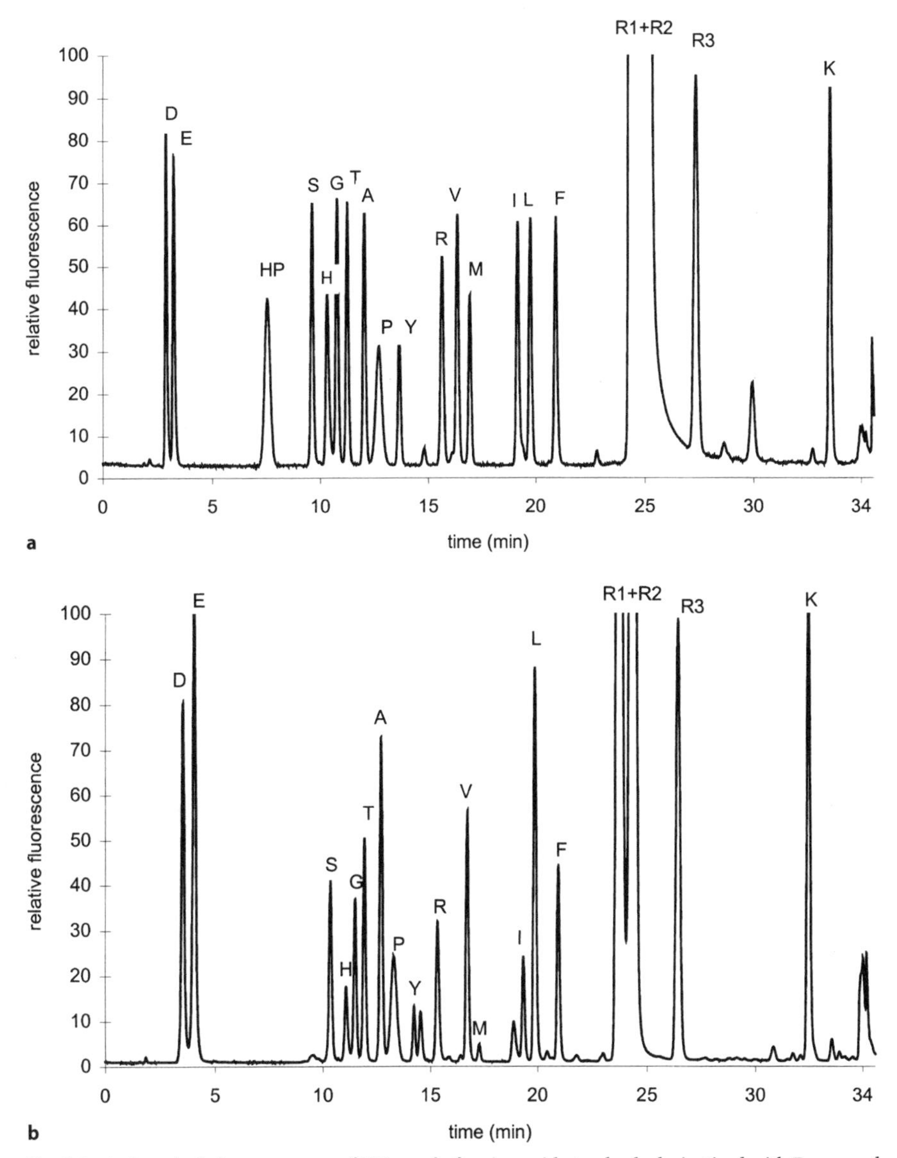

Fig. 7.4. A. A typical chromatogram of 125 pmol of amino acid standards derivatised with Fmoc, and separated according to the chromatography described. B. Chromatogram showing separation of a hydrolysate of PVDF-bound bovine serum albumin. *HP* hydroxyproline internal standard; *R1* Fmoc-hydroxylamine; *R2* Fmoc-OH; *R3* reagent peak present in blank derivatization, other peaks labelled with the standard one amino acid code

6.1 Hydrolysis Troubleshooting

Errors during hydrolysis usually fall into the categories of
(1) destruction of amino acids,
(2) incomplete hydrolysis, or
(3) lack of cleanliness during sample handling. Met and Tyr are the amino acids most sensitive to destruction during hydrolysis. They provide a sensitive way to evaluate hydrolysis quality by looking at their values in a calibration protein. The most likely cause for low Met and Tyr is poor quality HCl or the presence of oxygen from inadequate purging and evacuation of the hydrolysis vessel.

Incomplete hydrolysis will result in a low yield of most amino acids, and in particular the hydrophobic amino acids Ile, Leu and Val, whose amide bonds are more difficult to hydrolyse. A longer hydrolysis time may be used if incomplete hydrolysis is suspected, but this may cause losses for Ser, Thr and Tyr. A lack of acid vapour will also affect hydrolysis, so it should be checked that sufficient liquid HCl is left at the bottom of the hydrolysis vessel after evacuation and flushing with argon but before heating. Also check that the hydrolysis vessel is properly sealed, and that it remains sealed during the hydrolysis itself. If no HCl vapour is released after hydrolysis when opening the vessel in the fume cupboard, the hydrolysis vessel is likely to be leaking.

Contamination in amino acid analysis is a major issue but can be difficult to source. In our experience, high glycine levels are often carried over from 2-D gel buffers. Extensive washing of membranes before staining and hydrolysis should improve this. High levels of the amino acids glycine and serine indicate contamination from skin keratins, or from contaminated glassware or laboratory water. Dedicated glassware should be used to prepare all reagents and buffers used for amino acid analysis.

6.2 Fmoc Derivatization Troubleshooting

During the Fmoc derivatization procedure, in some cases certain amino acids may not be correctly or completely derivatised. This can lead to quantitation errors. An apparent low derivatization efficiency of His and Tyr may sometimes arise even though other amino acid levels appear normal. This is due to the cleavage reagent (Table 7.4) not being fresh. His and Tyr differ from other amino acids, in that they form di-Fmoc derivatives which must be completely converted by the cleavage reagent to the mono-substituted forms before they are quantitated.

An apparent low recovery of aspartic acid and glutamic acid may be seen whilst other amino acid levels appear normal. This can arise if the pH of the borate derivatization buffer is too low. If the pH of the borate buffer is less than 8.6, derivatization of these acidic amino acids will be very slow, and incomplete in the normal 90 second derivatization period.

6.3 Chromatography

Chromatography using the gradient given in Table 7.5 is very reliable. However, as the column ages, His may co-elute with Gly and Arg may elute later than usual. These problems can be corrected by a slight increase (0.5–1.0 mM) in ammonium dihydrogenorthophosphate concentration throughout the whole gradient.

7 Protein Identification by Database Matching

Numerous computer programs have been described to allow proteins to be identified by matching their amino acid compositions against protein databases (Eckerskorn et al. 1988; Jungblut et al. 1992; Shaw 1993; Hobohm et al. 1994; Galat et al. 1995; Wilkins et al. 1996a). The program with which we are most familiar is AACompIdent (Wilkins et al. 1996a,b; Golaz et al. 1996) that is available over the Internet at the address http://www.expasy.ch/ch2d/aacompi.html. This program matches the percent empirically measured AA composition of an unknown protein (e.g. those in Table 7.2) against the theoretical percent AA compositions of proteins in the SWISS-PROT and/or TrEMBL databases. A numerical score, which measures the degree of difference between the composition of the query protein and a protein in the database, is calculated for each database entry from the sum of the squared difference between the percent AA composition for all amino acids of the unknown protein and the database entry (Wilkins et al. 1996a). All proteins in the database are then ranked according to their score, from lowest (best match) to highest (worst match), and additional identification parameters (e.g. protein pI and mass) are applied. The best ranked proteins are shown in the results.

7.1 Using the AACompIdent Tool

Extensive documentation and instructions for the AACompIdent program are contained on the Internet page and are given in Wilkins et al. (1998a). Protocol 5 provides abbreviated details of how to best use this program.

Protocol 5. Use of the AACompIdent Program for Protein Identification

1. On a computer connected to the Internet, go to the AACompIdent tool at the address
2. Choose the relevant AA constellation to use in matching. For AA compositions determined by the methods described here, use Constellation 2.
3. Specify the e-mail address to which the results should be sent.
4. Enter analytical data that will be matched against the database. At a minimum, you must enter one or more species against which the compositional data should be matched (e.g. ESCHERICHIA COLI or PROKARYOTA or ALL), and the amino acid composition of the query protein in mol%.

5. Enter any other data that should be used to increase the search power. This might include (1) protein pI and mass estimated from 2-D gels, and the confidence limits for these values; (2) the SWISS-PROT name and amino acid composition data of a calibration protein that was analysed at the same time as the query protein; and/or (3) keywords to which the search should be restricted.
6. Click on the search button. The results should arrive by e-mail in a few minutes

7.2 Interpretation of AACompIdent Results

The AACompIdent output shows three lists, all of which have proteins ranked according to their amino acid difference score (Fig. 7.5). The first list contains the results of matching the AA composition of the query protein against all proteins from the species of interest that have the the specified keyword (if any), but without considering the specified pI and molecular weight. The second list is the result of matching the AA composition of the query protein against all proteins from all species in SWISS-PROT that have the specified keyword (if any), again without considering pI and Mw. The third list shows the results of matching the AA composition of the query protein only against the proteins from the species of interest

```
SpotNb BSA FMOC
==============
pI:    5.00     Range:    (3.00, 7.00)
Mw:   60000     Range: (30,000, 90,000)

The closest SWISS-PROT entries (in terms of AA composition)
for the species BOS TAURUS and the specified keyword:

Rank Score    Protein    (pI        Mw)  Description
===============================================
   1      5 ALBU_BOVIN    5.60     66433 SERUM ALBUMIN
   2     30 CNRC_BOVIN    5.54     98798 CONE CGMP-SPECIFIC 3',5'-CYCLIC
   3     30 PGF2_BOVIN    6.80     36742 PROSTAGLANDIN-F SYNTHASE 2 (EC 1.1.1.188)
   4     31 PLC1_BOVIN    6.42     23022 PLACENTAL LACTOGEN
   5     31 TRFL_BOVIN    8.67     76144 LACTOTRANSFERRIN

The closest SWISS-PROT entries (in terms of AA composition)
for the specified keyword and any species:

Rank Score    Protein    (pI        Mw)  Description
===============================================
   1      5 ALBU_BOVIN    5.60     66433 SERUM ALBUMIN
   2      7 ALBU_SHEEP    5.58     66328 SERUM ALBUMIN
   3     10 ALBU_RABIT    5.65     66020 SERUM ALBUMIN
   4     11 ALBU_HORSE    5.72     65752 SERUM ALBUMIN
   5     12 ALBU_RAT      5.80     65904 SERUM ALBUMIN

The SWISS-PROT entries having pI and Mw values in the specified range
for the species BOS TAURUS and the specified keyword:

Rank Score    Protein    (pI        Mw)  Description
===============================================
   1      5 ALBU_BOVIN    5.60     66433 SERUM ALBUMIN
   2     30 PGF2_BOVIN    6.80     36742 PROSTAGLANDIN-F SYNTHASE 2 (EC 1.1.1.188)
   3     38 SUPP_BOVIN    6.32     34017 PHENOL-SULFATING PHENOL SULFOTRANSFERASE
   4     39 ARRS_BOVIN    6.08     45275 S-ARRESTIN (RETINAL S-ANTIGEN) (48 Kda)
   5     41 PDI_BOVIN     4.73     55223 PROTEIN DISULFIDE ISOMERASE
```

Fig. 7.5. Sample output from AACompIdent, showing the correct identification of bovine serum albumin (BSA) by its amino acid composition. One microgram of PVDF-bound protein was analysed by Fmoc amino acid analysis. Compositional data used to obtain this result are shown in Table 7.2

that lie within the specified pI and Mw range and which also have the appropriate keyword (if any). The third list is the most restricted and potentially the most powerful search. In all lists, a score of 0 is a perfect match between the query protein and a protein in the database, with larger scores indicating increasing differences.

We have shown that a top-ranked protein is likely to be a correct identification if three conditions are met (Wilkins et al. 1996a). Firstly, the same protein, or type of protein, should be ranked top of all three lists. Secondly, the top-ranked protein from the third list should have a score less than 30 (indicating a "good fit" of the query protein with that particular database entry). Finally, the third list should show a large score difference (e.g. a factor of 2) between the top-ranked protein and the second ranked protein, indicating a unique matching of the query protein with the top-ranked database entry. The identification in Fig. 7.5 clearly meets with these three criteria, and would be classed as a correct identification. When these criteria are not met, additional parameters for the query proteins must be obtained to ensure that confident identification is achieved.

8
Identification by N-terminal Sequence Tags and Amino Acid Composition

The N-terminal sequence of a protein can be determined by automated Edman degradation. A sequence of 10 to 15 amino acids is usually sufficient to identify a protein by comparison with proteins in a database. However, when combined with other criteria such as pI, molecular weight and amino acid composition, a sequence tag of 3–5 amino acids can also lead to unambiguous protein identification (Wilkins et al. 1996b; 1998b). This is possible because sequence tags are highly specific. For example, there are 8,000 possible combinations of amino acids in a 3 amino acid sequence tag, 160,000 for a tag of 4 residues and 3,200,000 for 5 residues.

Whilst it may be desirable to sequence many proteins for 10 to 15 amino acids, Edman degradation remains slow and expensive. An alternative to extensive sequencing is that PVDF-bound proteins be sequenced for 4 amino acids to generate a sequence tag, following which the same protein samples are used for amino acid analysis. If proteins are N-terminally blocked or present in subpicomole quantities, no N-terminal sequence will be obtained. However, the subsequent amino acid analysis of the protein on the PVDF membrane will allow the amount of protein present (taking into account the apparent molecular weight from the blot) to be estimated. If there is more protein present in the sample than the minimum needed for sequencing, the protein is most probably blocked.

Sequence tag and amino acid composition data can be matched against databases using AACompIdent to achieve protein identification (Wilkins et al. 1996b). In one case of two closely related *E. coli* proteins, LIVJ and LIVK, amino acid composition data together with pI and molecular weight were not able to distinguish between them when submitted to AACompIdent (Ou et al. 1998). However, an N-terminal sequence tag in combination with amino acid composition data proved valuable, because the N-terminal sequence of LIVJ is EDIK and of LIVK is DDIK. An AACompIdent identification of LIVJ using sequence tag and amino acid compositional data is shown in Fig. 7.6.

```
SpotNb 16-160
===========
pI:    5.40     Range:    (4.90, 5.90)
Mw:   30,000    Range: (15,000, 45,000)

The closest SWISS-PROT entries (in terms of AA composition)
for the species ESCHERICHIA and the specified keyword:
Rank Score    Protein     pI          Mw    N-terminal Sequence
=======================================================
*  1      25 LIVJ_ECOLI     5.28      36744 edikvAVVGAMSGPVAQYGDQEFTGAEQAVADINAKGGIK
   2      28 LIVK_ECOLI     5.00      36911 DDIKVAVVGAMSGPIAQWGIMEFNGAEQAIKDINAKGGIK
   3      35 PSTS_ECOLI     6.92      34422 EASLTGAGATFPAPVYAKWADTYQKETGNKVNYQGIGSSG
   4      38 GABD_ECOLI     5.44      51720 MKLNDSNLFRQQALINGEWLDANNGEAIDVTNPANGDKLG
   5      41 SUCC_ECOLI     5.37      41393 MNLHEYQAKQLFARYGLPAPVGYACTTPREAEEAASKIGA

The closest SWISS-PROT entries (in terms of AA composition)
for the specified keyword and any species:
Rank Score    Protein     pI          Mw    N-terminal Sequence
=======================================================
   1      24 LIVJ_SALTY     5.37      36641 DDIKVAVVGAMSGPVAQYGDQEFTGAEQAIADINAKGGIK
   2      25 BRAC_PSEAE     5.24      36963 ADTIKIALAGPVTGPVAQYGDMQRAGALMAIEQINKAGGV
*  3      25 LIVJ_ECOLI     5.28      36744 edikvAVVGAMSGPVAQYGDQEFTGAEQAVADINAKGGIK
   4      26 BCHX_RHOSH     4.89      35576 MTDAPELKAFDQRLRDEAAEEPTLEVPQGEPKKKTQVIAI
   5      28 PUR2_YARLI     5.06      83759 MSLRILLVGNGGREHALAWKLAQSPLVERIFVAPGNGGTD

The SWISS-PROT entries having pI and Mw values in the specified range
for the species ESCHERICHIA and the specified keyword:
Rank Score    Protein     pI          Mw    N-terminal Sequence
=======================================================
*  1      25 LIVJ_ECOLI    5.28      36744 edikvAVVGAMSGPVAQYGDQEFTGAEQAVADINAKGGIK
   2      28 LIVK_ECOLI     5.00      36911 DDIKVAVVGAMSGPIAQWGIMEFNGAEQAIKDINAKGGIK
   3      41 SUCC_ECOLI     5.37      41393 MNLHEYQAKQLFARYGLPAPVGYACTTPREAEEAASKIGA
   4      42 GGT_ECOLI      5.21      39198 APPAPPVSYGVEEDVFHPVRAKQGMVASVDATATQVGVDI
   5      44 ARAF_ECOLI     5.61      33210 ENLKLGFLVKQPEEPWFQTEWKFADKAGKDLGFEVIKIAV
```

Fig. 7.6. Sample output from AACompIdent, showing the correct identification of the protein LIVJ with compositional data and protein sequence tag. The PVDF-bound protein sample was first subjected to five cycles of Edman degradation to generated the sequence tag, following which the same protein sample was used for Fmoc amino acid analysis. All data were used in the database searching procedure. Asterisks indicate if the protein carries the sequence tag anywhere in its sequence

8.1
Using the AACompIdent Tool with Sequence Tags and Compositional Data

To use amino acid and sequence tag data together for protein identification, fill out the AACompIdent form as in protocol 5 but do not immediately press the search button. Go to the end of the form, select the tagging option by clicking in the check box, and enter a protein sequence tag of up to 6 amino acids in single amino acid code into the tag box. Then specify if the sequence tag is N- or C-terminal, and press the search button. Results will be sent to you by e-mail.

When the sequence tag option is selected, the AACompIdent output will show 40 amino acids of each protein's predicted N- or C-terminal sequence instead of its description, and show an asterisk to the left of a protein's rank if the protein carries the sequence tag anywhere it its sequence (Fig. 7.6). If the tag is found in the displayed N- or C-terminal sequence, it will be shown in lowercase letters. We are confident that a protein from SWISS-PROT represents a correct identification if the query protein's empirically determined sequence tag of 3 or more amino acids is present at the expected N- or C-terminal position, and that this protein is ranked within the first 10 or so closest entries by amino acid composition.

9
Conclusions

In this chapter, we have discussed methods for the identification of proteins by their amino acid composition, and with the use of small sequence tags. As a final note, it is worthwhile discussing the variety of applications to which these methods have been recently applied, which might serve as useful references to the reader. Protein identification by amino acid composition, pI and molecular weight has been used for the identification of proteins from molecularly well-defined organisms such as *E. coli*, and yeast (Pasquali et al. 1996; Sanchez et al. 1996; Wilkins et al. 1996a). Identification with sequence tags and amino acid composition has been applied to proteins from human sera and tears (Wilkins et al. 1996b; Molloy et al. 1997a), and has been used to identify proteins across species boundaries (Cordwell et al. 1995; Galat and Rioux 1997; Molloy et al. 1997b). In some cases, amino acid compositional data has also been combined with peptide mass fingerprinting information, in both well-characterized organisms (Langen et al. 1997) and for cross-species identification (Wasinger et al. 1995; Wheeler et al. 1996). Finally, the large-scale trials of the ABRF also now include protein identification by amino acid composition, which serves as a useful reference to see what problems are generally experienced by laboratories when undertaking amino acid analysis for the purpose of protein identification (Mahrenholz et al. 1996; 1997).

Acknowledgements: These methods have been developed over many years and have benefited from the input of many people. Specifically, we acknowledge input from Jun Yan, Keli Ou, Yik Fung, Andrew Gooley, Ron Appel, Jean-Charles Sanchez, Denis Hochstrasser and Keith Williams. We thank Simon Gates for analytical data used in some tables and figures.

References

Bjellqvist B, Sanchez J-C, Pasquali C, Ravier F, Paquet N, Frutiger S, Hughes GJ, Hochstrasser DF (1993) Micropreparative 2-D electrophoresis allowing the separation of milligram amounts of proteins. Electrophoresis 14: 1375–1378

Cordwell SJ, Wilkins MR, Cerpa-Poljak A, Gooley AA, Duncan M, Williams KL, Humphery-Smith I (1995) Cross-species identification of proteins separated by two-dimensional gel electrophoresis using matrix-assisted laser desorption ionisation/time-of-flight mass spectrometry and amino acid composition. Electrophoresis 16: 438–443

Eckerskorn C, Jungblut P, Mewes W, Klose J, Lottspeich F (1988) Identification of mouse brain proteins after two-dimensional electrophoresis and electroblotting by microsequence analysis and amino acid composition. Electrophoresis 9: 830–838

Einarsson S, Josefsson B, Lagerkvist S (1983) Determination of amino acids with 9-fluorenylmethyl chloroformate and reversed-phase high-performance liquid chromatography. J Chromatogr 282: 609–618

Galat A, Rioux V (1997) Convergence of amino acid compositions of certain groups of protein aids in their identification on two-dimensional electrophoresis gels. Electrophoresis 18: 443–451

Galat A, Bouet F, Riviere S (1995) Amino acid composition of proteins and their identities. Electrophoresis 15: 1466–1486

Golaz O, Wilkins MR, Sanchez JC, Appel RD, Hochstrasser DF, Williams KL (1996) Identification of proteins by their amino acid composition: an evaluation of the method. Electrophoresis 17: 573–579

Haynes PA, Sheumack D, Kibby J, Redmond JW (1991a) Amino acid analysis using derivatization with 9-fluorenylmethyl chloroformate and reversed-phase high performance liquid chromatography. J Chromatogr 540: 177–185

Haynes PA, Sheumack D, Greig LG, Kibby J, Redmond JW (1991b) Applications of automated amino acid analysis using 9-fluorenylmethyl chloroformate. J Chromatogr 588: 107–114

Hobohm U, Houthaeve T, Sander C (1994) Amino acid analysis and protein database compositional search as a rapid and inexpensive method to identify proteins. Anal Biochem 222: 202–209

Hoogerheide J, Campbell C (1992) Determination of cysteine plus half-cystine in protein and peptide hydrolysates: Use of dithiodiglycolic acid and phenyl-isothiocyanate derivatization. Anal Biochem 201: 146–151

Jungblut P, Dzionara M, Klose J and Wittmann-Leibold B (1992) Identification of tissue proteins by amino acid analysis after purification by two-dimensional electrophoresis. J Protein Chem 11: 603–612

Langen H, Gray C, Roder D, Juranville JF, Takacs B, Fountoulakis M (1997) From genome to proteome: protein map of *Haemophilus influenzae*. Electrophoresis 18: 1184–1192

Latter GI, Burbeck S, Fleming J, Leavitt J (1984) Identification of polypeptides on two-dimensional electrophoresis gels by amino acid composition. Clin Chem 30: 1925–1932

Mahrenholz AM, Denslow ND, Andersen TT, Schegg KM, Mann K, Cohen SA, Fox JW, Yüksel KU (1996) Amino acid analysis–Recovery from PVDF membranes: ABRF-95AAA collaborative trial. In: Techniques in protein chemistry VII. Academic Press, San Diego, pp 323–330

Mahrenholz AM, Andersen TT, Bao Y, Cohen SA, Denslow ND, Hulmes J, Hunziker PE, Mann K, Schegg KM, West K (1997) ABRF Amino acid analysis survey: Identification of proteins electroblotted to PVDF. http://www.medstv.unimelb.edu.au/ abrf/researchcommittees/aaaarticles/aaaposter2.pdf

Meyer HE, Hoffmann-Posorske E, Heilmeyer LM Jr (1991) Determination and location of phosphoserine in proteins and peptides by conversion to S-ethylcysteine. Methods Enzymol 201: 169–185

Molloy MP, Bolis S, Herbert BR, Ou K, Tyler MI, van Dyk DD, Willcox MD, Gooley AA, Williams KL, Morris CA, Walsh BJ (1997a) Establishment of the human reflex tear two-dimensional polyacrylamide gel electrophoresis reference map: new proteins of potential diagnostic value. Electrophoresis 18: 2811–2815

Molloy MP, Herbert BR, Yan JX, Williams KL, Gooley AA (1997b) Identification of wallaby milk whey proteins separated by two-dimensional electrophoresis, using amino acid analysis and sequence tagging. Electrophoresis 18: 1073–1078

Ou K, Wilkins MR, Gooley AA, Tonella L, Sanchez J-C, Tyler M, Walsh BJ, Pasquali C, Hochstrasser DF, Williams KL (1998) Rapid identification and quantitation of *E. coli* proteins separated by 2-D electrophoresis. American Biotechnology Laboratory, in press

Ou K, Wilkins MR, Yan JX, Gooley AA, Fung Y, Sheumack D and Williams KL (1996) Improved high-performance liquid chromatography of amino acids derivatised with 9-fluorenylmethyl chloroformate. J Chromatogr A 723: 219–225

Pasquali, C., Frutiger, S., Wilkins, M.R., Hughes, G.J., Appel, R.D., Bairoch, A., Schaller, D., Sanchez, J.-C., Hochstrasser, D.F. (1996) Two-dimensional gel electrophoresis of Escherichia coli homogenates: the E. coli SWISS-2DPAGE database. Electrophoresis 17: 547–555

Sanchez JC, Golaz O, Frutiger S, Schaller D, Appel RD, Bairoch A, Hughes GJ, Hochstrasser DF (1996) The yeast SWISS-2DPAGE database. Electrophoresis 17: 556–565

Sanchez J-C, Rouge V, Pisteur M, Ravier F, Tonella L, Wilkins MR, Hochstrasser DF (1997a) Improved and simplified sample application using reswelling of dry immobilized pH gradients. Electrophoresis 18: 324–327

Sanchez J-C, Wilkins MR, Appel RD, Williams KL, Hochstrasser, D.F. (1997b) Identifying proteins for proteome studies: a two-dimensional gel electrophoresis approach. In: Crighton TE (ed) Protein function–a practical approach, 2nd Edition. pp 1–27

Shaw G (1993) Rapid identification of proteins. Proc Natl Acad Sci USA 90: 5138–5142

Wasinger VC, Cordwell SJ, Cerpa-Poljak A, Yan JX, Gooley AA, Wilkins MR, Duncan MW, Harris R, Williams KL, Humphery-Smith I (1995) Progress with gene-product mapping of the Mollicutes: *Mycoplasma genitalium*. Electrophoresis 16: 1090–1094

Wheeler CH, Berry SL, Wilkins MR, Corbett JM, Ou K, Gooley AA, Humphery-Smith I, Williams KL, Dunn MJ (1996) Characterisation of proteins from two-dimensional electrophoresis gels by matrix-assisted laser desorption mass spectrometry and amino acid compositional analysis. Electrophoresis 17: 580–587

Wilkins MR, Pasquali C, Appel RD, Ou K, Golaz O, Sanchez J-C, Yan JX, Gooley AA, Humphery-Smith I, Williams KL, Hochstrasser DF (1996a) From proteins to proteomes: large scale protein identification by two-dimensional electrophoresis and amino acid analysis. Bio/Technology 14: 61–65

Wilkins MR, Ou K, Appel RD, Sanchez J-C, Yan JX, Golaz O, Farnsworth V, Cartier P, Hochstrasser DF, Williams KL, Gooley AA (1996b) Rapid protein identification using N-terminal "sequence tag" and amino acid analysis. Biochem Biophys Res Commun 221: 609–613

Wilkins MR, Gasteiger E, Sanchez J-C, Williams KL, Appel RD, Hochstrasser DF (1998a) Protein identification and analysis tools in the ExPASy server. In: Link AJ (ed) 2-D Protocols for Proteome Analysis Humana Press. pp 531–552

Wilkins MR, Gasteiger E, Tonella L, Ou K, Sanchez J-C, Tyler M, Gooley AA, Williams KL, Appel RD, Hochstrasser DF (1998b) Protein identification with N- and C-terminal sequence tags in proteome projects. J Mol Biol 278: 599–608

Yamada H, Moriya H, Tsugita A (1991) Development of an acid hydrolysis method with high recoveries of tryptophan and cysteine for microquantities of protein. Anal Biochem 198: 1–5

Yan JX, Wilkins MR, Ou K, Gooley AA, Williams KL, Sanchez JC, Golaz O, Pasquali C, Hochstrasser DF (1996) Large scale amino acid analysis for proteome studies. J Chromatogr A 736: 291–302

Yan JX, Kett WC, Herbert BR, Gooley AA, Packer NH, Williams KL (1998) Identification and quantitation of cysteine in proteins separated by gel electrophoresis. J Chromatogr (in press)

CHAPTER 8

Identification by Amino Acid Composition obtained from Labeling

J. LABARRE[1] and M. PERROT[2]

1 Introduction

Now that the sequence of several micro-organism genomes has been completed, we are faced with the new challenge of characterizing the proteins which have been deduced from this sequence analysis. Two-dimensional (2-D) gel electrophoresis has already proven to be very instrumental in these functional studies. However, the important work of protein identification on 2-D gels is a prerequisite to these studies. Protein identification can be achieved by microsequencing and mass spectrometry. An alternative and more simple method consists of precisely measuring the isoelectric point (pI), the molecular mass (M_r) and a partial amino acid composition of the proteins to be identified. These values are then compared with theoretical pI, M_r and amino acid composition in protein databases to identify the protein matching the experimental data.

The amino acid composition of proteins separated by 2-D gel electrophoresis can be determined in two ways. The classical method consists of hydrolyzing proteins with a strong acid, and analysing the amino acid content by chromatographic techniques (see Chapt. 7). This method is simple but is restricted to relatively abundant proteins (about 300 ng of protein is needed for an accurate amino acid determination). To improve the sensitivity of this technique, labeling methods have been proposed (Latter et al. 1983, 1984). After labeling with a single amino acid, the radioactivity incorporated in a given protein is proportional to the amount of the corresponding amino acid in this protein. A prerequisite to this approach is that the label present in the marker amino acid should not be incorporated into other amino acids during the *in vivo* labeling.

Two different techniques have been described: (1) the single labeling method involves comparison of 2-D gels run with extracts separately labeled with each of the 20 amino acids; (2) the double labeling method relies on selective incorporation of two amino acids labeled with two different radioisotopes. In principle, the isotope ratio measured in a given protein spot on 2-D gel should be directly proportional to the ratio of these two amino acids for the corresponding protein. This method avoids the errors associated with 2-D gel comparison: differences in

[1] Service de Biochimie et Génétique Moléculaire, CEA-Saclay, 91191 Gif-sur-Yvette Cedex, France.
[2] Institut de Biochimie et de Génétique Cellulaires, UPR CNRS 9026, 1 rue Camille Saint Säens, 33077 Bordeaux Cedex, France.

cell labeling, in protein extraction and differential proteolysis between extracts. Good isotope couples are $^{14}C/^{35}S$, $^{35}S/^{3}H$ and $^{14}C/^{3}H$. In the two latter cases, the significant energy difference between ^{3}H and ^{14}C or ^{35}S allows the accurate quantitation of the two isotopes by scintillation counting. When radiolabeling is achieved with the two radioisotopes of similar energy ^{14}C and ^{35}S, the isotope quantitation relies on the decay of ^{35}S.

For each protein analysed, a program compares the experimental M_r, pI and amino acid ratios with the calculated values for each protein of a sequence database. A score is defined and used to rank matching proteins of the database, the top ranking protein being the best candidate for identification.

2 Choice of Labeling Amino Acids

2.1 Measurement of Amino Acid Interconversion

After labeling with a specific amino acid, the radioactivity in a given protein should be proportional to the amount of the corresponding amino acid in the protein. Thus, the first criterion in the choice of the marker amino acids is the absence of incorporation into other amino acids due to metabolic interconversions during the labeling. Such information can be obtained experimentally: after labeling with a marker amino acid, soluble proteins are extracted, hydrolyzed with strong acid and analysed for labeled amino acids. Using this method in the yeast *Saccharomyces cerevisiae*, Maillet et al. (1996) showed that 8 out of 15 amino acids analysed were metabolized weakly or not at all before incorporation into proteins (Table 8.1). The other amino acids tested were highly converted into other amino acids.

This method does not provide information about the metabolic fate of Cys, Met and Trp due to their destruction during acid hydrolysis. In these cases, interconversion rates can be estimated on 2-D gel by analysing the radioactivity incorporated in proteins devoid of Cys, Met or Trp after labeling with ^{35}S-Cys, ^{35}S-Met and ^{3}H-Trp respectively. For example, in yeast, after labeling with Trp or Cys, the very low labeled intensity of proteins known to be devoid of Trp or Cys indicates a poor interconversion of these amino acids. In contrast, labeling with ^{35}S-Met reveals a significant conversion of Met to Cys, since known proteins devoid of Met are labeled (Maillet et al. 1996).

2.2 Methods Used to Limit Interconversion

Experiments performed with yeast have shown that decreasing labeling times does not reduce amino acid interconversions (Table 8.1). However, interconversion rates can be greatly limited if an excess of the non-radioactive amino acid by-product is added to the medium (Maillet et al. 1996). For example, the presence of 1 mM unlabeled Leu in the culture medium blocks the interconversion of ^{14}C-Val to ^{14}C-Leu. Similarly, the metabolization of ^{35}S-Met to ^{35}S-Cys can be lim-

Table 8.1. Metabolic interconversion of amino acids in yeast

Labeled amino acid supplied in culture medium	Percent of radioactivity in amino acids incorporated in proteins			cytosolic pools (b)
	After 30 min	After 5 h	After 5 h (a)	
A (Ala)	**A** 43, L 29, V 18	**A** 39, L 33, V 20	(V,L) **A** 74, L 15	6.3
C (Cys)	**C** 85, M 15	**C** 85, M 15	(M) **C** > 95	n.d.
D (Asp)	**B** 32, E 9, A 5, V 7, I 14	**B** 28m E 23, I 13, R 14, A 10, P 7, V 5	n.d.	8.2
E (Glu)	**Z** 60, P 28, R 5, K 6	**Z** 40, P 16, R 14, K 27	(P,R,K) **Z** 81, P 13	24.4
F (Phe)	**F** 89	**F** 98	n.d.	0.2
G (Gly)	**G** 44, S 34, V 2, L 2	**G** 66, S 23	(S) **G** 60, S 13, V 6, L 10, A 7	2.8
H (His)	**H** 87	**H** 90	n.d.	0.2
I (Ile)	**I** 82	**I** 92	n.d.	0.5
K (Lys)	**K** 93	**K** 91	n.d.	0.5
L (Leu)	**L** 92	**L** 98	n.d.	0.5
M (Met)	**M** 67, C (33)	**M** 67, C (33)	(C) **M** >95	n.d.
P (Pro)	**P** 99	**P** 95	n.d.	n.d.
R (Arg)	n.d.	**R** 78, P 15	n.d.	0.6
S (Ser)	**S** 37, G 11, A 5, V 7, L 10	**S** 31, G 23, A 12, V 11, L 16	(G) **S** 44, G 23, A 10, V 7, L 3	2.6
T (Thr)	**T** 64, I 25	**T** 50, I 33	(I,L,V) **T** 82	1.2
V (Val)	**V** 47, L 41	**V** 52, L 42	(L) **V** 90	1.7
W (Trp)	n.d.	W > 90	n.d.	n.d.
Y (Tyr)	**Y** 85	**Y** 98	n.d.	0.3

B, Asp + Asn; Z, Glu + Gln, n.d., not done.
(a) In the third experiment acids in brackets were added to the medium at a final concentration of 1 mM.
(b) From Messenguy et al. (1980).

ited by addition of unlabeled Cys. These results are probably due to the isotope dilution of the labeled by-product.

Another method to avoid Met interconversion consists of using Met which carries the label (^{3}H or ^{14}C) on the methyl group. Metabolism of such Met results in the transfer of the methyl group (and the label) to the methylation pathway but not to the Cys biosynthetic pathway.

These interconversion data have been established in *Saccharomyces cerevisiae* but are not available with other organisms. Maillet et al. (1996) observed that amino acids that are not metabolized have the smallest cytosolic pools (Table 8.1). A possible explanation is that amino acids with small pools are rapidly incorporated in proteins, making them available for too short periods to be significantly converted. If it applies to other organisms, one could use such correlation between amino acid pool size (Messenguy et al. 1980) and amino acid interconversion to help in the decision of which amino acid to use in specific labeling experiment.

We should mention here that in principle, it is also possible to label with any amino acid if the metabolization rates are known and incorporated in the search program (see Sect. 8.2).

2.3 Other Criteria

Other criteria must be considered in the choice of radioactive amino acids:

1. They must be discriminant: usually, the less used amino acids (Trp, Cys, Met, His) have the most unequal distributions among proteins and are likely to give the most useful information.
2. The amino acids must be incorporated into proteins with high efficiency. The most efficient labeling is generally obtained with isotopes with high specific activity such as ^{35}S- and ^{3}H-labeled amino acids. U-^{14}C-amino acids have the lowest specific activity and must be used in large quantity for long-term labeling. The efficiency of amino acid transport systems should be taken into account to obtain high labeling yield. Indeed, amino acid permeases may be absent or inhibited in specific experimental conditions. Addition of excess amino acids to limit interconversion may also reduce transport of the marker and the labeling efficiency. As a general rule, rich media or media supplemented with unlabeled amino acids should be avoided.
3. The cost of labeled amino acids should be examined: ^{35}S-Met and ^{3}H-Met are the cheapest amino acids. ^{35}S-Cys and other ^{3}H-labeled amino acids are of intermediary cost. U-^{14}C-amino acids are the most expensive.
   In conclusion, according to these criteria, at least 10 different amino acids should be selected to perform a minimum of 8 different labelings.

Protocol 1: Measurement of Amino Acid Interconversion

1. Label cells with the ^{14}C-amino acid (20 to 100 μCi) in conditions that will be used in the double labeling procedure. Collect cells by centrifugation, wash with 200 μl H_2O, and resuspend the pellet in 200 μl H_2O.
2. Disrupt cells by sonication (in case of bacteria easy to lyse) or with a Mini BeadBeater (Biospec Products) for micro-organisms resistant to sonication. Precipitate proteins by adding 400 μl of 50 % trichloracetic acid. Keep on ice 1 h. After centrifugation, remove the supernatant and solubilize the pellet with 50 μl NaOH 10 N.
3. Hydrolyze proteins at 100 °C for 12 h in 6 M HCl. Evaporate HCl under vacuum. Add about 400 μl H_2O and evaporate under vacuum. Add a minimum volume (10 to 20 μl of 0.01 N HCl)
4. Samples can be analysed for ^{14}C-amino acids by HPLC (Column Adsorbosphere o-phthaldialdehyde (OPA) HS 5 μm, 100 × 4.6 mm; Alltech) equipped for OPA derivatization, essentially as described by Godel et al. (1984). Alternatively, samples can be analysed by cellulose thin layer chromatography: two different solvent systems (butanol 1/acetic acid/water 90/15/33 or isoamylalcohol/acetic acid/pyridine/water 20/5/40/20) can be used. Radiolabeled amino acids are then quantified by phosphorimager technology. Similar results are obtained by both methods.

3 Single Labeling Method

The single labeling approach was first described by Latter et al. (1983, 1984) for the analysis of proteins from a transformed human cell line and used by Garrels et al. (1994) to identify yeast proteins. The method consists in comparing several gels of cell extracts, labeled with one of the 20 ^{14}C- or ^{35}S-labeled amino acids. Spots from 2-D gels are quantified by microdensitometry of autoradiograms or by phosphor-imager technology to determine the relative amounts of radioactivity incorporated in analysed as well as in reference proteins. Reference proteins are proteins previously identified on 2-D gels by other techniques (microsequencing, mass spectrometry). Their theoretical amino acid sequences are known and they are used as internal standards to calibrate for amino acid composition. This simple method allowed the tentative identification of 17 abundant protein spots of transformed human cell lines (Latter et al. 1984). It is especially suited for the identification of spots which are particularly poor or rich in some amino acids (Giometti and Anderson 1981; Latter et al. 1983). However, this approach is inappropriate as a general spot identification method since it involves the comparison of different gels and is therefore subject to artifacts due to differences in cell labeling and protein extraction:

(1) during cell labeling, the physiological conditions are modified by the presence of the different amino acids in the culture medium (for example, Leu in case of ^{14}C Leu labeling) and may alter the expression of numerous enzymes. For example, in yeast, the addition of 0.5 mM Leu in minimal culture medium represses the expression of Leu1p, Leu2p, Ilv5p and Gdh1p and induces the expression of several amino acid biosynthetic enzymes (Maillet et al. 1996).

(2) Protein extraction for 2-D gel electrophoresis is not strictly identical from one sample to another: some proteins are not recovered with reproducible yield, probably due to problems of solubility or proteolysis during protein extraction.

4 Double Labeling Method Based on ^{35}S Decay

To avoid the artifacts mentioned above, the double labeling technique is an elegant method to obtain a ratio of amino acids in a single gel. The two radioisotopes should be chosen for their ability to be easily distinguished and quantified. The first way is to use two radionuclides with very different half-lives : ^{14}C and ^{35}S with respective half-lives of 5730 years and 87.5 days. Accurate measurement of radioactivity decay in spots allows the determination of the two radioisotopes and then the amino acid ratios (Garrels et al. 1994).

Table 8.2. Labeling methods

Labeling method	Analysis	Advantages	Disadvantages	References
Simple ^{14}C-^{35}S	Phosphorimager	Easy technique	Inaccurate Expensive[a]	Latter et al. (1984)
Double ^{14}C-^{35}S	Phosphorimager	Easy technique	Inaccurate Expensive[a]	Garrels et al. (1994)
Double ^{14}C-^{3}H	Liquid scintillation counting	Accurate Sensitive	Complex Expensive[a]	Maillet et al. (1996)
Double ^{35}S-^{3}H	Liquid sctintillation counting	Accurate Sensitive	Complex	Maillet et al. (1996)

[a] Due to the cost of U-^{14}C-labeled amino acids.

4.1 Labeling

Practically, a series of 8 to 10 double labelings is performed. In each case, cells are simultaneously labeled with a ^{35}S-amino acid (Cys or Met) and a U-^{14}C-amino acid. The labeling should satisfy the following three points:

1. A high incorporation of both radioisotopes. A difficulty appears with the use of ^{14}C-amino acids, due to their low specific activity (even when they are uniformly labeled). In this case, a longer labeling period may be necessary to obtain efficient incorporation into proteins.
2. An average isotopic ratio $^{35}S/^{14}C$ in proteins ranging from 1 to 3. These proportions are desired for the highest accuracy in subsequent quantitation of the two radioisotopes.
3. Incorporation of both isotopes throughout the labeling period. If the isotopes are not concomitantly incorporated, some errors will appear for at least two categories of proteins: proteins with very high turnover and proteins induced or repressed during the labeling period.

For example, in yeast, ^{35}S-Met is very quickly incorporated into proteins (500 μCi of ^{35}S-Met is incorporated in 10^7 cells within 10 min), whereas similar radioactivity amounts of the ^{14}C-amino acid are slowly incorporated (more than 4 h).

As a consequence, after a few hours, proteins with high turnover have a decreased ^{35}S content. Moreover, some proteins may be induced or repressed during the labeling period. For these two classes of proteins, their $^{35}S/^{14}C$ ratios are respectively underestimated or overestimated. The best way to obtain equal incorporation of both isotopes throughout the labeling period is to dilute ^{35}S-Met with unlabeled Met.

4.2 Determination of Amino Acid Ratios

After extraction and 2-D gel electrophoresis, gels are exposed and the spots to be analysed are quantified by phosphorimager technology. The radioactivity R_0 is precisely measured at time 0. This value corresponds to Eq.(1):

$$R_0 = A_C + A_S, \quad (1)$$

where A_S and A_C are respectively the ^{35}S and ^{14}C values at time 0. The quantitation is repeated several times during 3 to 4 months in order to obtain a decay curve (Garrels et al. 1994). Extrapolation of this curve to infinite times gives the A_C value. A_S is then deduced as the difference between R_0 and A_C values.

An alternative method consists of precisely measuring the remaining radioactivity R_1 after one ^{35}S decay period [Eq. (2)]:

$$R_1 = A_C + (A_S/2). \quad (2)$$

A_S and A_C are then easily deduced from R_0 and R_1 values by the following equations:

$$A_S = 2 \times (R_0 - R_1); \quad (3)$$
$$A_C = 2 \times R_1 - R_0. \quad (4)$$

The reference proteins are used to demonstrate that the deduced isotope ratios are proportional to their theoretical amino acid ratios. The calibration curves are then used to determine the amino acid ratios for the proteins analysed.

This method is simple and has been used successfully by Garrels et al. (1994) for the probable identification of 34 yeast proteins. The main drawback of this ^{35}S-based method is its intrinsic imprecision. First, each isotope value depends on two radioactive measurements. As deduced from Eqs (3) and (4), the uncertainty equations are:

$$\Delta A_s = 4 \times \Delta R \quad (5)$$
$$\Delta A_c = 3 \times \Delta R \quad (6)$$

Second, these ^{35}S and ^{14}C data are not independent : if one radioisotope is overestimated, the other isotope is underestimated [consequence of equation (1)] and the error on the isotope ratio is then increased.

For example, an error of 4 % on R_0 and R_1 determinations may lead to an error of 16 and 12 % on A_S and A_C determinations and 30 % on $^{35}S/^{14}C$ ratios.

To increase accuracy, the experiment can be independently repeated two or three times. Another way to improve precision is to increase the time between the first and the last quantitative exposures (6 to 9 months). For example, after 9 months (three times ^{35}S decay period), the uncertainty Eqs (5) and (6) become close to:

$$\Delta A_S \approx 2 \times \Delta R \text{ and } \Delta A_C \approx \Delta R.$$

With the same example of 4 % error on R_0 and R_3 determinations, the error is 8 and 4 % on A_S and A_C determinations and 13 % on $^{35}S/^{14}C$ ratios. However, these improvements increase the length of the experiment before one can get conclusive results.

5 Double Labelings Method Using Scintillation Counting.

5.1 Method

$^{3}H/^{35}S$ and $^{3}H/^{14}C$ double labeling methods are based on the energy emission difference between ^{3}H and ^{35}S or ^{14}C. This difference allows a simultaneous quantitation of each radioisotope by scintillation counting.

Cells are simultaneously labeled with a ^{3}H-labeled amino acid and a ^{35}S or a U-^{14}C-labeled amino acid. As in the previous technique (Sect. 4.1), the amount of each radioisotope should be chosen to obtain:

(1) a high radioactive incorporation,
(2) an isotope ratio $^{3}H/^{35}S$ or $^{3}H/^{14}C$ ranging from 1 to 3 (these values represent the optimum proportions for accurate scintillation counting of both radioisotopes) and
(3) the incorporation of both markers throughout the labeling period. The most efficient labeling is obtained with $^{3}H/^{35}S$ because each isotope is incorporated within 30 or 40 min. It is then possible to analyse proteins of low abundance which are transiently induced under specific conditions. These proteins can be efficiently labeled with both isotopes during their short induction period (about 30 min). This method was used by Godon et al. (1998) to identify proteins induced under heat shock and oxidative stress.

After 2-D gel electrophoresis, the spots to be analysed and reference proteins are extracted, treated with a solubilizer and a counting scintillant, and counted for ^{3}H and ^{14}C or ^{35}S isotopes. As mentioned above, the reference proteins are used to ensure that the experimental isotope ratios are proportional to the theoretical amino acid ratios (Fig. 8.1). The regression curve is then used to determine the amino acid ratios of the proteins analysed. This method has allowed the identification of more than 100 protein spots on yeast 2-D maps (Maillet et al. 1996; Godon et al. 1998).

The advantages of this method rely on the performances of liquid scintillation counting: accuracy and sensitivity. The standard deviation for the amino acid ratio measurement ranges from 5 to 20 % and spots of 100 ng or less can be analysed. However, this technique is time consuming, because for each double labeling, each spot must be independently extracted, treated and counted. An easier approach would be to quantify the two isotopes by imager technology. Unfortunately, phosphorimagers are not yet able to detect efficiently ^{3}H emission which complicates the quantitation of double labeled samples by this method. Two other recently developed techniques, the β-imager and the μ-imager technologies, allow the simultaneous and accurate quantitation of ^{3}H and ^{35}S or ^{14}C radioisotopes when the labeled sample has been previously transferred onto membrane (Sandkamp, 1997). Over the next years, these technologies should improve the double labeling analysis.

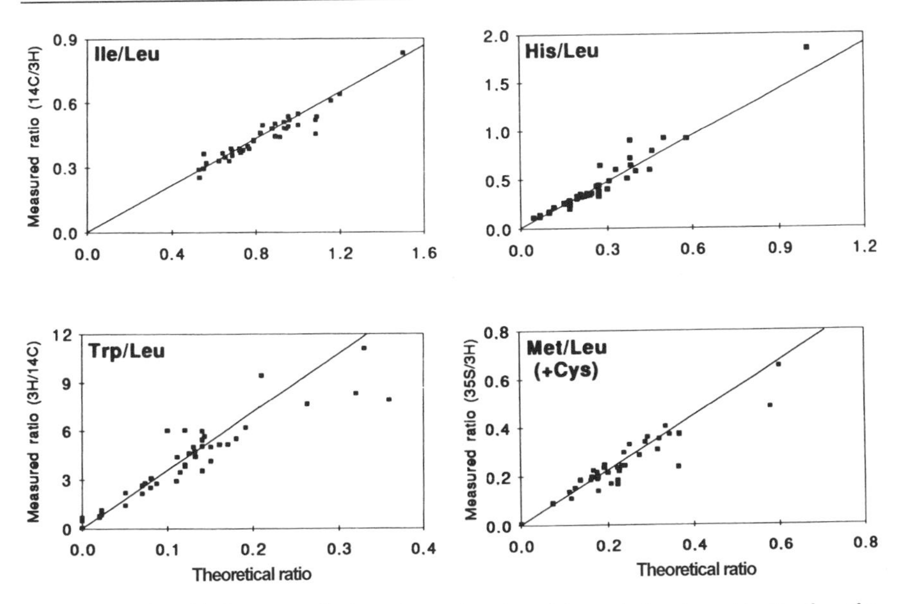

Fig. 8.1. Measured $^{14}C/^{3}H$ ratio or $^{35}S/^{3}H$ ratio as a function of theoretical amino acid ratios for reference proteins. The lines were drawn to minimize the average error. Standard deviations were 6 % for Ile/Leu, 15 % for His/Leu, 19 % for Trp/Leu, and 21 % for the Met/Leu experiment supplemented with 1 mM Cys

Protocol 2

1. Cells are grown to mid log phase in minimum medium (without amino acid). Labelings are performed on 2 ml cultures at about 0.6 OD_{600}.
2. For the ^{35}S-Met/^{3}H-labeling, add to the culture medium:
   - 500 μCi to 1 mCi of the ^{3}H-labeled amino acid
   - 200 μCi to 400 μCi of the ^{35}S-labeled Met
   - 200 μM to 1 mM unlabeled Cys to reduce ^{35}S-Met to ^{35}S-Cys conversion
   - 1 to 10 nM of unlabeled Met. This concentration should allow complete ^{35}S-Met incorporation throughout the labeling period. This concentration may differ from one microorganism to another.
3. For the ^{35}S-Cys/^{3}H-labeling, add to the culture medium:
   - 500 μCi to 1 mCi of the ^{3}H-labeled amino acid
   - 200 to 400 μCi of the ^{35}S-labeled Cys.

   If Cys is not converted to Met, addition of unlabeled Met is not required. Otherwise, since Cys transport systems are generally less efficient than Met permeases, addition of unlabeled Cys may be inappropriate.
4. After 30 min to 1 h incubation, cells are collected by centrifugation, washed with 200 μl water and frozen as pellet at –80 °C.
5. Perform 8 to 10 different double labelings, with either ^{35}S-Met or ^{35}S-Cys and different ^{3}H-amino acids not interconverted: for example, ^{35}S-Met/^{3}H Leu, ^{35}S-Met/^{3}H-Lys, ^{35}S-Met/^{3}H-Phe, ^{35}S-Met/^{3}H-Tyr, ^{35}S-Met/^{3}H-Trp, ^{35}S-Met/^{3}H-His, ^{35}S-Cys/^{3}H-Leu, ^{35}S-Cys/^{3}H-His.

6. Prepare cell extracts for 2-D gel electrophoresis by a standard method. Run gels with large quantities of labeled protein extracts: 50–100 μCi of ^{3}H, 30–50 μCi of ^{35}S. This corresponds approximately to 150–250 μg of proteins.
7. After electrophoresis, stain gels with Coomassie Brillant Blue R 250. Silver staining is prohibited because it may interfere with scintillation counting. The gel is dried on Whatman paper and exposed for autoradiography.
8. Number and pinpoint the spots on a transparency placed on the autoradiogram. Point at least ten background controls in different locations of the map. Separate the transparency from the autoradiogram and overlay it precisely on the gel to punch out round sections (1.5–2.2 mm diameter) at the location of each spot using a sharpened cannula (the transparency serves as tracing-paper and allows precise extraction of all the spots, even those invisible on Coomassie Blue stained gels). After extraction, reexpose the gel for autoradiography to check the quality of spot excision.
9. Distribute the spots in 4 ml plastic scintillation counting vials, rehydrate the spots with 50 μl H_2O for 30 min and keep them for 15 h at room temperature in 0.5 ml NCS (Nuclear Chicago Solubilizer, Amersham). After addition of 2 ml BCS-NA (Biodegradable Counting Scintillant Non-Aqueous, Amersham), samples are gently shaken (by inverting the vials 8 to 12 times) and kept for 7 to 10 more days at room temperature in the dark for stabilization. Samples are gently shaken again before counting.
10. Count the vials in a scintillation counter which measures both isotopes and their ratios. Each sample is counted for at least 10 min. Verify that the average $^{3}H/^{35}S$ values range from 1 to 3. Use control vials to determine the background level. Usually, the average background is between 200 and 400 dpm for ^{3}H and between 30 and 60 dpm for ^{35}S. If the dpm value is not three times higher than the background level, the accuracy is too low and the measure should be discarded.
11. For each ^{35}S-Met/^{3}H-labeling, select reference proteins with known N-terminal sequences. These proteins have a well-defined number of Met residues. Plot the $^{35}S/^{3}H$ isotope ratio of these reference proteins against their theoretical amino acid ratio: N_{Met}/N_{3H} (where N_{met} and N_{3H} are the numbers of Met and ^{3}H-labeled amino acid residues in the proteins). Draw the regression line. The experimental values should be in good agreement with the theoretical amino acid ratios. Verify that the regression coefficient is close to 1. If some values appear far away from the curve or are aberrant, examine the cause of error.
12. The calibration curve is used to deduce the amino acid ratios of proteins analysed.

6 Experimental Determination of pI and M_r

The isoelectric point and molecular weight are deduced from the migration pattern of the protein. The location of reference proteins in the first dimension is plotted as a function of the calculated pI values of the corresponding polypeptides. Similarly, the location of each reference protein spot in the second dimension is plotted as a function of the log of the calculated values of their M_r. For each polypeptide analysed, M_r and pI are then determined using the standard curves. The uncertainty in predicting the protein migration parameters with this procedure is less than 0.3 pH units for pI and less than 15 % for M_r (Boucherie et al. 1995).

7 Construction of the Database

Software is available on the Internet (AACompIdent, PropSearch and MultiIdent) to search databases for proteins matching the experimental parameters pI, M_r and amino acid data. Unfortunately, the amino acid data available in these programs concern the amino acid composition and not the amino acid ratios. Consequently, it is necessary to construct a protein database which includes data of protein amino acid ratios and to design a program to search this database.

Theoretically, the database should include all protein sequences of the model micro-organism. However, the database may be limited by selecting proteins encoded by genes of high codon bias index (CBI). The CBI of a gene is correlated to the abundance of the corresponding protein (Bennetzen and Hall 1982). As proteins visible on 2-D gel electrophoresis belong to the most abundant proteins, they are supposed to be encoded by genes with high CBI. For example, on yeast 2-D gels, we have identified by different methods the product of 219 genes (see YPM server available through the www network). Only three of them have a CBI < 0.1 and the corresponding spots are not visible under standard conditions. Accordingly, from about 6000 yeast open reading frames, the database can be restricted to 2700 sequences with CBI > 0.1. A cutoff at CBI > 0.2 would limit the database to 1300 proteins, but would probably eliminate a few proteins which are visible on the gel.

Similarly, when one uses a 2-D gel electrophoresis technique limited to acidic and neutral pI, the basic proteins (with calculated pI > 8 or 9) can also be eliminated from the database. Conversely, in some cases, it is crucial to extend the database: the N-terminal Met cleavage of the database proteins is not known. Therefore, using data obtained from Met labeling may lead to important errors in calculation of amino acid ratios involving Met, especially in proteins with low Met content. Consequently, the proteins of the database for which the cleavage of the N-terminal Methionine is unknown can be duplicated in the database, one sequence retaining the N-terminal Met and the other with this residue removed.

8 Search in a Protein database

8.1 General Method

The program is designed to search the database for proteins that match the experimental values pI, M_r and amino acid ratios. This is generally achieved by calculating a distance (or score) between the query protein and each protein of the database. Proteins are then ranked according to their scores. The closest protein to the query is the best candidate for identification.

Practically, the software should calculate:

(1) the pI, the M_r and the amino acid ratios (rA) of each protein of the database taking into account known post-translational modifications (N- or C-terminal cleavage and N-acetylation, phosphorylation, glycosylation, etc .);

(2) a distance between the query protein (m) and each protein (x) of the database. This distance is the addition of a distance in isoelectric point (dI), a distance in molecular weight (dM) and 8 to 10 distances in amino acid composition (dA):

$$d(x) = dI(x) + dM(x) + \Sigma\,(dA(x))$$

where

$$dI(x) = |pI(m) - pI(x)|$$
$$dM(x) = |1 - M_r(x)/M_r(m)|$$
$$dA(x) = |1 - rA(x)/rA(m)|$$

The proteins having the shortest distances (best scores) are selected. The best candidate corresponds to the lowest distance (d1). For each search, the 5 to 10 first candidates should be listed with their scores and their parameters (pI, M_r, CBI and amino acid ratios). It is also possible to list all the partial distances dA(x) (Table 8.3). This will ease the search of errors when scores are too high (see Sect. 9.1).

8.2 Improvement of Distance Calculation

The calculation of the distance can be improved in several ways:

1. Euclidian distance. An Euclidian distance $d = (\Sigma\,(dA(x))^2)^{1/2}$ may be preferred (Hobohm et al. 1994). In practice, it gives essentially the same results as the linear distance.
2. Weights. In principle, all dA can be weighted:

   $$d(x) = \Sigma\, a_i(dA_i(x))$$

   where a_i is the weight.

   Partial distances with the smallest standard deviation should be favored with higher weights than the others. For example, weights can be defined as inverse of the standard deviation.
3. pI and M_r windows. The uncertainty in predicting the pI and M_r is less than 0.3 pH units for pI and less than 15 % for M_r (Boucherie et al. 1995). To avoid elimination of proteins which do not migrate exactly at their predicted place on 2-D gels, a window of +/- 0.3 pI units and +/- 15 % M_r can be defined:

   $$dI(x) = 0 \text{ if } |pI(m) - pI(x)| < 0.3;$$
   $$dI(x) = |pI(m) - pI(x)| - 0.3 \text{ if } |pI(m) - pI(x)| > 0.3;$$
   $$dM(x) = 0 \text{ if } |1 - M_r(x)/M_r(m)| < 0.15;$$
   $$dM(x) = |1 - M_r(x)/M_r(m)| - 0.15 \text{ if } |1 - M_r(x)/M_r(m)| > 0.15.$$

Table 8.3. Examples of results for six different searches in the yeast database. For each spot (A, B, C, D, E and F), the experimental data pI, M_r, and the amino acid ratios Leu/Met (L/M), Lys/Met (K/M), His/Met (H/M), Phe/Met (F/M), Tyr/Met (Y/M), Trp/Met (W/M), Leu/Cys (L/C) and His/Cys (H/C) are indicated on the first line (bold characters). On the five next lines, are listed the five best scored proteins (shortest distances), with their theoretical pI, M_r, and amino acid ratios expressed as the relative difference (in percent) with the experimental data. In this experiment, dL was defined as the sum of standard deviation (dL = 1.4)

Gene	Score	CBI	pI	Mr	L/M	K/M	H/M	F/M	Y/M	W/M	L/C	H/C
A	**Experimental data**		**5.68**	**41000**	**0.259**	**0.399**	**0.902**	**0.464**	**1.013**	**1.090**	**0.161**	**0.935**
GLN1	0.759	0.68	5.75	41616	-14	3	-8	-4	-23	5	-46	3
PPC1	1.958	0.42	5.88	60982	-63	-5	-12	-19	9	23	-105	1
RIP1	2.155	0.43	5.91	20098	-16	6	17	6	51	31	-52	24
PPC1*	2.170	0.42	5.88	60982	-72	-11	-18	-26	4	18	-108	0
GLY1	2.280	0.52	5.70	42725	-74	7	-3	24	13	64	-147	-2
B	**Experimental data**		**4.60**	**67000**	**1.298**	**1.606**	**0.617**	**1.750**	**1.027**	**0.320**	**1.437**	**0.805**
SSA1	0.886	0.83	4.70	69677	13	3	-14	-1	8	26	18	-7
SSA2	0.972	0.89	4.66	69380	13	5	-14	6	8	26	18	-7
SSA4	1.792	0.24	4.74	69561	-1	-11	-39	-14	-34	11	-10	-50
SSB2	1.979	0.88	5.09	66504	13	19	-8	17	22	31	-11	-36
SSA3	2.061	0.20	4.76	70456	21	20	-30	14	-13	31	0	-63
C	**Experimental data**		**5.76**	**30000**	**1.280**	**2.520**	**2.679**	**3.348**	**3.640**	**5.960**	**0.414**	**0.973**
HOR2*	1.026	0.38	5.72	27813	0	1	10	16	21	33	-23	-14
GPP1*	3.029	0.70	6.09	30438	20	18	23	40	37	52	-32	-31
HOR2	3.375	0.38	5.72	27813	29	29	36	40	43	52	-17	-9
MAK31*	3.631	0.18	5.00	25223	-25	44	10	20	41	22	3	28
CAP1	3.892	0.16	4.96	30698	-41	1	-34	3	-32	50	-102	-101
D	**Experimental data**		**4.62**	**28500**	**0.633**	**1.216**	**0.385**	**0.589**	**0.413**	**0.010**	**3.676**	**ND**
EGD2	1.423	0.72	4.49	18619	-5	-10	-4	-13	-29	10	40	-
YLR221C	2.468	0.12	4.54	24655	-42	-29	-56	-27	-45	10	18	-
VMA4	3.091	0.43	4.98	26381	-70	-7	100	15	-94	10	2	-
YTA1*	3.303	0.24	4.67	48255	-21	46	25	12	7	99	53	-
YLR192C*	3.501	0.37	4.77	29563	5	9	11	3	45	99	46	-
E	**Experimental data**		**5.20**	**23500**	**0.924**	**0.923**	**1.557**	**1.269**	**1.067**	**1.290**	**0.793**	**1.827**
HAM1	1.686	0.19	5.28	22092	23	-2	31	-31	17	31	-26	4
YDR399W	1.751	0.36	5.30	25190	31	-8	23	29	-20	7	22	26
UBC1	1.894	0.12	4.87	24178	4	4	-3	30	0	31	-58	-44
YPR127W*	1.904	0.33	5.57	38600	-3	0	-7	-9	-27	5	30	38
TFS1*	1.998	0.15	6.06	24357	39	-3	8	-10	-20	7	32	12
F	**Experimental data**		**4.60**	**40000**	**0.690**	**0.595**	**0.717**	**0.767**	**0.800**	**0.380**	**0.544**	**0.684**
APA1	0.913	0.43	4.67	36330	13	9	-12	8	8	12	10	-11
RBK1	1.043	0.19	4.95	36983	3	-7	-12	-1	-11	-17	-9	-22
YCR035C*	1.400	0.14	4.64	44058	-10	12	20	44	-5	0	-19	18
YHR029C	1.585	0.14	5.36	32563	39	-1	16	-4	0	-5	26	2
YCR035C	1.640	0.14	4.64	44058	-1	20	27	49	4	8	-16	19

* Proteins of the duplicated database without N-terminal Met.

4. Calculation of dA(x). dA(x) is calculated as $|1-rA(x)/rA(m)|$ (see above). However, the double labeling technique may give unprecise measurements of rA(m) when this value is very low (< 0.15). In this case, the dA(x) distance can be highly overestimated. To avoid this problem, the definition of dA(x) can be modified as follow:

   $dA(x) = |rA(m)-rA(x)|/rA(m)$ if $rA(m) > 0.15$;
   $dA(x) = |rA(m)-rA(x)|/0.15$ if $rA(m) < 0.15$.

5. Interconversions. Metabolized amino acids may be used for labeling. In this case, the amino acid interconversions must be taken into account in the program. For example, in yeast, after labeling with ^{35}S-Met, 66 % of the radioactivity is recovered in Met and 33 % in Cys. In this case, for each protein, the isotope ratio $^{35}S/^{3}H$ is theoretically proportional to the following ratio: $(0.66 \times N_{Met} + 0.33 \times N_{Cys})/N_{3H}$, where N_{Met}, N_{Cys} and N_{3H} are respectively the numbers of Met, Cys and ^{3}H-labeled amino acid residues in the proteins.

9 Result Analysis

9.1 Definition of the Limit Distance

Once protein scores are obtained, it is necessary to examine whether the top ranking protein represents the correct identification. A simple approach is to arbitrarily consider the top ranking candidate as the correct identification. However, this may not be true especially when the quality of the experimental data is poor. To gain confidence in the identifications, a general method consists of defining a limit distance (dL) and searching as follow: if $d1 < dL$, the first candidate is considered as identified; if $d1 > dL$, no identification is provided (though it is highly probable that the correct protein is one of the first candidates). dL can be chosen after a research test with experimental data of reference proteins. The best dL value is the one giving the highest number of correct identifications with a minimum number of errors. The sum of standard deviations of amino acid ratios corresponds generally to an optimized dL value (Maillet et al. 1996).

When data are of good quality, the scores of the second, third and next candidates d2, d3, d4, etc are similar and clearly higher than d1. Consequently, as proposed by Wilkins et al. (1996), another criterion can be considered: $d2/d1 >$ limit ratio (for example = 1.5) and can help the analysis when d1 is close to dL. However, this criterion should not be used for very homologous proteins.

9.2 Test with Reference Proteins

It is important to test the reliability of the search with experimental data of reference proteins. This test is representative when several reference proteins are available (30 to 50). This analysis allows us: to adjust the weights of the distance coefficients; to define the limit distance dL; to measure a proportion of correct results. This proportion should be higher than 90 %, to obtain sufficient confidence in the search.

9.3 Homologous Proteins

Very homologous proteins have close amino acid compositions. This method easily identifies protein families but does not discriminate easily between members of the same family. Nevertheless, correct identifications can usually be obtained even between proteins with more than 90 % sequence identity (Maillet et al. 1996).

9.4 Failures in Protein Identification

A significant proportion (15 to 30 %) of the proteins analysed cannot be identified because d1 $>$ dL and a small fraction (2 to 8 %) of the identifications are inaccurate. These failures may be due to:

(1) Unprecise determination of amino acid ratios.
(2) Overlapping of spots. If two or more proteins of similar abundance migrate at the same place on 2-D gels, the resulting amino acid ratios will not correspond to any of these proteins.
(3) Unknown post-translational modifications (phosphorylation, N-acetylation, glycosylation, N- or C-terminal cleavage) or aberrant migrations which change the apparent pI, M_r or even the amino acid ratios in case of proteolytic cleavage. Interestingly, most of the spots for which identification failed, migrate as small proteins, suggesting that some of them could be cleavage products from higher molecular weight proteins.
(4) Absence of the corresponding protein in the database. This is possible when the genome sequence of the model micro-organism is not completed or when the database is restricted to sequences of high CBI.
(5) The theoretical possibility that two completely different proteins match the same partial amino acid composition.

9.5 Examples of Result Analysis

Table 8.3 reports examples of results for six different searches in the yeast database. For each spot (A, B, C, D, E and F), the experimental data pI, M_r, and the amino acid ratios Leu/Met (L/M), Lys/Met (K/M), His/Met (H/M), Phe/Met (F/M), Tyr/Met (Y/M), Trp/Met (W/M), Leu/Cys (L/C) and His/Cys (H/C) are indicated on the first line (bold characters). On the five next lines, are listed the five best scored proteins (shortest distances), with their theoretical pI, M_r, and amino acid ratios expressed as the relative difference (in percent) with the experimental data. In this experiment, dL was defined as the sum of standard deviation (dL = 1.4). Genes with stars (*) correspond to proteins of the duplicated database without N-terminal Met.

A: According to the criterion, dL < 1.4, this spot is identified as Gln1p (confirming previous identification). Notice an important error (46 %) in the determination of L/C ratio.
B: The five first candidates belong to the same Hsp 70p family. However, Ssa1p and Ssa2p have much better scores (lower than 1.4). We conclude that this spot corresponds to one of these two very homologous proteins (this spot was previously identified as Ssa2p).
C: According to the criterion, dL < 1.4, this spot is identified as Hor2p* devoid of its N-terminal Met. Notice that score of Hor2p with N-terminal Met is very high (3.375). Without the duplication of the database which incorporated N-terminal cleavages, the identification would not have been possible.
D: Due to the absence of H/C data, the dL limit is 1.25 in this case. The best score d1 (1.423) is higher but close to dL (1.25). The second criterion can then be considered. The second score d2 (2.468) being much higher than d1 (1.423), we identified this spot as Egd2p. Notice that a significant part of the high score of Egd2p is due to an aberrant migration in the second dimension.
E: No identification is provided because d1 $>$ dL and d1 is similar to d2.
F: This case is rare; though d1 $<$ dL, no identification is provided because the two first candidates have similar scores and do not belong to the same family. However, Apa1p is more probable because its CBI (0.43) is markedly higher than the CBI of Rbk1p (0.19).

10 Conclusion

The amino acid analysis by the labeling method is a very good first step for a global identification of proteins on 2-D gels. However, its predictive nature can result in a few errors (about 5 %) in protein identification. This method is especially suited for the systematic proteome analysis in micro-organisms whose genome has been sequenced. It has already been successfully used in the identification of more than a hundred yeast proteins (Garrels et al. 1994; Maillet et al. 1996; Godon et al. 1998). However, the important investment it requires makes this technique not suitable for punctual identifications.

The method described here can be the first step in more specific proteomic analysis. For example, it is possible to study protein expression under different culture conditions and in genetically engineered strains lacking or overexpressing specific regulators of gene expression. Here also, the double labeling method could be a very instrumental tool for precise quantitation of protein expression (Godon et al. 1998).

Acknowledgement: The authors wish to thank M. Toledano, D. Spector and I. Maillet for critical reading of the manuscript.

References

Bennetzen JL , Hall BD (1982) Codon selection in yeast. J Biol Chem 257: 3026–3031
Boucherie H, Dujardin G, Kermorgant M, Monribot C, Slonimski P, Perrot M (1995) Two-dimensional protein map of *Saccharomyces cerevisiae*: construction of a gene-protein index.Yeast 11: 601–613
Garrels JI, Futcher B, Kobayashi R, Latter GI, Schwender B, Volpe T, Warner JR, McLaughlin CS (1994) Protein identifications for a *Saccharomyces cerevisiae* protein database. Electrophoresis 15: 1466–1486
Giometti CS, Anderson NL (1981) A variant of human nonmuscle tropomyosin found in fibroblasts by using two-dimensional electrophoresis. J Biol Chem 256: 11840–11846
Godel H, Graser T, Foldi P, Pfander P, Furst P (1984) Measurement of free amino acids in human biological fluids by high-performance liquid chromatography. J Chromatogr 297: 49–61
Godon C, Lagniel G, Lee J, Buhler J M, Kieffer S, Perrot M, Boucherie H, Toledano M, Labarre J (1998) The H_2O_2 stimulon in *Saccharomyces cerevisiae.* J Biol Chem 273: 22480–22489
Hobohm U, Houthaeve T, Sander C (1994) Amino acid analysis and protein database compositional search as a rapid and inexpensive method to identify proteins. Anal Biochem 222: 202–209
Latter GI, Metz E, Burbeck S, Leavitt J (1983) Measurement of amino acid composition of proteins by computerized microdensitometry of two-dimensional electrophoresis gels. Electrophoresis 4: 122–126
Latter GI, Burbeck S, Fleming J, Leavitt J (1984) Identification of polypeptides on two-dimensional electrophoresis gels by amino acid composition. Clin. Chem. 30: 1925–1932
Maillet I, Lagniel G, Perrot M, Boucherie H, Labarre J (1996)
Rapid identification of yeast proteins on two-dimensional gels. J Biol Chem 271: 10263–10270
Messenguy F, Colin D., ten Have JP (1980) Regulation of compartmentation of amino acid pools in *Saccharomyces cerevisiae* and its effects on metabolic control. Eur J Biochem 108: 439–447
Sandkamp B (1997) Application de détecteurs à comptage direct à la séparation simultanée de différents isotopes dans le domaine de l'autoradiographie. Thèse de doctorat de l'université Pierre et Marie Curie, Paris VI
Wilkins MR, Pasquali C, Appel RD, Ou K, Golaz, O, Sanchez JC, Yan JX, Gooley AA, Hughes G, Humphery-Smith I, Williams KL, Hochstrasser D (1996) From proteins to proteomes: large scale protein identification by two-dimensional electrophoresis and amino acid analysis. Bio/Technology 14: 61–65

CHAPTER 9

Identification of Proteins by Amino Acid Sequencing

F. Lottspeich[1]

1
Introduction

Purpose of Protein Sequencing in Proteomics

Since 1975 when the publications of O'Farrel and Klose appeared, the biochemical society dreamed of knowing the identity of separated proteins (Klose 1975; O'Farrel 1975). However, in 1975 the protein chemical methods were far to insensitive to handle the minute protein quantities that could be applied to 2-D gels. Furthermore, the separated proteins were embedded in a polyacrylamide gel matrix which is almost incompatible with any protein chemical analysis method. Several steps of methodological invention and improvements had to take place before the diagnostic 2-D gel electrophoresis evolved into the modern proteome science. First of all, the amino acid sequence technology had to be improved to a sensitivity level where sequence information from a few picomoles of protein could be generated, which will be described in this chapter. However despite this improvement in sequencing sensitivity the identification of the many proteins within a 2-D gel would not have been possible without development on the sample preparation side i.e. electroblotting of proteins onto chemically inert membranes (Vandekerckhove et al. 1985, Aebersold et al. 1986) and cleavage of proteins directly in gel (Aebersold et al. 1987, Eckerskorn and Lottspeich 1989). In recent years the modern and rapid mass spectrometry (MS) methods like electrospray MS and matrix-assisted laser desorption/ionization (MALDI) MS have almost completely taken over the identification of gel-separated proteins. These protein identification methods exploiting exclusively mass spectrometric techniques are rapid and powerful if the sequences of proteins investigated are already stored in a database. However, major limitations arise when dealing with proteins from organisms not sequenced at the DNA level. Here still the classical protein chemical methods like amino acid sequence analysis have to be applied. In general, today the challenges in protein characterization are rather in automation and optimization than in fundamental problems.

[1] Max-Planck-Institute for Biochemistry, 82152 Martinsried, Germany.

2 N-Terminal Sequence Analysis by Edman Chemistry

In 1949 the Swedish scientist Pehr Edman introduced the sequential degradation of proteins and peptides (Edman 1950). The chemistry of the cyclic reaction is shown in Fig 9.1 and comprises three steps, coupling, cleavage and conversion.

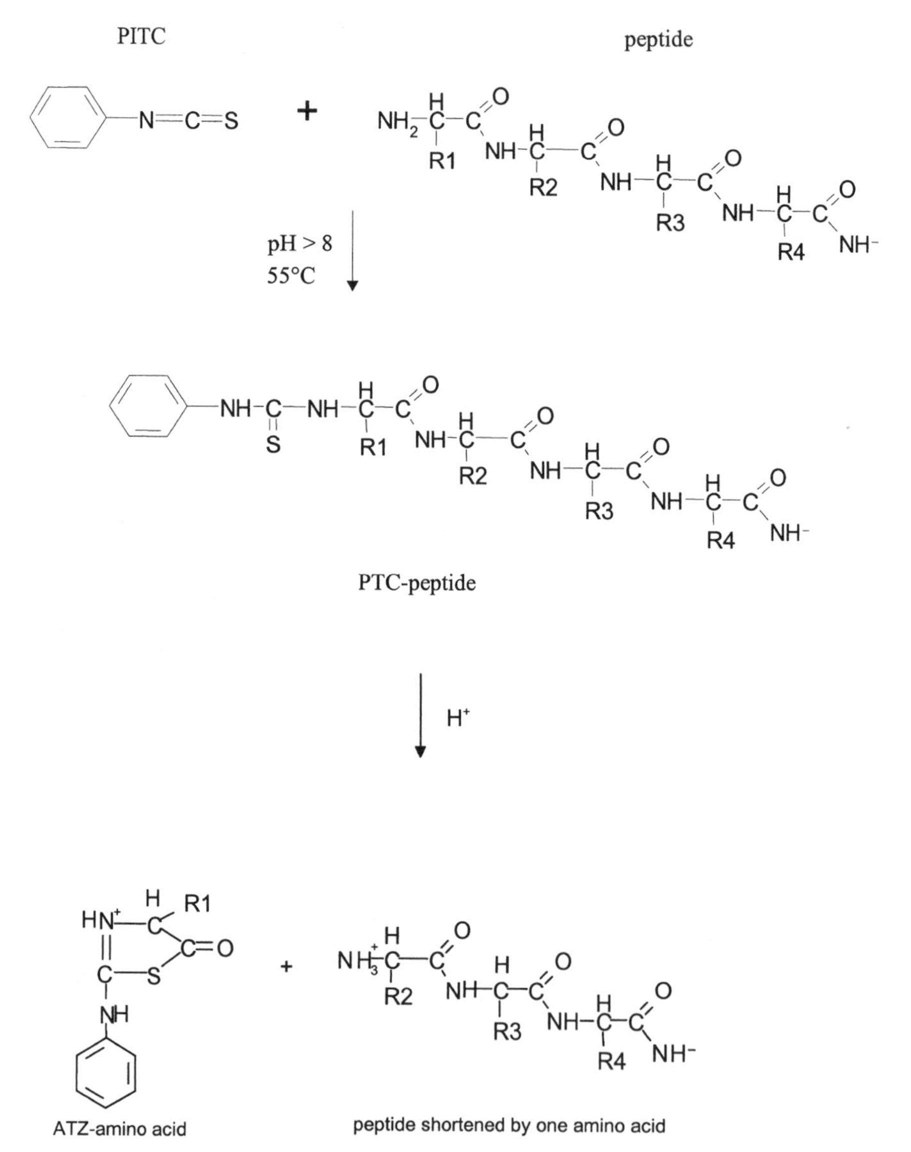

Fig. 9.1. Chemical reactions of the Edman degradation. *PITC* phenylisothiocyanate; *PTC* phenylthiocarbamyl; *ATZ* anilinothiazolinone

Until today this method, named the "Edman degradation", remained the only technique to obtain *de novo* amino acid sequences routinely.

The first step, the so-called coupling, is the addition of the Edman reagent phenylisothiocyanate to the free unprotonated N-terminal amino group of the peptide, yielding a phenylthiocarbamylpeptide. This reaction requires alkaline conditions of about pH 9 and an elevated temperature of 40–55 °C for about 15–30 min. A higher pH increases the coupling reactivity but increases also the alkali-catalyzed hydrolysis of phenylisothiocyanate to aniline. Aniline then reacts with excess phenylisothiocyanate yielding diphenylthiourea, the most common byproduct of Edman degradation. The coupling reaction is almost quantitative and very specific for amino groups. Thus, also the ε-amino groups of lysine are reacted to ε-phenylthiocarbamyllysine. Then, the excess of phenylisothiocyanate and some of the diphenylthiourea is extracted with an apolar organic solvent, usually ethyl acetate, leaving the rather hydrophilic phenylthiocarbamylpeptide in the reaction compartment. After coupling and washing the reaction conditions are changed to acidic with, e.g., neat trifluoroacetic acid. In this step, which is called the "cleavage", the nucleophilic attack of the sulphur on the carbonyl group of the first peptide bond results in a cleavage of the first peptide bond. The products of this reaction are a peptide shortened by one amino acid and an anilinothiazolinone amino acid derivative.

The anilinothiazolinone amino acid is sufficiently hydrophobic that it can be separated from the rather hydrophilic peptide by extraction with an apolar solvent, e.g. ethyl acetate or chlorobutane. The anilinothiazolinone derivatives of the amino acids are not entirely stable to allow reliable chromatography identification. Therefore, in the so-called conversion reaction, the anilinothiazolinone amino acids are rearranged to more stable derivatives, the phenylthiohydantoin (PTH) amino acids. The PTH amino acids are then identified by their retention times in reversed phase high performance liquid chromatography (HPLC) by comparison with a standard chromatogram containing all the common PTH-amino acid derivatives. After the extraction of the anilinothiazolinone amino acid the residual shortened peptide, which exhibits a new free N-terminal amino group and is still left in the reaction compartment, can be subjected to a new cycle of coupling and cleavage.

All the reactions of the Edman degradation perform with very high efficiency, resulting in an overall yield from step to step (i.e. the repetitive yield) of over 95 % routinely. The repetitive yield is directly connected with the length of the sequence that can be obtained. With a repetitive yield of 90 % one can expect to get sequence information of maximal about 15 to 20 steps. With the repetitive yield of 95 % a sequence length of about 30 to 40 amino acid residues may be obtained. In special cases and under optimal conditions even higher repetitive yields are possible leading to long sequences of up to more than 70 amino acid residues.

To reach high repetitive yields, several precautions have to be taken:

- Any contamination of the sample with salts, detergents or dyes has to be avoided, because these compounds may interfere with the reaction efficiency of the Edman reactions and consequently leads to the accumulation of by-products and out-of-frame sequences, which prohibits more extended sequence runs.

- All reactions have to be carried out under exclusion of oxygen since a sulfur to oxygen exchange in the phenylisothiocyanate or in phenylthiocarbamylpeptide results in a complete stop in the Edman degradation. Oxygen in the phenylcarbamylpeptide is not capable of performing the nucleophilic attack towards the carbonyl bond and consequently no cleavage can occur.
- The reactions have to be carried out in specialized dedicated and fully automated instruments, the sequencers, where any manual interference is avoided and only specially purified high quality reagents and solvents should be used.

The main difference between the instruments on the market is the design of the reaction compartment, where the sample is applied and must remain throughout all the degradation steps. Today two versions of instruments for the automated version of the Edman degradation are used. In one version, the gas phase sequencer (Hewick et al. 1981), the sample is non-covalently attached to a glass fiber disc or a piece of a chemically inert membrane consisting usually of polyvinylidenfluoride (PVDF). A volatile base (usually N-methylpiperazine/water/methanol) is used during coupling for the application of the alkaline pH which is delivered in gaseous form by bubbling an argon stream through an aqueous solution of the base. This type of delivery was chosen, since a liquid aqueous base would dissolve the protein and wash it out of the reaction compartment. By addition of phenylisothiocyanate the coupling is performed. The excess of reagent and reaction by-products are extracted and after drying the cleavage is performed by delivering trifluoroacetic acid. In the original version of the gas phase sequencer also the trifluoroacetic acid delivery was performed in gaseous form. Today, a precisely defined liquid delivery is preferred which allows a more rapid reaction and thus shorter cycle times. Therefore, these modern types of gas phase sequencers are also named pulsed liquid phase sequencers.

A completely different technical way to apply the Edman chemistry is realized in the biphasic column sequencer (Totty et al. 1993). In this instrument the reaction compartment consists of two small chromatographic columns. One of these columns is filled with a hydrophobic reversed phase material which very efficiently binds protein material under aqueous (i.e. polar) conditions via hydrophobic interactions. The second column is filled with a hydrophilic silica gel, where proteins are well retained under apolar conditions applied during the extraction steps with organic solvents to remove reaction by-products or the anilinothiazolinone amino acid. The protein is applied usually directly on the top of the reversed phase column, and can be easily washed free of salts and other contaminating polar impurities with e.g. 0.1 % trifluoroacetic acid. Then, assembly of the reaction compartment is finished by connecting the second column to the top of the reversed phase column. The protein remains in this type of reaction compartment during the whole degradation cycle in the area between the two columns. To ensure this local fixation the flow direction in this instrument can be delivered to both ends of the combined biphasic column reactor. Aqueous solutions will always be delivered through the silica gel column in direction reversed phase column, organic solvents always through the reversed phase column towards the silica gel column. Also with this instrument repetitive yields of over 95 % can be realized.

Despite the fact that the Edman reaction performs with very high yield, the reactions are well understood and the instruments are well developed, a number of problems and limitations especially at the low picomole level may make amino acid sequence analyses still difficult or even sometimes impossible.

- The sample has to be homogeneous. In an individual cycle of the Edman degradation the terminal amino acid residues of all the protein species present are cleaved off, are extracted together and converted to PTH amino acids. If a single pure protein is present almost no difficulties arise in the interpretation of the HPLC chromatogram to identify and quantify the amino acid. If, however, in the case of protein mixtures several PTH amino acids are present in a single degradation cycle interpretation may become problematic. If, for example two proteins are present in a mixture the assignment of the amino acids to the corresponding main and minor sequence is only possible when the molar amount of the two proteins and consequently also the amount of the PTH amino acids differ significantly. The interpretation often becomes uncertain when the ratio of the protein amounts is less than 2:1. The difficulties arise because the PTH amino acids exhibit widely different response factors, depending on which amino acid, reaction condition and instrument is present.
- Small peptides can be partly lost during the washing procedures with the organic solvents. The degree of this wash-out loss is not foreseeable since it depends on hydrophobicity and peptide length, both normally not known during sequence analysis. Especially in mixtures of peptides the sequence interpretation and correct assignment of a particular PTH amino acid to the corresponding sequence may become very difficult, due to the unpredictable different wash-out loss of different peptides.
- Polar contaminations are poorly extracted by the organic solvent washing and thus influence the efficiency of the Edman degradation for many cycles.
- Apolar contaminations also reduce the quality of the Edman degradation reactions, but are usually removed during the washes and extraction steps in the first degradation cycle. Consequently they show up in the PTH chromatogram of the first degradation cycle, but do not influence further cycles. This behavior is the reason why often in otherwise good amino acid sequence degradations the first step cannot be determined.
- For reasons not well understood about 50% of the protein applied to a sequencer will appear to be not sequencable. One possible explanation of this phenomenon called "initial yield" is that the protein layer in tight hydrophobic contact with the surface of the reaction compartment is not amenable to the Edman chemistry.
- The main problem in amino acid sequence analysis is N-terminal blocked proteins, i.e. proteins which have no free alpha-amino group available for coupling with the Edman reagent phenylisothiocyanate. In nature probably over 50% of the proteins are blocked. This is caused by either a posttranslational modification (acetylation, formylation, etc.) or a cyclization of an N-terminal glutamine residue to pyroglutamic acid. Removal of the blocking group of a protein is rarely successful, since the enzymes available for this purpose are usually active for only small peptides and not with proteins. Therefore for the protein chemical analysis, characterization and identification of an N-terminally

blocked protein the generation of internal sequences has almost routinely to be done. Then, to determine the nature of the blocking group, the blocked N-terminal peptide has to be isolated and sequenced by other methods, usually using mass spectrometry.

Mainly because of the frequent existence of blocked proteins a rough knowledge of the protein amount used for the analysis is absolutely necessary. When getting a sequence one has to know if the sequence obtained is derived from the main constituent of the sample or if eventually the sequence obtained is derived from a protein mixture where one (and maybe even the major one) is N-terminally blocked and therefore not amenable to the Edman chemistry. Only a careful estimation of the protein amount present can give an indication if the sequence obtained belongs really to the (major) protein believed to be analyzed. This quantitative estimation of the peptide amount is in practice very difficult, since especially at the microscale no accurate protein determination method exists and the initial yield in amino acid sequencing (see above) is protein dependent. Therefore, the presence of a single sequence never guarantees a pure protein.

In summary, the reactions of the Edman amino acid sequence degradation are well understood, and excellent instruments are commercially available. The sensitivity has improved dramatically, from several nanomoles starting material in the early 1970s to now less than 1 pmol. With this high sensitivity the tiny amounts of protein material present in 2-D gel spots can be sequenced routinely. However, the interpretation of the PTH identification especially with sequence analyses of very little protein material or with mixed samples still requires a great deal of expertise and experience. The sample preparation techniques for amino acid sequence analysis (electroblotting or cleavage in gel/on membrane) are now well established and have become realistic high throughput methods. The main limitations of chemical amino acid sequence analysis are blocked proteins and the fact that it is a rather expensive and slow method. Therefore, it is not by any means a high throughput technique. Nevertheless, today the Edman degradation is extremely valuable and still unreplacable since it is the only technique which is capable of sequencing from intact proteins and – despite the great success of mess spectrometry – it is the most efficient technique in generating "de novo" sequences of peptides in routine.

3 Other Possibilities for Generation of Amino Acid Sequences

3.1 Ladder Sequencing

Since the development and the automatization of proteomics there has been an urgent need for high throughput sequencing methods. One solution is to start with a defined peptide and generate of a mixture of fragments of this peptide by any kind of limited cleavage. Ideally, in this peptide mixture all fragments of the peptide differing only by one single amino acid are present (Fig. 9.2). These so-called peptide ladders can be generated from the N-terminal as well as from the

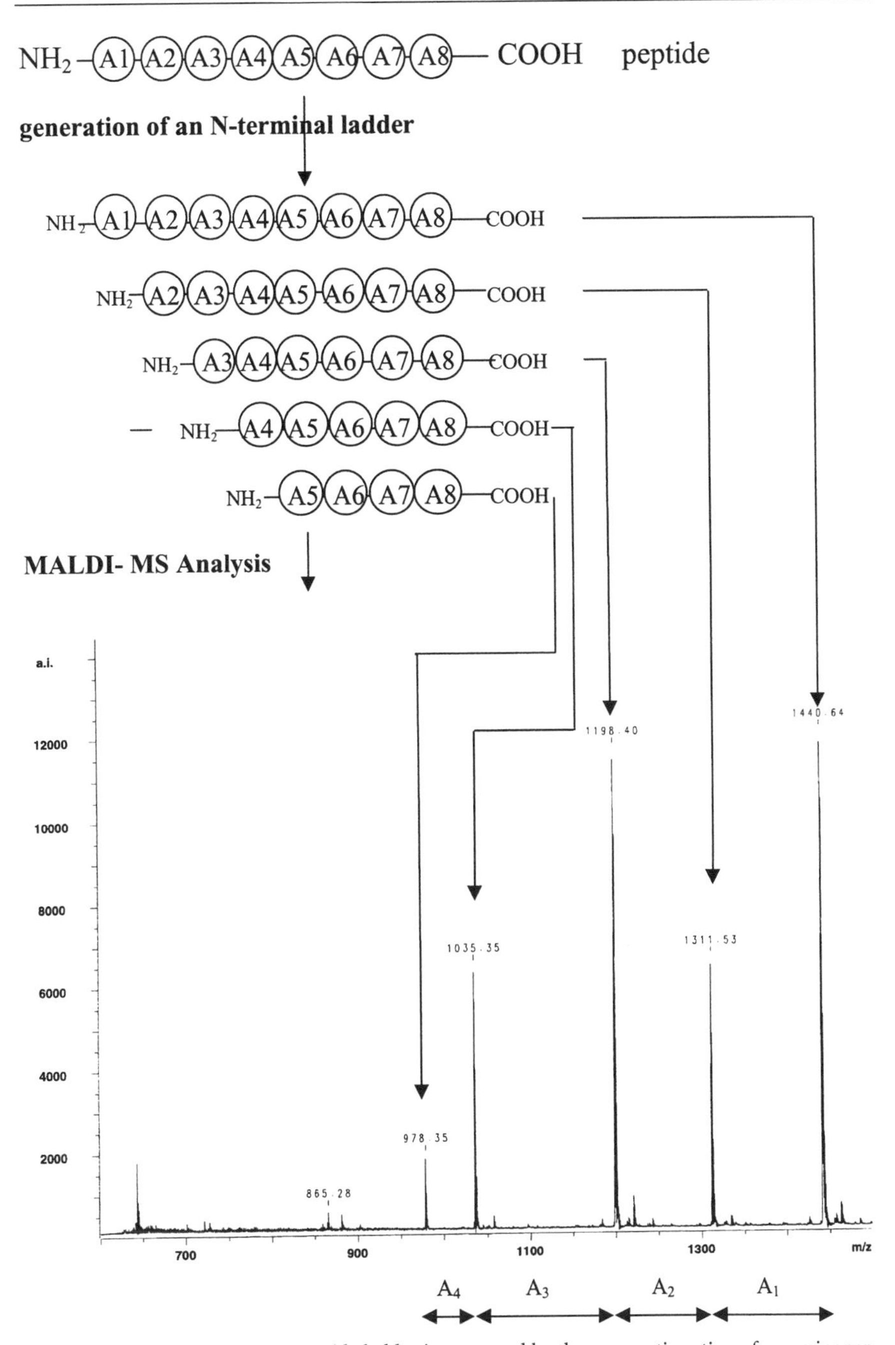

Fig. 9.2. Ladder sequencing. A peptide ladder is generated by the enzymatic action of an aminopeptidase by successive and limited cleavage of the N-terminal amino acids. The peptide mixture is analyzed by MALDI MS

C-terminal end of the peptide by chemical methods (Chait et al. 1993) or by time-controlled limited degradation using various exopeptidases (amino- and carboxypeptidases). The identification of the peptide mixture is then usually done by MALDI mass spectrometry which is rapid, sufficiently accurate and can easily handle peptide mixtures due to the simple spectra generated. From the differences in the masses the amino acid sequence can be deduced (Fig. 9.2).

Several problems prevented the method from becoming popular and widely used in high throughput approaches:

- The isobaric amino acids leucine and isoleucine cannot be distinguished by this method. Lysine and glutamine which differ in molecular mass only by 0.036 may otherwise be differentiated by the best instruments, but can easily be distinguished if the ε-amino group of lysine is chemically modified, e.g. by phenylisothiocyanate as it was done in the first application where the ladder was generated by several rounds of Edman degradation with a low amount of a "blocking" coupling reagent phenylisocyanate (Wang et al. 1993).
- MALDI mass accuracy obtained from peptides larger than 40 amino acid residues is usually not sufficient to discriminate between several possible amino acids. Therefore, ladder sequencing today is restricted to peptides of this size. However, with improved mass resolution and mass accuracy of MALDI mass spectrometry the applicable mass range probably will increase. There is also the possibility to apply electrospray mass spectrometry which has a better mass accuracy for larger proteins. Improved deconvolution programs will certainly overcome today's limitation, i.e. the interpretation of the complicated mixture spectra.
- An ideal ladder would contain all peptide species produced in almost equivalent amounts. In practice this is hardly achieved. Kinetic problems with the enzymatic or chemical cleavage lead to uneven distribution of the fragments. Therefore, some gaps in the sequence appear and only very short useful sequence data might be obtained. In enzymatic ladder sequencing usually a number of overlapping ladders can be obtained with a careful adaptation of the reaction times and conditions.
- It is not a very sensitive method. Since the amount of the original peptide is divided up into many fractions, each of the peptide fragments is present only in a much lower amount. The better the ladder generated the less of the individual peptide species is available for analysis. The high sensitivity of MALDI compensates somewhat for this fact, but in practice ladder sequencing can be performed only with higher femtomole amounts of starting peptide – very comparably with high sensitivity Edman sequencing.
- It has been shown that in MALDI analysis of protein mixtures severe and almost unpredictable suppression of individual peptides may occur. This results in gaps in the amino acid sequence.

In summary ladder sequencing is a promising tool. It is readily susceptible to automatization and therefore suited to high throughput approaches. Data interpretation is rather simple. The drawbacks are a limited sensitivity, some problems in generation of the ladders and a certain molecular weight restriction according to the mass determination methods used. However, since it generates "de novo" sequences much more rapidly and simpler compared with classical Edman or tandem mass spectrometry methods, in the future, ladder sequencing might be an alternative to high throughput protein sequencing.

3.2 C-Terminal Amino Acid Sequence Analysis

Two types of approaches to generate C-terminal amino acid sequences exist: chemical methods and the enzymatic methods. The chemical methods are all based on the original Schlack-Kumpf chemistry (Schlack and Kumpf 1926) where a thiocyanate reagent is coupled to the C-terminal amino acid residue. Since the carboxyl group per se exhibits only little reactivity it has to be activated by acetic acid anhydride/acetic acid under elevated temperatures to form a mixed anhydride. This can then be reacted with a thiocyanate compound to yield a peptidyl thiohydantoin. The C-terminal amino acid residue is then cleaved off as an amino acid thiohydantoin and is identified by reverse phase HPLC. However, despite a formal similarity with the N-terminal Edman degradation, side reactions and severe reactivity problems of certain amino acids drastically decrease the efficiency of the chemical C-terminal analysis. In general, only very short sequences will be obtained and a rather large peptide amount (100 pmol to more than 1 nmol) is needed. As a consequence, in a proteome approach chemical C-terminal amino acid sequencing will be hardly applied.

Originally, enzymatic C-terminal amino acid sequencing was performed by measuring the amount of amino acids released during a time course of an enzymatic digestion of proteins or peptides using carboxypeptidases. The large amount of peptide material needed for this approach and problems with kinetics of the enzymatic degradation hinders this method from being generally successful. Since the introduction of MALDI mass spectrometry the enzymatic approaches for C-terminal sequencing using carboxypeptidases are increasingly used in a mode of ladder sequencing (see above). In this case, it is not the amino acids released from the peptide that will be measured but rather the residual shortened peptides generated by amino acids cleaved off the C-terminal end. To improve the general applicability different carboxypeptidases and various digestion times have to be used.

3.3 Mass Spectrometry

The peptide bonds between the individual amino acids are more susceptible to cleavage than the other bonds resulting in different types of fragments. If the charge remains after cleavage on the N-terminal part of the peptide the fragments are said to belong to the a, b, or c series. If the charge remains on the fragments of the C-terminal series they will be named x, y or z fragments (see Fig. 9.3). If the complete ion series is obtained and can be determined the peptide sequence can be deduced since the adjacent signals of one series differ by the mass of one amino acid residue. However, the degree of cleavage and the preferred cleavage sites very much depend on the amino acid distribution within the

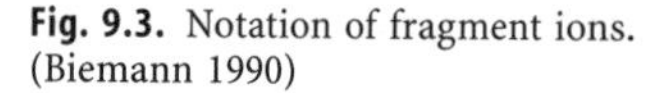

Fig. 9.3. Notation of fragment ions. (Biemann 1990)

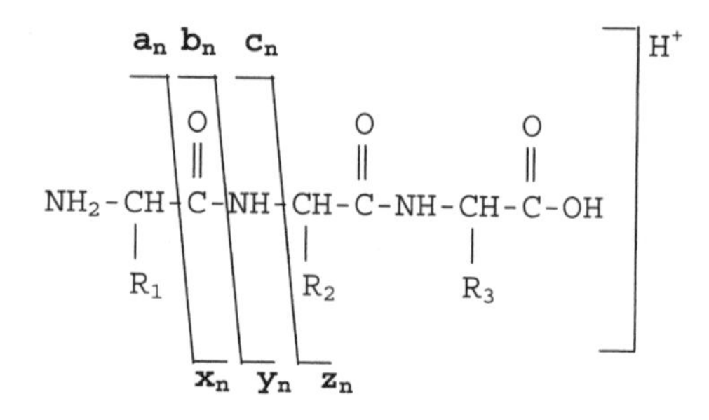

peptide and so far cannot be predicted. A complete set of fragments is almost never obtained, the intensity of the fragment ions is very different or sometimes the fragmentation information is simply missed. Intense fragments often belong to the b and y series, corresponding to a cleavage in the middle of the peptide bond, but the certain assignment of a fragment ion to a specific series is usually very difficult.

In MALDI mass spectrometry the occurrence of metastable decomposition of peptide ions during the flight time in the field free drift path was observed (Spengler et al. 1992). The resulting so-called post source decay (PSD) ions have different masses but the same velocity as the original molecular ion and therefore cannot be detected using a linear detector. However, the PSD fragments can be separated according to their masses in the electric field of a reflector. By adjusting the reflector voltages the metastable ions can be brought into the focus of the detector and thus their masses can be determined. The PSD spectra can then easily be used to confirm peptide sequences in protein databases, where calculated MSMS spectra are compared with the experimental spectra (e.g. http://prospector.ucsf.edu/, or as a good start site for further programs: http://expasy.hcuge.ch/www/tools.html#proteome). This can be done almost online with the spectra-acquisition and together with the peptide molecular mass gives highly significant data and therefore is a powerful tool for fast protein identification. The sensitivity of this method is in the low femtomole to attomole range and far below the Edman sequencing possibility. However, the interpretation of PSD mass spectra obtained from "new" sequences, i.e. sequences not stored in a database, is rather difficult, extremely time consuming, and almost never gives complete unambiguous amino acid sequences. Therefore PSD mass spectrometric sequencing of unknown peptides cannot be used in a high throughput approach.

With electrospray ionization, mainly used as nano-electrospray or capillary liquid chromatography-electrospray, mass and sequence information is usually obtained with triple quadrupole instruments. In the first quadrupole a single peptide ion is selected and introduced in a second quadrupole, the collision chamber, which is filled with a gas like helium or nitrogen. The peptide ions interact with the gas molecules and will be fragmented. The peptide fragments can then be analyzed in a third quadrupole. Similar to PSD mass data, also the uninterpreted collision induced (CID) mass spectra together with the molecular weight of the parent ion are used in rapid protein identification by comparison with calculated spectra from databases. The interpretation of "de novo" sequences is still complicated, but due to the different and higher energetic mode of fragmentation usually somewhat simpler than with PSD spectra. Due to this easier interpretation CID mass spectra are preferably used for the determination of posttranslational modifications.

The identification methods relying on mass sectrometry will be discussed in more detail in the next chapter (Chap. 10)

4
The interface from Two-Dimensional Gel to Amino Acid Sequencing

Usually in amino acid sequence analysis the pure salt and detergent-free sample is dissolved in a volatile solvent, spotted onto a chemically inert membrane and assembled in the reaction cartridge of a protein sequencer. With samples separated by 2-D gels the situation appears more difficult. During running and staining of the gel N-terminally blocking agents may be present or can be generated. After the separation the proteins are associated with residual SDS, buffer salts and staining dyes, all incompatible with direct N-terminal sequencing. Furthermore, the proteins are embedded in a polyacrylamide-gel matrix which is not stable during the chemical reactions of the Edman degradation. Consequently, if amino acid sequence analysis is to be applied to 2-D gel-separated proteins, two main strategies are available. In the first one, direct amino acid sequencing is performed on the whole protein. This cannot be done on the gel directly after electrophoresis and staining because of chemical incompatibilities. Consequently, the protein must be first transferred on a chemically inert membrane (see below and Chap. 6). This approach is most sensitive and the least laborintesive available with Edman chemistry. However N-terminal blocking is a major limitation, especially with intracellular proteins from eukaryotes. In this case, another approach must be used. The protein is first degraded into peptides, either with enzymes or chemically (e.g. cyanogen bromide). This step can be performed either on proteins transferred on membranes or directly in gel. The peptides are then eluted and separated (generally by reverse phase chromatography). The purified peptides are collected and then submitted to Edman sequencing by application of the peptide in the reaction chamber. This strategy is almost universal but very labor-intensive. Furthermore, it requires several picomoles of the starting protein.

In addition to this strategy choice, several sample preparation steps and certain precautions have to be considered in order to correctly apply Edman degradation to 2-D gel separated proteins.

4.1
Two-Dimensional Gel Separation

During the electrophoretic separation a blocking of the free amino groups of the proteins has to be avoided. The main risks for such an N-terminal blockage are carbamylation due to cyanate in urea and oxidations due to peroxides in detergents or to residual persulfate present in gels. Cyanate is slowly produced from urea at high pH and elevated temperatures. Therefore, to circumvent this problem the urea should be of high purity, the solution should be freshly prepared, and elevated temperatures at high pH are to be avoided. The second cause for N-terminal blocking are mainly aged detergents. Here again, in using high quality substances and solutions freshly prepared the risk of chemical modification of the proteins is negligible. A further considerable risk is the contamination of the sample with other proteins. In fact, today the limitation in sequence analysis of electrophoretically separated proteins is often the appearance of human keratin in the sample which was introduced during electrophoresis and during gel- and sample-handling. The problem can be minimized by wearing powder-free gloves and not touching the gel and the buffers.

4.2
Alternatives in Interfacing from Gel to Sequence Analysis-Elution, Electroblotting or Cleavage in Gel

4.2.1
Elution

At first thought the most straightforward solution to recover the electrophoretically separated proteins from the polyacrylamide gel seems to simply cut out the desired protein and then elute it from the gel with the help of suitable solvents. However, in practice – at least with low amounts of protein material – this approach is not feasible. The main cause lies in the visualization step after the gel electrophoresis, the staining procedure. The most common approaches for staining are Coomassie Blue staining which binds unspecifically to proteins and silver staining which relies on binding of silver to amino acid side chains (e.g. carboxyl and sulfhydryl groups). Both these staining procedures involve fixing the proteins within the gel to inhibit diffusion within and out of the gel matrix. In Coomassie Blue R250 staining the proteins are simultaneously fixed and stained in the Coomassie Blue staining solution (containing acetic acid as fixing agent). The complete gel turns blue and the gel has to be destained, usually with 10 % acetic acid/15 % methanol. Using the Coomassie Blue G250 staining procedure the proteins have to be fixed with isopropanol/acetic acid before the staining. The Coomassie Blue G250 is less soluble than R250 and is used in a colloidal state rather than in solution. It binds preferentially to proteins and not to the gel matrix. Therefore, no destaining is necessary.

In the Coomassie staining methods the proteins are fixed and during the fixation the SDS is removed from the protein. This is denatured and becomes entwined in the gel matrix. To now recover the intact proteins in good yield from such stained gels requires rather harsh conditions (acidic/organic mixtures). Thereby also dyes, residual SDS and polyacrylamide oligomers will be extracted, which all prohibit the direct amino acid sequence analysis. As a consequence the protein has to be desalted. The desalting step, however, is usually accompanied by substantial protein loss and therefore is not suited for micro amounts.

A slightly more favorable situation arises if non-denaturing staining methods are applied, like SDS precipitation, zinc-staining, or staining with Ponceau S. These methods do not fix the proteins in the gel and therefore the proteins are less denatured and easier to elute compared with Coomassie staining. These stains are less sensitive or in the same sensitivity range as Coomassie. They have the same disadvantage in eluting together with the protein of interest a lot of contaminating material. Prior to amino acid sequence analysis the contaminations have to be removed in a desalting step which is prone to protein loss.

Due to all these problems, the usual approach to obtain N-terminal amino acid sequence information from a gel-electrophoretically separated intact protein is electroblotting.

4.2.2 Electroblotting

One of the major steps toward modern protein biochemistry was the introduction of electroblotting of electrophoretically separated proteins onto chemically inert membranes (Vandeckerkhove et al. 1985, Aebersold et al. 1986, Eckerskorn et al. 1988). With this technique it became possible to transfer low microgram and submicrogram amounts of protein onto a solid support which binds the protein non-covalently via hydrophobic interactions. This binding is sufficiently stable for extensive washing steps with aqueous solvents to be applied and thus the protein can be freed from all the disturbing polar contaminants, mainly SDS and salts. For the transfer the proteins must not be stained, since stained proteins are almost SDS-free and in close interaction with the gel matrix, which hinders the migration of the protein out of the gel onto the membrane. In reality, almost no protein can be transferred from a stained gel. In contrast, the Coomassie Blue dye alone is moving in the electrical field leaving the protein in the gel. The Coomassie Blue will bind to the hydrophobic membrane, suggesting that a protein band is present.

For subsequent amino acid sequence analysis of proteins several types of hydrophobic membranes, like high-affinity PVDF membranes and modified glass fiber membranes, are commercially available (Eckerskorn and Lottspeich 1993). The well-known nitrocellulose membrane, frequently used for immunoblotting, cannot be used for amino acid sequencing, since it will dissolve during the Edman degradation conditions.

The immobilized protein can be visualized usually by Coomassie Blue staining. The stained protein spot can then be cut out and directly applied in the reaction cartridge of a protein sequencer. The technique is rather robust and today the most sensitive possibility to get amino acid sequence information of a protein (less than 1 pmol protein may be sufficient to get an short N-terminal sequence). Care should be taken, that all solutions and reagents are freshly prepared and of high purity.

Despite the simplicity and the high sensitivity of sample preparation of electrophoretically separated proteins via electroblotting, mainly two limitations are connected with this approach. More than 50 % of the natural existing proteins are N-terminally blocked and therefore a sequence analysis will fail. Secondly, the direct amino acid sequence analysis will give only a very limited number (10–30) of amino acid residues from the N-terminal end of a protein, which is usually not sufficient for cloning purposes or to find posttranslational modifications. Both limitations can be circumvented by the cleavage approach, which can be performed on membrane-transferred proteins or directly in gel (see below). In this case, nitrocellulose membrane can also be used. The main advantage of cleavage on membrane compared to in gel is the much cleaner reaction conditions. However, problems arise from the difficult recovery od peptides from the membrane and also from the blotting process itself (see Chap. 6).

4.2.3 Cleavage in Gel

The limitations of the blotting method led to the development of the most frequently used approach, the cleavage of proteins directly within the gel matrix. The gel can be stained with almost any staining method available, provided the amino groups are not modified during the procedure. Almost all non-covalent staining procedures like Coomassie, India ink, amido black, Ponceau S, Zn-acetate, etc. in principle do no harm to the primary structure of the proteins and thus are compatible with subsequent sequencing. Care has to be taken that all staining solutions contain high-purity reagents and that no contamination is introduced during handling of the gels. A very common contamination found in the analysis of gel-separated proteins is human keratin. Another very common atrefact lies in the solutions used for destaining of organic dyes (e.g. Coomassie). These solutions contain very frequently a mixture of acid (generally acetic acid) and water-miscible alcohols (methanol to isopropanol). Consequently, esters are formed in these mixtures (methyl, ethyl or isopropyl acetates). The problem arises from the esters, which can react with amino groups to yield amides. This will result in N-terminal artefactual blocking or in lysine acetylation, thereby preventing cleavage at lysine by trypsin or endoproteinase LysC. The very simple remedy to this situation is to destain with solution containing either acid or alcohol, but not a mixture of the two. These solutions perform correct but slower destaining than the acid-alcohol, mixtures.

In the more sensitive silver staining procedures the proteins are usually fixed with formaldehyde or even crosslinked with glutaraldehyde. Both are highly reactive amino-group blocking reagents and should be avoided if subsequent Edman sequencing is planned. A modified version of the silver staining without aldehyde fixing (Blum et al 1987) can be used.

However, in routine Coomassie Blue staining is used most commonly, since protein amounts of down to 100–200 ng can well be detected with this staining and this is in the order of magnitude which is at least necessary to successfully obtain amino acid sequence information.

The stained protein spots are excised, washed and the protein enzymatically cleaved. For sequencing purposes the endoproteinase LysC cleavage is preferred over trypsin, due to the longer fragments produced. Buffer conditions for the optimal cleavage can be applied. Salts are easily removed during the following HPLC separation of the peptides. The peptides may be collected in vials and then applied for sequencing. For peptide amounts below 10 pmol the HPLC separation should preferably be performed by nano-LC on <320 μm reversed phase columns and the effluent should be collected directly onto a PVDF membrane using special robotic instruments. The lack of transfer-steps results in a very good yield of the peptides.

5 Conclusion

In the last few years especially, the sample preparation and the interface from gels to analytical techniques have become routine at a very high quality level. Today, with some precautions, sufficient sequence information for identification or characterization can be obtained from almost any protein which is present in a gel down to an amount of about 1 pmol. Even more sensitive is the approach of electroblotting the 2-D separated proteins onto chemically inert membranes. However, some limitations concerning blocked proteins apply. If sufficient material is available, the analysis of internal fragments is highly recommended, where usually a high sequence coverage and eventually the elucidation of posttranslational modifications can be obtained.

6 Perspectives

Due to its rapid development and high sensitivity mass spectrometry of peptides has superseded a great deal of analyses from the classical chemical amino acid sequencing method. Protein identification of proteins known by DNA sequencing and stored in databases can now be performed much easier using solely mass spectrometry techniques fulfilling the requirements of high throughput techniques used in proteomics. However, for direct analysis of larger intact proteins and confident amino acid sequence determination of unknown peptides Edman amino acid sequencing is still unavoidable. In future, rapid and high throughput chemical or enzymatic methods in combination with mass spectrometry might overcome the two major limitations of Edman degradation, cost and speed.

References

Aebersold R, Teplow DB, Hood LE, Kent SB (1986) Electroblotting onto activated glass: high efficiency preparation of proteins from the analytical sodium dodecyl sulfate-polyacrylamide gels for direct sequence analysis. J Biol Chem 261: 4229–4239

Aebersold R, Leavitt J, Hood LE, Kent SH (1987) Internal amino acid sequence analysis of proteins separated by one- or two-dimensional gel electrophoresis after in situ protease digestion on nitrocellulose. Proc Natl Acad Sci USA 84: 6970–6974

Biemann K. (1990) Applications of tandem mass spectrometry to peptide and protein structure. In: (Burlingame, A.L., McCloskey, J.A., eds.) Biological mass spectrometry

Blum H, Beier H, Gross HJ (1987) Improved silver staining of plant proteins, RNA and DNA in polyacrylamide gels. Electrophoresis 8: 93–99

Chait BT, Wang R, Beavis RC, Kent SBH (1993) Protein ladder sequencing. Science 262: 89–92

Eckerskorn C, Mewes W, Goretzki HW, Lottspeich F (1988) A new siliconized glass fiber as a support for protein chemical analysis of electroblotted proteins. Eur J Biochem 176: 509–519

Eckerskorn C, Lottspeich F (1989) Internal amino acid sequence analysis of proteins separated by gel electrophoresis after tryptic digestion in the polyacrylamide matrix. Chromatographia 28: 92–94

Eckerskorn C, Lottspeich F (1993) Structural characterization of blotting membranes and the influence of membrane parameters for electroblotting and subsequent amino acid sequence analysis of proteins. Electrophoresis 14: 831–838

Edman P (1950) Method for determination of the amino acid sequence in peptides. Acta Chem Scand 4: 283–290

Hewick RM, Hunkapiller MW, Hood LE, Dreyer J (1981) A gas-liquid-solid-phase peptide and protein sequencer. J Biol Chem 256: 7990–7997

Klose J (1975) Protein mapping by combined isoelectric focusing and electrophoresis of mouse tissues. A novel approach to testing for induced point mutations in mammals. Humangenetik 26: 231–243

O'Farrell PH (1975) High resolution two dimensional electrophoresis. J Biol Chem 250: 4007–4021

Schlack P, Kumpf W (1926) On a new method for the determination of the constitution of peptides. Hoppe Seyler 's Z Physiol Chem 154: 125–170

Spengler B, Kirsch D, Kaufmann R, Jaeger E (1992) Peptide sequencing by matrix assisted laser-desorption mass spectrometry. Rapid Commun Mass Spec 6:105–108

Totty NF, Waterfield MD, Hsuan JJ (1993) Accelerated high-sensitivity microsequenceing of proteins and peptides using a miniature reaction cartridge. Protein Science 1:1255–1224

Vandeckerkhove J, Bauw G, Puype M, Van Damme J, Van Montagu M (1985) Protein blotting on polybrene coated glass-fiber sheets: a basis for acid hydrolysis and gas phase sequencing of picomole quantities of protein previously separated on sodium dodecyl sulfate polyacrylamide gel. Eur J Biochem 152: 9–19

Wang R, Chait B, Kent S (1993) Protein ladder sequencing: a conceptually novel approach to protein sequencing using cycling chemical degradation and one step readout by matrix-assisted laser desorption mass spectrometry. In Angletti R (ed) : Techniques in protein chemistry IV., Academic Press, San Diego, pp. 409–418.

CHAPTER 10

Identification of Proteins by Mass Spectrometry

G. L. Corthals[1], S. P. Gygi[2], R. Aebersold[2] and S. D. Patterson[3]

1 Introduction

Biological mass spectrometry (MS) is now an indispensable tool for rapid protein and peptide structural analysis, and the widespread use of MS is a reflection of its ability to solve structural problems not readily or conclusively determined by conventional techniques. All mass spectrometers (MS) have three essential components that are required for measuring the mass of individual molecules that have been converted to gas-phase ions prior to detection. The components are an ion source, a mass analyzer and a detector. Ions produced in the ion source are separated in the mass analyzer by their mass (m) to charge (z) ratio, and (usually) detected by an electron multiplier. MS data are recorded as "spectra" which display ion intensity versus the m/z value. The two techniques that have

[1] The Garvan Institute of Medical Research, St Vincent's Hospital, University of New South Wales, Sydney, Australia.
[2] Department of Molecular Biotechnology, University of Washington, Seattle, Washington, USA.
[3] Amgen Inc, Protein Structure, Amgen Centre, Thousand Oaks, California, USA.

Table of abbreviations:

MS	mass spectrometers and mass spectrometry
ESI	electrospray ionization
MALDI	matrix-assisted laser desorption/ionization
MS/MS	tandem mass spectrometry
CID	collision induced dissociation
TOF	time-of-flight
SDS-PAGE	sodium-dodecyl-sulfate polyacrylamide gel electrophoresis
IEF	isoelectric focusing
1-D	one-dimensional electrophoresis
2-D	two-dimensional electrophoresis
kDa	kilo Dalton
mu	mass units
RP	reversed-phase
HPLC	high pressure liquid chromatography
U	atomic mass unit
CBB	coomassie brilliant blue
PSD	post-source decay
%T	total percent acrylamide and crosslinker per weight volume
PVDF	polyvinylidenedifluoride
LC-MS	Liquid chromatography mass spectrometry
CE	capillary electrophoresis

become preferred methods for ionization of peptides and proteins are electrospray ionization (ESI) and matrix-assisted laser desorption/ionization (MALDI), due to their effective application on a wide range of proteins and peptides (Karas and Hillenkamp 1988; Fenn et al. 1989). Although different combinations of ionization techniques and mass analyzers exist, MALDI usually uses a time-of-light (TOF) tube as a mass analyzer while ESI is traditionally combined with quadrupole mass analyzers capable of tandem mass spectrometry (MS/MS). Instruments capable of MS/MS have the ability to select ions of a particular *m/z* ratio from a mixture of ions, to fragment selected ions by a process called collision induced dissociation (CID) and to record the precise masses of the resulting fragment ions. If this process is applied to the analysis of peptide ions, in principle the amino acid sequence of the peptide can be deduced. Both ionization techniques give best results with salt- and detergent-free samples, although MALDI is more tolerant of sample contaminants. Besides the improvement in ionization performance, certain key events have contributed to popularize the use of ESI and MALDI:

(1) the development of microscale capillary reversed-phase high performance liquid chromatography (capillary LC, and LC-MS) (Karlsson and Novotny 1988; Kennedy and Jorgenson 1989) that can be directly coupled to an ESI interface;
(2) the development of computer algorithms that correlate MS and MS/MS generated data with database information (Henzel et al. 1993; James et al. 1993; Mann et al. 1993; Pappin et al. 1993; Yates et al. 1993; Eng et al. 1994; Mann and Wilm 1994; Clauser et al. 1995);
(3) MS software development allowing instrument control and versatility;
(4) development of post-source decay (PSD), a MALDI-TOF process whereby peptide sequence information can be obtained (Kaufmann et al. 1994);
(5) pre clean-up up of small (1 to 5 l) volumes (Zhang et al. 1995), useful for both ESI and MALDI;
(6) development of very low ESI flow rates (10–25 nL/min), or "nanoelectrospray", enabling long analysis times of as little as 1 µl of sample (Wilm and Mann 1996). The long analysis time permits many experiments on the same sample, and is particularly useful for optimizing MS/MS fragmentation patterns; and
(7) liquid extraction of peptides produced by in-gel proteolytic digestion (Shevchenko et al. 1996b).

Currently most proteins are identified by MS using ESI-MS or MALDI-TOF-MS. Both instruments are capable of analysis at comparably high sensitivities. Since the ability to identify a protein depends on the quality of the sample, this chapter explains some of the principles of MS and MS/MS, and highlights some useful techniques that may be required for protein identification by MS.

2 Protein Preparation Methods Compatible with MS

MS generated data is mostly as good as the sample that was prepared for analysis, immanently linking the successful identification of proteins to the quality of the sample and the sample preparation methods used. Fractionation of complex protein mixtures prior to MS is typically done by gel electrophoresis, as it provides purified samples efficiently and reproducibly. Both one-dimensional sodium-dodecyl-sulfate polyacrylamide gel electrophoresis (SDS-PAGE) (Laemmli 1970) and two-dimensional gel electrophoresis (2-DE, (isoelectric focusing) IEF/SDS-PAGE) (Klose 1975; O'Farrell 1975; Gorg et al. 1988) are rapidly becoming preferred methods for sample preparation. For high sensitivity analysis, gel-separated proteins have certain advantages over proteins prepared by other technologies:

(1) gels can be dried and stored for later analysis;
(2) the same gel electrophoresis conditions are suitable for the separation of most proteins;
(3) quantitative and qualitative sample control is afforded through visualization of the sample prior to MS analysis.

Furthermore, 2-DE gels are particularly useful for separating complex samples as they provide valuable added information missed by column separations of peptide mixtures such as:

(1) molecular mass;
(2) isoelectric point;
(3) relative abundance; and
(4) post-translational modification (isoforms) of the intact protein.

While for the majority of protein mixtures 2-DE may be useful, it should be noted that there are particular limitations inherent in the current 2-DE technology:

(1) proteins with high (> 8) and low pI's (< 4) are difficult to resolve, although recent progress in gel technology has extended the useful separation range of 2-DE gels (Gorg et al. 1997);
(2) very large proteins may not be resolved. It appears that proteins above approximately 100 kDa become underrepresented in gel patterns, and proteins heavier than 150 kDa are usually not seen at all, even though such proteins are visible in 1-D SDS-PAGE gels. It is not clear whether heavy proteins are lost in the first dimension or during transfer to the second dimension;
(3) there is growing evidence that hydrophobic proteins may not be amenable to 2-DE.

This is illustrated by the fact that hydrophobic proteins are rarely observed (Zewert and Harrington a, b, 1992; Zewert and Harrington 1993; Wilkins et al. 1998). Every project aimed at analysis of gel-separated proteins by MS has therefore to assess the advantages and limitations of the separation technique used. 2-DE provides the highest resolution, but may not display the all proteins in the sample. 1-D SDS-PAGE is a more general protein separation technique, but individual bands may contain multiple polypeptides. The implications of heterogeneous protein samples for protein identification are discussed later in the section on correlative data analysis (section 7).

MS analysis of proteins is not limited to samples separated by polyacrylamide gel electrophoresis. Essentially, any separation method is suitable, provided the peptides can be isolated in, or transferred into a solvent which is compatible with

MS analysis. To eliminate or minimize the need for protein separation McCormack et al. (1997) have shown that protein identification is feasible directly from proteolytic products of protein mixtures. The method was successfully used for the identification of components in inducible protein-protein complexes and immunoprecipitated complexes. All the proteins in the sample were co-digested and the resulting complex peptide mixture was separated by reversed-phase high pressure liquid chromatography (RP-HPLC) and analyzed by ESI-MS. Since the many peptides generated by the digestion of a relatively simple protein complex could easily overwhelm the resolution of a one-dimensional chromatography system, A. J. Link et al. ("in press", Nature Biotechnology) have extended this approach to include multidimensional peptide separation by column chromatography. In this approach two chromatography systems with orthogonal separation properties are connected on-line with a tandem MS. Compared to one-dimensional peptide separation this approach has several advantages, including simplified sample handling, process automation and increased number of identified proteins. In spite of these advances the vast majority of proteins currently analyzed by MS are isolated by gel electrophoresis.

2.1 Visualization/Staining Methods for Gel Separated Proteins

Many protein stains are compatible with subsequent MS analysis of gel – separated proteins. Among them, Coommassie Brilliant Blue (CBB) and silver staining are preferred because they are rapid and simple and do not interfere with MS analysis. The lower limit of detection for CBB and silver staining are, respectively, $\sim$ 0.1 μg and $\sim$ 1 ng. For a 50-kDa protein these limits of detection translate into 20 pmol and 20 fmol, respectively. These stains are useful for detecting both gel-separated and membrane-bound proteins. Silver staining in particular has had an enormous impact for MS applications. Following the demonstration by Shevchenko and colleagues (Shevchenko et al. 1996b; Wilm et al. 1996) that modified silver-staining of proteins in gels did not interfere with enzymatic digestion and subsequent peptide extraction, many MS labs switched to silver staining for low-level protein analysis. The key to successful peptide extraction from silver-stained gels is not to use fixatives such as gluteraldehyde or formaldehyde (Shevchenko et al. 1996b; Rabilloud et al. 1998). The procedures used prior to MS analysis therefore are a variation on previous, commonly used protocols (Merril et al. 1979, Rabilloud 1992). Silver staining and MS-compatible protocols are discussed in Chapter 5, Section 4. Equally compatible with MS analysis and similar to gel silver staining in sensitivity is imidazole-SDS-zinc reverse staining (Fernandez-Patron et al. 1995). Courchesne et al. (1997) have identified proteins by MALDI-MS and post-source decay MALDI-MS (see Sections 6.3.1 and 6.3.2.) at 125 and 250 fmol respectively, using this reverse staining method. Recently, Steinberg reported the use of the fluorescent dyes SYPRO Orange and SYPRO Red for direct detection of gel-separated proteins, with similar sensitivity to that of silver staining (Steinberg et al. a, b, 1996). To enhance sensitivity of the Amido Black stain which is commonly used for detection of proteins electroblotted on membranes prior to proteolysis (Aebersold et al. 1987), van Oostveen et al. (1997) developed

Table 10.1. A rapid silver stain method used for nitrocellulose or polyvinylidene difluoride membrane-bound proteins. The stain is useful for detection of 500 nmol to 100 fmol protein. This procedure was previously published by van Oostveen et al. (Ostveen et al. 1997)

Make the following solutions:
40 % sodium citrate (w/v), 20 % ferrous sulfate (w/v), and 20 % silver nitrate (w/v)
For preparation of a 25 ml staining solution:
Vortex the following solutions (the solution should become yellow)
1.25 ml sodium citrate (sol. 1)
1 ml ferrous sulfate (sol. 2)
22.5 ml H_2O
Then add ***dropwise*** *until a dark-brown precipitate becomes visible*
0.25 ml silver nitrate (sol. 3)
Stain/destain:
Add 10 ml of stain to membrane
Develop for 10 min max
Stop stain and clear background with 3 × 5 min H_2O washes

a colloidal silver staining procedure for nitrocellulose staining. The procedure is extremely rapid (< 10 min), has a detection limit of approximately 10 ng protein per band, and is compatible with MS analysis of extracted peptides (Oostveen et al. 1997; see Table 10.1). There are also other protein stains for membrane bound proteins such as Amido Black (100 ng) and Ponceau S (100 ng) (Sanchez et al. 1992), although, like CBB, these stains are not very sensitive.

3 Enzymatic Digestion of Proteins

3.1 In-Gel Digestion

The objective of digestion protocols prior to MS identification of proteins is to obtain sufficient enzymatic or chemical cleavage to successfully extract peptides from the gel matrix in a form that is directly compatible with MS analysis. Over the past 15–20 years there has been an incremental change from passive elution (Hager and Burgess 1980) to electroelution (Hunkapiller et al. 1983) and methods where staining was totally omitted (Ward and Simpson 1991; Ortiz et al. 1992) to the currently applied methods of in-gel digestion of proteins (Rosenfeld et al. 1992; Jeno et al. 1995). In-gel digestion methods were originally developed for fractionation of peptides by RP-HPLC for N-terminal sequence analysis; in these applications the use of small amounts of detergents such as SDS and Tween-20 could be tolerated (Kawasaki and Suzuki 1990, Kawasaki et al. 1990, Rosenfeld et al. 1992, Hellman et al. 1995). However, the increased use of MS for peptide analysis resulted in the modification of these methods either to use MS-compatible buffers (e.g., (Kirchner et al. 1996)) or to eliminate them entirely by not including them in any prior separation procedures (Moritz et al. 1995). These methods are now exclusively used for the MS analysis of silver-stained proteins (Shevchenko et al. 1996b, Wilm et al. 1996).

Trypsin is preferentially used to digest proteins for MS analysis. Trypsin cleaves C-terminal to the two basic amino acids, Lys and Arg (not at X-Pro), generating a wide distribution of peptide masses useful for MS analysis (typically 500–2500 u). Of particular importance for ESI-MS, the basic side chains of the C-terminal Lys and Arg residues allow for the attachment of a second proton in addition to the proton typically attached to the N-terminal amino group. Therefore, doubly charged peptides are generated frequently. The positioning of the charges at either the N- or C-terminal end of the peptide ion simplifies the inter-

Table 10.2. In gel digestion of proteins following silver staining. The short method is adapted from Shevchenko et al. (1996b), and the long method from Moritz et al. (1996)

Trypsin digestion – short method

1. After washing the gel 3 × 5 min, excise protein spot/band whilst paying attention to leaving the spot as small as possible to limit inclusion of acrylamide contamination (detected by MS). Dehydrate the gel pieces in CH_3CN for approximately 10 min. Remove CH_3CH and SpeedVac until dry. We use a positive and a negative control, which is useful to confirm possible contaminations and can be used to qualify the digestion. For a positive control either a standard can be run with the sample and treated as normal or a known protein from the gel can be used. A negative control is a piece of gel where no protein is present, but that was passed by the running buffer during electrophoresis.
2. Reswell gel pieces in small microfuge tubes for 45 min in buffer containing trypsin and 50 mM NH_4HCO_3, everything at 4 °C (approx. 5 μl/mm^3 gel). The gel pieces should be reswollen without excess buffer. Suggested amount of trypsin is 12.5 ng per μl buffer for proteins that have been silver stained. Digest overnight at 37 °C (or at least for 3 h).
3. Centrifuge gel pieces and collect supernatant if possible. Further extract peptides by 1 × 20 min wash with 20 mM NH_4 HCO_3 (centrifuge then collect), and three washes with 5 % acetic acid in 50 % CH_3CN (20 min between changes) at room temperature.
4. SpeedVac pooled supernatants until dry. Store at minus 20 °C, but do not keep sample in long storage (months), as peptides become difficult to extract after storage; direct use is preferred.

Note do not use more than 1 μg trypsin per sample for MS analysis.

Trypsin digestion – long method (higher coverage)

1. After washing the gel 3 × 5 min, excise protein spot/band and dehydrate in CH_3CN for approx. 10 min. Remove CH_3CN and SpeedVac until dry.
2. Reswell gel pieces in small microfuge tubes with 10 mM DTT, 0.2 M Tris-HCl, pH 8.4, 2 mM EDTA. Reduce proteins for 120 min at 40 °C.
3. Alkylate cysteine by adding 2 % 4-vinylpyridine (v/v) to the gel pieces in the reduction buffer. Vortex briefly and incubate for 60 min in the dark at room temperature. Note: the solution will become white after addition of the 4-vinylpyridine.
4. Stop alkylation by adding 2 % β-mercaptoethanol (v/v). Incubate for 60 min.
5. Wash gel pieces with 50–100 μl aliquots of 100 mM NH_4HCO_3 for 10 min. Dehydrate with CH_3CN reswell in 100 mM NH_4HCO_3, and shrink again with CH_3CN.
6. Remove liquid phase and SpeedVac.
7. As from point 2 in short method. Reswell gel pieces in small microfuge tubes for 45 min in buffer containing trypsin and 50 mM NH_4HCO_3, everything at 4 °C. Digest overnight at 37 °C (or at least for 3 h).
8. Centrifuge gel pieces and collect supernatant and further extract peptides by 1 × 20 min wash with 20 mM NH_4 HCO_3 (centrifuge then collect), and three washes with 5 % acetic acid in 50 % CH_3 CN (20 min between changes) at room temperature.
9. SpeedVac pooled supernatants until dry. Store at minus 20 °C; direct use is preferred.

Note do not use more than 1 μg trypsin per sample for MS analysis.

pretation of CID spectra obtained by MS/MS. In-gel enzymatic digestion of proteins and subsequent liquid extraction of peptides is now commonly used for MS analysis of proteins at picomole to low femtomole concentrations (Shevchenko et al. 1996b). A protocol for in-gel tryptic digestion of proteins that is rapid and generally applicable is described in Table 10.2. It is essentially the same as described previously (Shevchenko et al. 1996a, Wilm et al. 1996), although minor changes have been made. A similar method (Table 10.2), which is longer and involves reduction and alkylation of cysteine residues, is useful when a higher coverage of the protein sequence by the peptides recovered is required. Artifactual modification of peptides or non-specific enzymatic cleavage reduces the chance of protein identification, because the match-up of peptide masses with protein or nucleotide sequences in databases becomes difficult without precise knowledge of the peptide modification. This is particularly relevant for "small" proteins where fewer peptides are generated than with "large" proteins. In the long protocol, reduction and alkylation is performed prior to digestion. We have found the alkylation protocol using 4-vinylpyridine described by Moritz et al. (1996) to yield more superior in-gel alkylation of cysteine-containing peptides than protocols based on iodoacetamide (Jeno et al. 1995; Shevchenko et al. 1996b; Wilm et al. 1996). If the peptides are analyzed by ESI-MS, one must be aware that S-pyridylethylation has a tendency to add a proton at the site of the alkylated cysteine residue. Thus, triply charged tryptic peptides are frequently observed. Both digestion protocols in Table 10.2 are useful for all forms of PAGE and gels of varying thickness up to 1.5 mm. We have successfully used trypsin with 15 % T (total percent acrylamide and crosslinker per weight volume) acrylamide gels. A higher concentration of %T may be compatible with in-gel tryptic digestion. The porosity of the gel should be such that the relatively small (25-kDa) trypsin molecule can still penetrate the gel. Otherwise only proteins migrating at the gel surface will be digested. The digestion protocols are particularly useful for overnight digestion. Typically, 2 days are required for protein identification: on day 1 samples are prepared for overnight digestion and on day 2 samples are extracted and lyophilized, reconstituted for MS analysis and analyzed. A wide range of proteases are commercially available that may be useful to specifically digest a protein of interest (see Table 10.3).

3.2 Membrane Digestion

Enzymatic digestion of proteins electroblotted to nitrocellulose was first introduced for the generation of peptides for internal chemical sequencing (Aebersold et al. 1987). The same method essentially unchanged was shown to be compatible with ESI-MS (Hess et al. 1993) and MALDI-TOF (Zhang et al. 1994) analysis of the generated peptides. Numerous variations with respect to the procedure and the membrane have been described (Patterson 1994; Patterson and Aebersold 1995).

Digestion of proteins on membranes differs from in-gel protein digestion by the added electroblotting step that may lead to protein loss, at least for some proteins. "On" membrane digests are, however, fast, simple to perform and generally

Table 10.3. Proteolytic enzymes and their cleavage specificities useful for protein digestion

Enzyme	Cleavage site	Exception	Reference
Arg-C	RX Some KX	Some RX	Poncz and Dearborn (1983)
AspN	DX X-cysteic acid Some XE	–	Bentz et al. (1990)
Chymotrypsin	FX, YX, WX, LX Some MX, IX, SX, TX, VX, HX, GX, AX.	XP	Spackman et al. (1960)
Clostripain	R	–	–
Cyanogen-Bromide	M	–	–
Elastase	GX, AX, SX, VX, LX, IX	XP	Grunnet and Knudsen (1983)
Glu-C	EX, DX in phosphate buffers EX ammonium bicarbonate buffers	XP	Tomasselli et al. (1986)
IodosoBenzoate	W	–	–
Lys-C	KX Some NX	–	Perides et al. (1987)
Pepsin	FX, LX, EX	XV, XA, XG	Konigsberg et al. (1963)
Proline-Endopept	PX	–	–
Pronase	Most peptide bonds	–	Garner et al. (1974)
Staph-Protease (V8)	EX	–	–
Trypsin	KX, RX	XP	Lill et al. (1984)
Trypsin-K	KX	XP	–
Trypsin-R	RX	XP	–

X represents any amino acid.

yield peptide samples which are less prone to contamination than those generated form in-gel digestion. Membrane digestion on nitrocellulose or polyvinyldifluorine (PVDF) is often used for the analysis of ^{32}P-labeled proteins, as their exposure to film is easily detected on the exposed membrane and the samples generated are directly compatible with 2-D phosphopeptide analysis on cellulose plates (Boyle et al. 1991). For membrane digestion we use a method recently reported by van Oostveen et al. (1997) where membrane bound proteins are enzymatically digested after colloidal silver staining. For other methods we refer to a review by Patterson (1994).

4
Sample Clean-up for ESI-MS and On-line Liquid Chromatography

ESI-MS has a low tolerance to salts, detergents, non-volatile buffers and other contaminants, because they interfere with the ionization process and may obscure relevant signals in the mass spectrometer. Removal of these contaminants from peptide samples is therefore essential and can efficiently be performed using reversed-phase chromatography resin. Peptides remain bound to the resin while contaminants are washed away. This process can be performed on-line with the MS [liquid chromatography-MS (LC-MS)] or off-line (for off-line sample clean-up, see next section). The coupling on-line of microscale RP-HPLC columns and ESI-MS has been a major contribution to the success of ESI-MS. In addition to sample clean-up, LC-MS has the following advantages:

(1) peptides are concentrated during separation, thus improving the level of sensitivity;
(2) peptides in peptide mixtures are separated. Therefore peptides in individual fractions are analyzed sequentially by MS and MS/MS;
(3) post-column flow-splitting allows for the recovery of a large fraction of the purified peptide sample for further analysis, without significant reduction in the detection sensitivity in the ESI-MS instrument (Hess et al. 1993).

Arbitrarily, microcolumns with an internal diameter greater than 300 μm have been termed microbore columns and columns with an internal diameter smaller have been termed capillary columns. Typical inner diameters for capillary columns are 50 or 100 μm, respectively (Hunt et al. 1991; Griffin et al. 1991). Generally microbore columns are useful for the analysis of Coommassie Blue-stained proteins (microgram level sensitivity) and capillary columns are used for low-level protein detection of e.g. silver-stained proteins (nanogram level sensitivity). Microbore columns and capillary columns of excellent quality are commercially available. It is also possible to pack excellent capillary columns with common laboratory equipment. Several suitable methods have been published for constructing in-house capillary columns by slurry packing. Table 10.4 describes a procedure based on Karlsson and Novotny (1988) and Kennedy and Jorgenson (1989).

ESI-MS detection of proteins and peptides is concentration dependent. Accordingly, a post-column flow-splitter is frequently used, which directs a small fraction of the eluting analyte directly into the MS, while the remainder of the fraction can be collected for further analysis (Hess et al. 1993; Klarskov et al. 1994). If the sample is introduced into the MS with a microspray needle (1–50 μm tip) up to 98 % of the analyte can be recovered from microbore columns without significant reduction in ESI-MS detection sensitivity (Ducret et al. 1996). Post-column flow-splitting is particularly useful for phosphopeptide analysis in which collected ^{32}P-radiolabeled fractions can be quantified by scintillation counting or

Table 10.4. Instructions for slurry packing a microcapillary 75 μm (inner diameter) column

1. A 20 to 30 cm pieces of fused (silica) capillary[a] is cut (inner diameter 75 μm) and rinsed with 2-propanol briefly and air-dried. Next, one end of the capillary is tapped into a small vial containing underivatized silica until approx. 1 mm has been packed. The silica is sintered in a blue flame very briefly so that a porous opening is left. The frit can be checked under a low-power microscope before packing if desired.
2. The column packing (e.g. Magic C18 packing materials[b], 200-Å pore, 5 μm particle) is packed into the capillary with a pressure bomb at a constant pressure of approx 1000 psi until 10 cm is reached. This is achieved by adding ~ 100 μg packing material to 0.5 ml 2-propanol bomb on a stirring plate with the capillary dipping into the liquid slurry. The column is checked during packing under a low-power microscope.
3. After packing the column is washed with 0.5 % acetic acid and finally washed with methanol before storage in 20 % methanol.

[a] Fused (silica) capillary – Polymicro Technologies, Tucson, Arizona.
[b] Magic is from Michrom Bioresources, Auburn, California.

further analysis by 2-D phosphopeptide mapping. For chromatography on capillary columns the flow rate of the mobile phase is reduced to sub microliter per minute flow rates to maintain adequate column and linear flow rates. Solvent gradients at nanoliter per minute flow rates are usually generated by pre-column flow-splitting of solvents generated at microliter per minute flow rates. There are two alternatives to achieve nanoliter per minute solvent gradient without the need for flow-splitting. In the first, a gradient is pre-formed and stored in an ancillary loop and then pumped though the column after the sample has been loaded (Davis et al. 1995). In the second, a microgradient device capable of directly generating nanoliter per minute solvent gradients is directly coupled to the column (Ducret et al. 1998a). In contrast to microbore columns, post-column flow-splitting from capillary columns is technically difficult. Hunt and co-workers demonstrated, however, that even from capillary columns significant fractions of the separated analytes can be recovered (Hunt et al. 1992). For further information Yates and co-workers have written an excellent guide to preparation and operation of capillary LC columns coupled with ESI-MS (Yates et al. 1998).

5 Sample Clean-up for MALDI-MS

Although MALDI-MS is considerably more tolerant of buffers and salts compared with earlier forms of MS, it is still advisable to remove as many contaminants from a sample as possible as these species will be competing with the peptides/proteins for protons in the gas phase, resulting in suppression of ionization for the species of interest. Lecchi and Caprioli (1996) have elegantly demonstrated the dramatic effects of a high sample purity on the detection sensitivity in MALDI-TOF analysis of peptides (Lecchi and Caprioli 1996). MALDI-MS is performed on samples which have been mixed with matrix (small organic molecules see Section 6.3.1), resulting in co-crystallization of the analyte with the matrix. Crystallization results in some separation of peptides/proteins from salts and buffers as these tend to be excluded from the crystals. Therefore, the simplest method to decrease the concentration of residual salts is to rinse the crystals with cold acidified water (e.g., 0.1 % trifluoroacetic acid) (Beavis and Chait 1990). A few groups have developed alternate 'on-probe' clean-up approaches which essentially rely on the exclusion of salts and other contaminants from the crystals by increasing the number of 'nucleation' sites (sites where crystals form). Others have employed different surfaces for desorption and sample clean-up (e.g., (Zaluzec et al. 1994), an approach that will not be covered further here). Xiang and Beavis (1994) accomplished sample clean-up on the probe by forming, on the metal probe surface, a layer of matrix crystals which they then crushed. Non-adherent crystals were removed and the sample was added in a solvent which resulted in partial dissolving of the preformed surface. The surface could then be washed vigorously with water. Vorm et al. (1994) also decoupled matrix and sample handling by the fast evaporation of matrix dissolved in acetone. This resulted in smaller crystals of a more uniform surface. The sample was then applied in a low concentration organic solution, and following crystallization could be washed thoroughly. This approach has the added bonus that the mass accuracy is improved in some instruments because the surface is very uniform (Vorm et al. 1994). However, it is also possible to employ a more traditional clean-up methodology based-on solid-phase extraction. It should be pointed out that these procedures are equally suitable for sample clean-up prior to nanospray ESI-MS (Wilm and Mann 1996).

A number of groups simultaneously developed methods for the clean up of small quantities of peptide mixtures which often, but not always, were derived from in-gel digests (e.g., Zhang et al. 1995; Shevchenko et al. 1996b; Courchesne and Patterson 1997; Gevaert et al. 1998; Jensen et al. 1998). All of these approaches employ reverse-phase material to bind the peptides, allowing salts, buffers and other polar gel-related contaminants to be washed away (or significantly reduced in concentration). The bound peptides can then be eluted in a small volume (from sub mircoliters to low mircoliters) of high concentration organic, acidified solvent [e.g., 70 % acetonitrile (v/v) in 0.1 % formic acid (v/v)], thereby affecting both concentration and clean-up of the sample.

6 Mass spectrometers

MS provides the ability to accurately measure the mass of almost any molecule that can be ionized in the gas-phase. With the introduction of the two 'soft' ionization methods, ESI and MALDI, a decade ago (Karas and Hillenkamp 1988; Fenn et al. 1989), the possibility of mass analysis of large (> 10 000 kDa) biologically derived polymers was realized. A mass spectrometer consists of three components:
(1) an ionization source,
(2) a mass analyzer, and
(3) a detector; and measures the *m/z* ratios of ions under vacuum. Various combinations of ionization sources and mass analyzers have been made. However, this chapter will describe the three most common types of MS commercially available, employing two ionization sources (ESI and MALDI) and three mass analyzers (ion trap, quadrupole and TOF).

6.1 Quadrupole Ion Trap MS

The quadrupole ion trap MS (ITMS) is now established as a compact, cost-effective and highly sensitive detector for HPLC and capillary electrophoresis (Fig. 10.1). ITMS can provide both molecular weight and structural analysis of biological macromolecules. Many reviews describe the history and development of ITMS technology as well as applications (Cooks et al. 1991; McLuckey et al. 1994; Schwartz and Jardine 1996; Jonscher and Yates 1997). Peptide ionization by ESI occurs by spraying analytes at atmospheric pressure from the tip (1–5 μm) of a fine capillary (50–100 μm) at high voltages (low kVs). As the highly charged droplets evaporate, peptide ions with one or more charged protons are ejected into the gasphase. These ions are then drawn into the MS which operates under

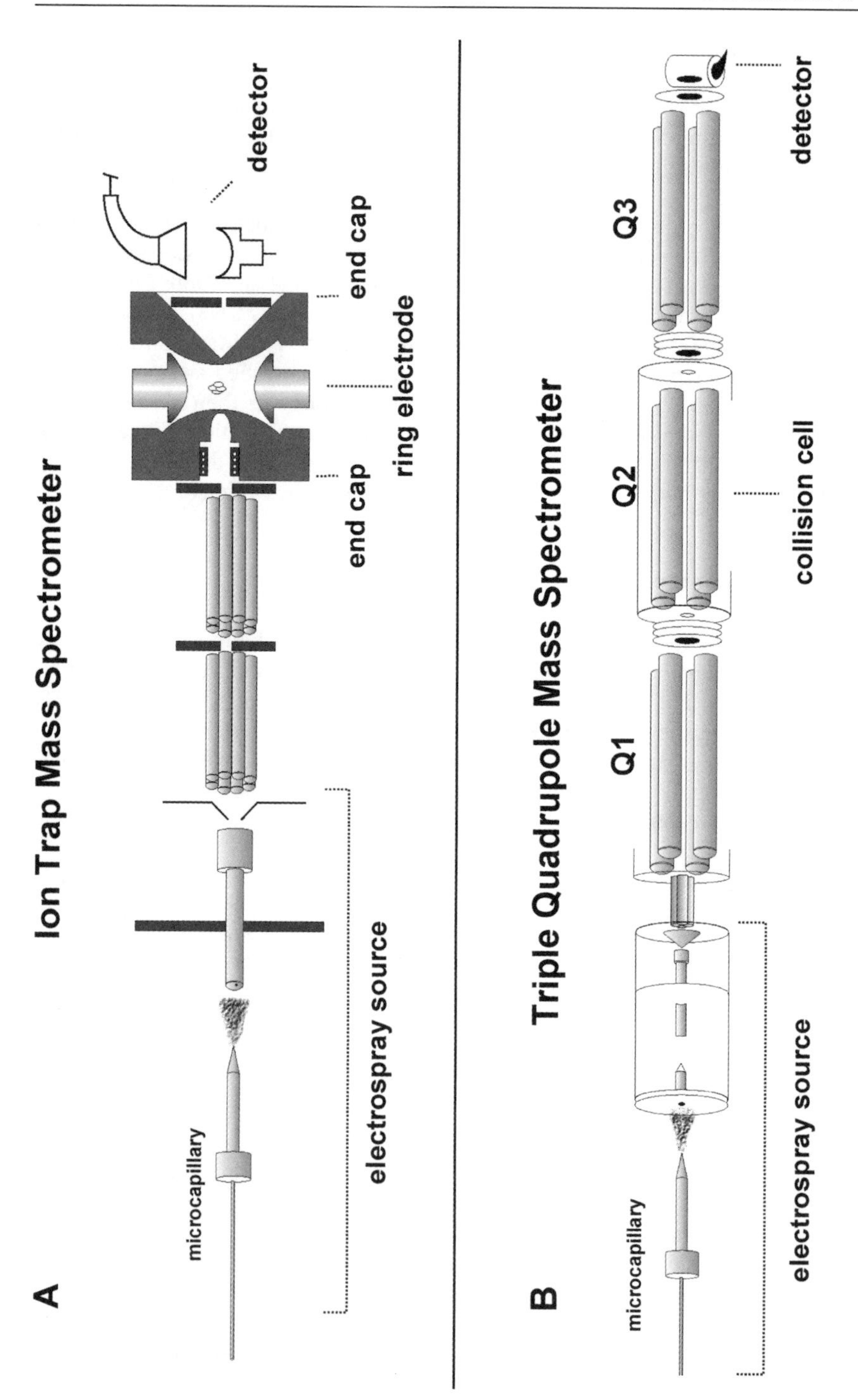

Fig. 10.1. An ion trap and triple quadrupole (*Q1, Q2, Q3*) mass spectrometer

vacuum, where the ions are trapped by a radio frequency (RF) trapping field. Ions enter and exit the ion trap through holes in the two end-cap electrodes. Helium gas present in the trap collisionally cools ions and forces them into the center of the ion trap. Ions above a certain minimum *m/z* remain trapped, cycling in a sinusoidal motion. The magnitude of the RF voltage determines the frequency and motion of the ions in the trap. To generate a mass spectrum, the RF voltage is ramped up linearly while applying a small voltage across the two end-cap electrodes, causing ions of successive *m/z* values to become unstable and be ejected axially from the trap in a process termed 'resonance ejection'. As ions are ejected from the trap they are detected by an off-axis conversion dynode with an electron multiplier detector. The detector is configured 'off-axis' to reduce background noise from neutral species that may be deflected onto the dynode. A schematic of the ITMS is shown in Fig. 10.1A.

Ion trap mass spectrometers are capable of performing CID and indeed can perform $(MS)^n$ experiments on a selected ion species (Jonscher and Yates 1997). Acquiring an MS/MS spectrum in an ion trap consists of selecting a single ion species by ejecting the ions of a larger *m/z* ratio and those present of a lower *m/z* ratio than the target ion. The trapped ion species is then energized and made to collide with neutral gas molecules of argon or helium. A spectrum is then recorded in the usual manner. Many successive MS/MS steps can be performed in an ion trap as long as a signal persists. The ability to perform $(MS)^n$ experiments is particularly useful for the detailed structural analysis of components, such as peptides containing post-translationally modified amino acids.

6.2
Triple Quadrupole MS

For protein identification purposes the triple quadrupole (TQ) MS is primarily used as a device for the generation of peptide CID spectra. This type of instrument is, however, extremely versatile and can perform specifically specialized experiments such as neutral loss scan, precursor ion scan and through in-source fragmentation, MS/MS/MS. These capabilities are particularly useful for the analysis of modified peptides and proteins. Typically, ions are produced by ESI at atmospheric pressure and introduced into the vacuum of the MS through a series of differentially pumped zones. The design is based on the properties of the quadrupole mass filter. A quadrupole mass filter consists of four parallel metal rods arranged to allow ions to pass lengthwise between the rods. Two opposing rods have both DC and AC voltages applied, while the other two rods have the same potential applied, but of opposite polarity. Oscillation of the applied voltages affects the trajectory of ions traveling down the flight path centered between the four rods. For given DC and AC voltages, only ions of a certain *m/z* ratio pass through the quadrupole filter. All other ions are ejected from their original path and eliminated. The variation of the potential applied to the quadrupole rods over time, and the consequential selection of ions of specific *m/z* ratios, is referred to as scanning. A mass spectrum is obtained by monitoring the ions passing through the quadrupole filter as the voltages on the rods are varied.

A TQ-MS contains three quadrupoles, termed Q1, Q2, Q3 (Fig. 10.1B). The function of Q1 and Q3 is to filter ions of specific *m/z* ratio by scanning the potentials applied to the quadrupole rods. Q2 serves as the collision cell and is constructed to transmit ions without selection. Q1 is a conventional quadrupole analyzer, set to transmit ions of selected *m/z* value at a given time. Q2, the collision cell, contains inert collision gas. An RF voltage is applied transmitting fragmentation products of an appropriate charge. Q3 records the *m/z* ratios of the fragment ions that originate from the fragmentation of the precursor ion selected by Q1. The advantage of the TQ-MS for acquiring MS/MS spectra is the fine control that the user can maintain over the entire process. The collision gas pressure and the energy imparted to the selected ion species can be manipulated to optimize fragmentation. A consequence of the sophisticated instrument control language, as part of the instrument in the most advanced TQ-MS instruments, is that they are characterized by a high degree of flexibility in selecting parameters and optimizing procedures.

6.3 MALDI-MS

6.3.1 Linear MALDI-MS

MALDI-MS-TOF is probably the simplest type of MS both conceptually and by design. Basically, samples are ionized and their *m/z* ratio is measured in a TOF mass analyzer (Fig. 10.2). The entire mass range is observed with each laser shot. The primary data can be summed and smoothed to enhance signal:noise if necessary. Masses can be measured with an accuracy of ~ ± 0.01–0.05 % up to ~ 25 kDa, and ± 0.05–0.3 % up to 300 kDa. MALDI is tolerant of a significant concentration of buffers and salts commonly employed in biological research such as phosphate, urea, non-ionic detergents, and some alkali metal salts. Such properties make MALDI-MS ideal for many aspects of biological research, particularly for the measurement of peptide masses and intact protein molecular weights, with measurements of antibodies (982000 ± 2000 u) having been made by Nelson et al. (1994).

In MALDI-MS-TOF molecules are ionized following a series of events starting with the mixing of the analytes with a small molecular aromatic 'matrix' compound usually dissolved in acidic organic solvent (see Table 10.5) on a metallic slide or probe. After drying, the sample:matrix co-crystal is inserted into the vacuum chamber of the MS which is held under a high vacuum of 10^{-5} to 10^{-8} Torr. A high voltage (e.g., +20 kV to generate positive ions) is applied to the sample slide/probe and a laser directed at the sample is fired, resulting in the desorption event which occurs if the matrix crystals absorb the photon energy of the wavelength of the laser. Energy deposition into the matrix molecules results in their emission of the absorbed energy as heat. This rapid heating of the crystals causes sublimation of the matrix crystals with the subsequent expansion of the matrix and analyte into the gas phase. This process ionizes the sample by protonation/deprotonation, cation attachment/cation detachment, or oxidation/reduc-

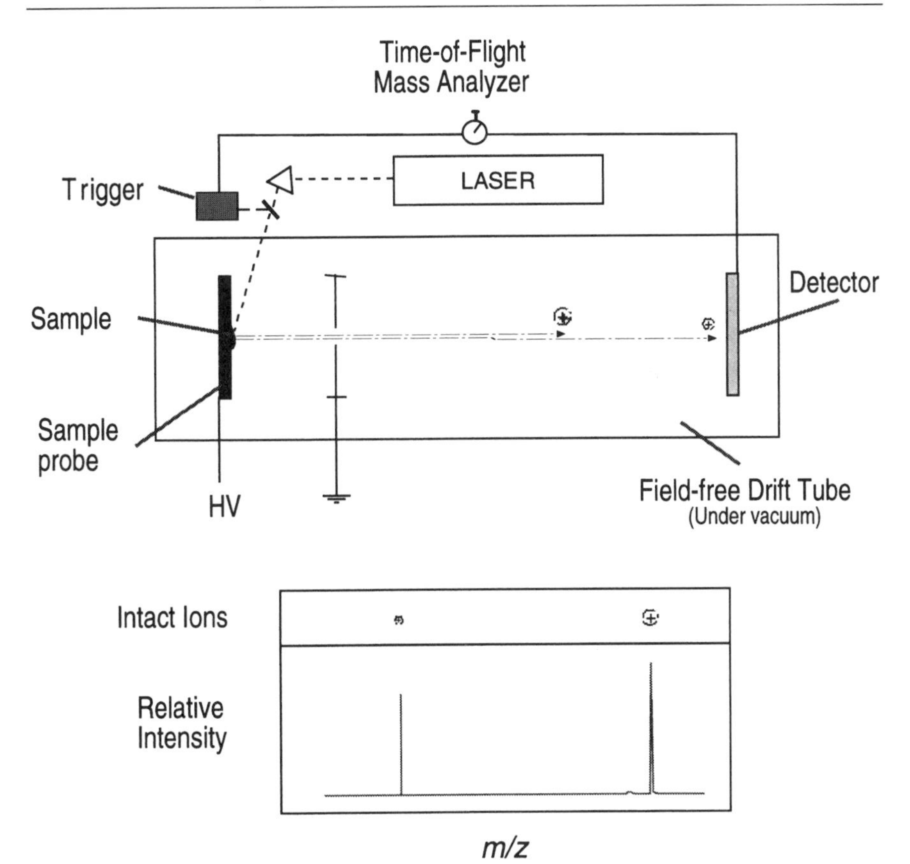

Fig. 10.2. A MALDI-MS instrument

Table 10.5. Common MALDI-MS matrices

Matrix	Stucture	Observed MH+[a]	Samples
Sinapinic acid: 3,5 dimethoxy sinapic acid	MeO, CH==CHCO2H, HO, OMe	225 207	Proteins, large peptides
4HCCA: α-cyano-4-hydroxybenzoic acid	CH=C(CN)CO2H, HO	190 172 379	Peptides, proteins (multiply charged), post-source decay
DHB: 2,5-dihydroxy benzoic acid	OH, CO2H, OH	155 137	Peptides proteins, carbohydrates, nucleic acids, polymers

[a] The first *m/z* listed is the intact protonated molecule; remaining ions listed are the result of losses of water, or dimerization.

tion in the gas phase. These ionized molecules are repelled from the surface (which is at high voltage) and accelerated towards a series of lenses at close to ground, thereby directing them into a field-free drift region. All of the ions have essentially the same final kinetic energy as they enter the field-free drift tube. Therefore, their arrival time (or time-of-flight) at the linear detector will vary depending upon their mass. A detector that has been triggered by the laser pulse records the time-of-flight from ionization to the end of the flight tube. The flight time of all molecules ionized by an individual laser pulse is then recorded (Fig. 10.2). Smaller ions fly faster than larger ions, and their *m/z* ratios practically can be calculated from their flight time using compounds of known mass as calibrants (usually a matrix ion and a peptide or protein).

Inherent to the ionization process is a small spread of kinetic energy resulting in a somewhat low resolving power [~ 450–800 u FWHM [full-width, half-maximum)] for peptides and (~ 50–400 u) proteins. At this resolution the isotope distribution of peptides cannot often be determined. Because the initial energy spread is mass dependent, the peak broadening becomes worse for larger ions. Two popular means of improving the resolution are through the use of an ion mirror (a reflectron) and/or 'time-lag focusing'. A schematic representation of a MALDI-MS-TOF with a reflectron is shown in Fig. 10.3. The reflectron has an applied voltage slightly higher than the accelerating voltage at the source. This results in a slowing of the ions as they enter the reflectron until they stop and are

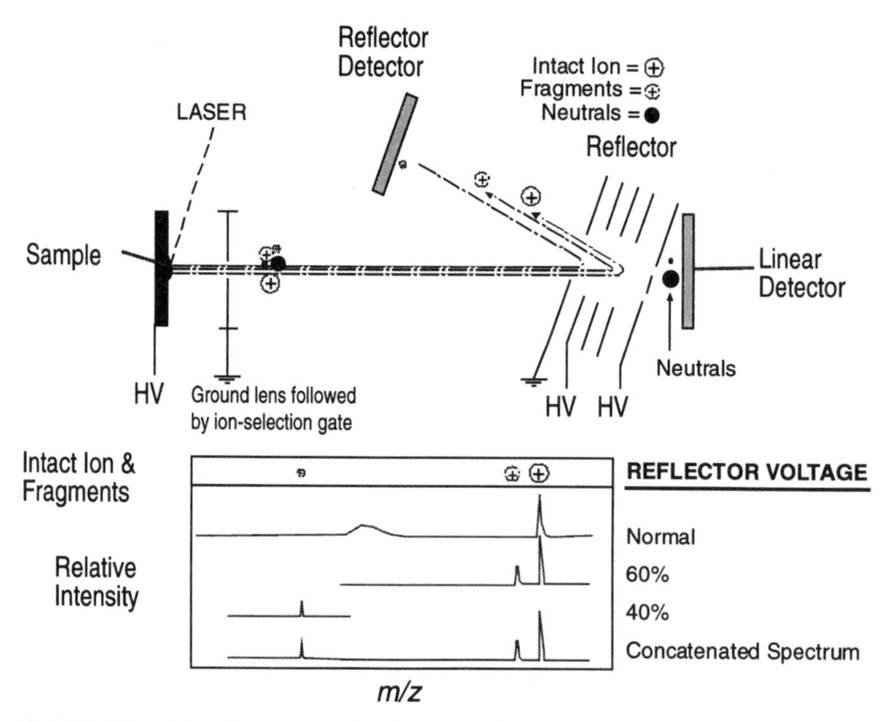

Fig. 10.3. A MALDI-MS instrument with a dual stage reflectron

then reaccelerated back out of the reflectron to a second detector at an angle. The reflectron therefore essentially acts as an energy-focusing device correcting for the initial energy spread of the ions. This occurs because ions with a slightly lower kinetic energy (slower ions) do not penetrate the reflectron as far and therefore turn around faster, thus catching up with the ions of slightly greater kinetic energy which have penetrated the reflectron further. Therefore, ions of a given *m/z* ratio are 'focused' spatially, having flight times that are very close together, resulting in a higher mass resolution.

Another means of correcting for the initial spread of kinetic energies during the MALDI process is to use a technique developed by Wiley and McLaren (1953) known as 'time-lag focusing', or more recently as 'delayed extraction' (Vestal et al. 1995). As the names imply, MALDI ions are created in a field-free region and allowed to spread out before an extraction voltage is applied to accelerate the ions into the drift tube. Not only does this result in a much decreased energy spread of the ions, but it may also limit peak broadening due to metastable decomposition from the collision of ions during extraction. Mass resolution has been shown to increase to ~ 2000–4000 u for peptides in a linear instrument and when the technique is applied to instruments fitted with a reflectron this further increases to ~ 3000–6000 u (Brown and Lennon 1995; Vestal et al. 1995).

6.3.2 PSD-MALDI-MS

Primary structural information can be generated in a MALDI-MS-TOF instrument through the fragmentation of selected peptides. These sequence-specific fragment ions are generated by a process known as post-source decay (PSD). Ions formed by MALDI can undergo metastable decay, i.e., ions may acquire excess energy from multiple collisions with matrix ions in the source during the desorption/acceleration process, resulting in sufficient energy to fragment the ion (Kaufmann et al. 1994). Because the metastable fragmentation has occurred 'post' source/ionization, the fragment ions will all have the same velocity as their precursor, but only a fraction of its kinetic energy. Therefore, the precursor and fragment ions (both charged and neutral) will arrive at the linear detector at approximately the same time and same *m/z* (Fig. 10.3 and 4). As mentioned above, this can result in some peak broadening. However, the fragment ions can be resolved by exploiting their kinetic energy differences through the use of an ion mirror (reflectrant) as described above. A dual stage (Fig. 10.3) and curved-field reflectrons (Fig. 10.4) have been successfully applied. The dual stage reflectron focuses ions of a limited kinetic energy range; therefore the voltage of the reflectron is decreased successively over 7–14 steps (bringing fragment ions of lower and lower kinetic energy into focus) until the last spectrum is acquired at about 5–10 % of the initial voltage. The spectra are then 'stitched' together with appropriate software and calibrated to yield the complete fragment ion spectrum. An alternative approach is the use of a curved-field reflectron described by Cornish and Cotter (1994) in which all of the fragment ions can be resolved in a single experiment due to all of the ions being focused in a long reflector which has a series of lenses at decreasing voltages (Fig. 10.4). Both types of instruments

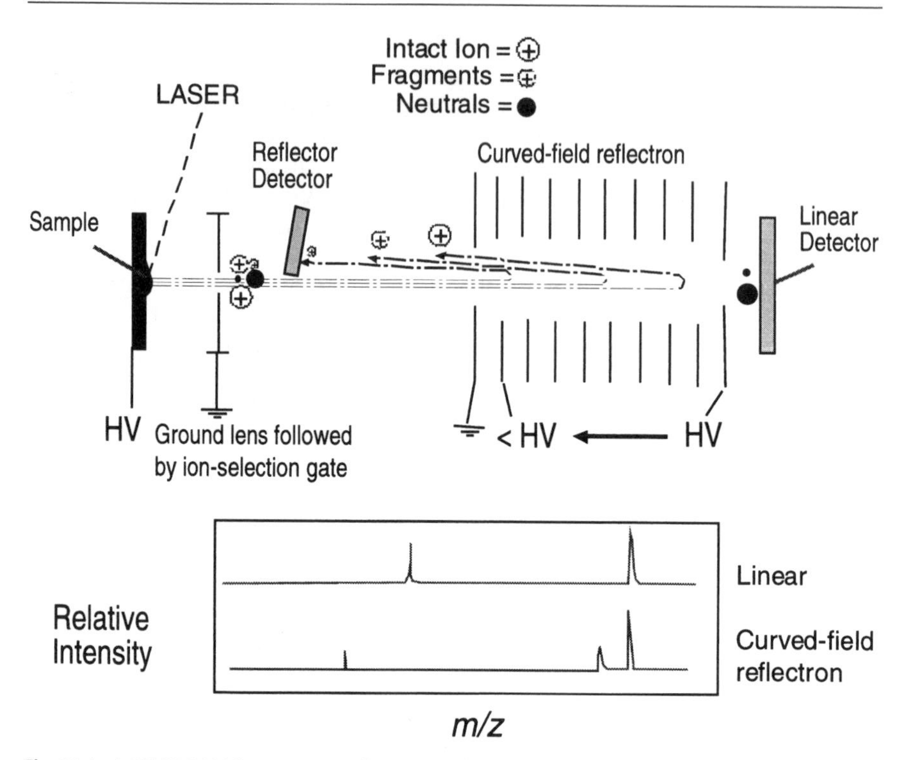

Fig. 10.4. A MALDI-MS instrument with a curved-field reflectron

have an ion gate just after the focusing lenses that enables low-resolution (±2.5%) selection of a precursor ion from a mixture. Therefore, sequence-specific fragment-ion information potentially can be derived from a number of peptides in an unfractionated mixture, as shown in Fig. 10.5, provided that the peptide mixture is of limited complexity. The matrix most often employed for PSD-MALDI-MS is 4HCCA (Table 10.5) as it is considered to be a relatively 'hot' matrix resulting in greater fragmentation than other matrices.

MS fragmentation of peptides primarily occurs at the amide bond (—CO-NH–) between two amino acid residues. The most commonly observed fragmen-

→

Fig. 10.5. PSD-MALDI-MS analysis using a curved-field reflectron-MALDI-MS of peptides derived from an in-gel LysC digest of 1 pmol of phosphorylase b (loaded on the gel). Four of the peptides observed in the linear MALDI-MS spectrum (see *inset* to panel **A**, where the selected peptides are labeled) from 1/5th of the 70% v/v MeCN/1% v/v HCCOH eluate from a microcolumn were subjected to curved-field reflectron (CFR)-MALDI-MS (**A** = MH^+ 845.8; **B** = MH^+ 1102.3; **C** = MH^+ 659.2; **D** = MH^+ 1290.8). The peptide sequence of each of the fragmented peptides is shown. Some of the fragment ions are labeled with their observed masses and the corresponding fragment-ion designation (according to Biemann (1990), with internal fragments represented by their sequence and whether they are a or b series, and immonium ions as "Imm"). (Courchesne et al. (1997) reproduced with permission from Wiley-VCH)

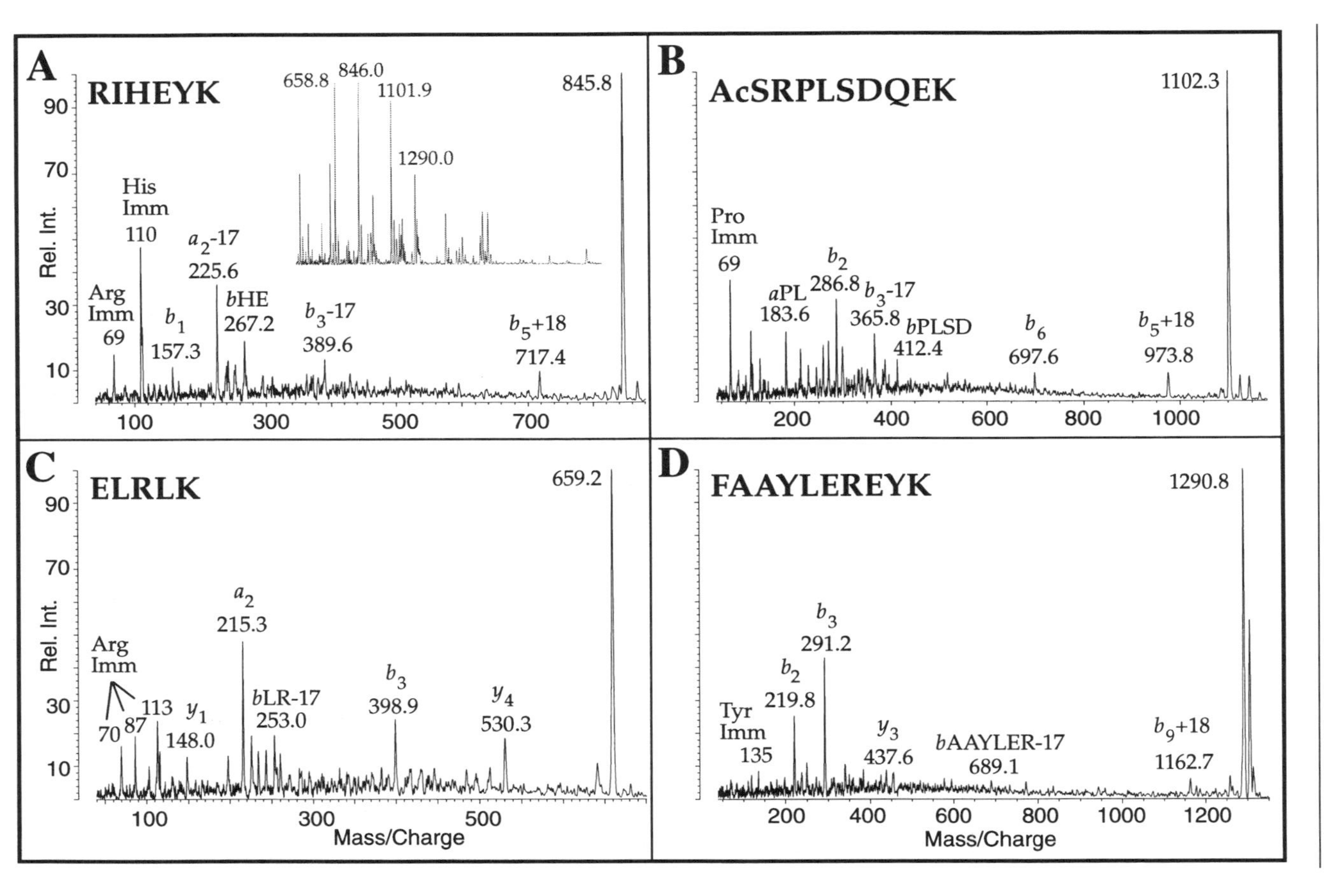
A
RIHEYK
Rel. Int.
90
70
30
10
Arg Imm 69
His Imm 110
b1 157.3
a2-17 225.6
bHE 267.2
b3-17 389.6
b5+18 717.4
658.8
846.0
1101.9
1290.0
845.8
100
300
500
700
B
AcSRPLSDQEK
Pro Imm 69
aPL 183.6
b2 286.8
b3-17 365.8
bPLSD 412.4
b6 697.6
b5+18 973.8
1102.3
200
400
600
800
1000
C
ELRLK
Rel. Int.
90
70
30
10
Arg Imm
70
87
113
y1 148.0
a2 215.3
bLR-17 253.0
b3 398.9
y4 530.3
659.2
100
300
500
Mass/Charge
D
FAAYLEREYK
Tyr Imm 135
b2 219.8
b3 291.2
y3 437.6
bAAYLER-17 689.1
b9+18 1162.7
1290.8
200
400
600
800
1000
1200
Mass/Charge

tation ions are *b* and *y* series ions: *b* series are sequence-specific fragment ions derived from the N-terminus and *y* series are from the C-terminus. PSD-MALDI-MS fragment ion spectra are composed of *b* and *y* series ions, as well as abundant internal fragments, neutral losses, immonium ions and *a* series ions (immonium ions are low mass ions of single amino acids of 27 u lower than that of the amino acid residue (HN=CH-R); *a* series are identical to the *b* series, but with the loss of CO (28 u). Neutral losses of water (–18 u) can be formed from fragment ions containing hydroxyl amino acids, Ser or Thr, and ammonia (–17 u) can be lost from fragment ions containing Gln, Lys, or Arg. Strong internal fragments are often generated during PSD-MALDI-MS-TOF if Pro is present in the peptide, resulting in unidirectional internal fragments starting at Pro and extending C-terminally. However, unlike most other forms of gas-phase fragmentation, PSD-MALDI-MS is not well controlled – the only parameters that one can alter are the laser fluence and the matrix. In addition, only singly charged ions are available for fragmentation, and these require more energy for fragmentation than the doubly or triply charged ions that are generated by ESI sources commonly coupled to quadrupole or ion trap mass spectrometers. As a result, many peptides are not amenable to fragmentation by PSD-MALDI-MS-TOF. Another constraint to fragmentation of singly charged ions applies to post-translationally modified peptides. The bond between the peptide and the modification (e.g., *O*-linked carbohydrate and sometimes phosphate on Ser or Thr) may be weaker than the peptide bonds themselves, therefore the predominant fragment ion observed may result from the loss of that modification with little or no peptide bond fragmentation. Consequently, TQ or IT MS is preferred for characterization of post-translationally modified peptides as well as de novo sequencing.

7 Protein Identification by Correlating MS Data with Sequence Databases

The identification of proteins is considerably simplified if the protein is represented in a sequence database. In this case the identity is established by correlating experimental data with sequence databases. If the sequence is not contained in a sequence database it has to be determined by de novo sequencing, which is slower, less sensitive and requires more operator input (section 7.3). Traditionally, partial protein sequences were determined to allow synthesis of degenerate oligonucleotide primers as a means to clone the gene coding for the protein, typically by a polymerase chain reaction (PCR)-based strategy. For many proteins isolated from species for which little or no sequence information is known, this is still the method of choice for the determination of complete gene and protein sequence. For those working with species for which the complete genome is known (*e.g., Saccharomyces cerevisiae, Eschericia coli* or *Haemophilus influenza*) protein identification has become considerably simpler as essentially all possible protein sequences are represented in the genomic sequence database. Another source of extensive nucleotide sequence information is the result of expressed sequence tag (EST) sequencing efforts in human and mouse (Adams et al. 1991). ESTs are short sequences of 250–400 base pairs generated by random single-pass sequencing of cDNA libraries. The wealth of nucleotide sequence information for open reading frames or anonymous stretches of coding sequence in EST databases provides researchers with a new possibility for protein identification from mammalian cells. In this new era of extensive nucleotide sequence information, the 'universe' of potential protein sequences in a given organism can be deduced from the six-way translations of the nucleotide sequence. Therefore, correlative approaches can be employed to determine what gene or gene fragment codes for the isolated protein. Once identified, the complete nucleotide sequence of the coding gene, or a considerable portion thereof, is already known.

Two types of MS data have been used for protein identification by correlation with sequence databases;

(1) the accurate mass of peptides (within 5 ppm resolution) derived by specific cleavage of the isolated protein,
(2) CID spectra from individual peptides isolated after proteolysis of the target protein. The methods to generate these data have been described earlier in this chapter. An important caveat to these correlative analyses is that the protein sequence, or protein sequence translated in six frames from nucleotide sequence, is contained within the database being searched.

7.1 Peptide-Mass Searching

Following the introduction of MALDI-MS-TOF and ESI-MS, which allowed accurate mass measurements of peptides, a number of groups described algorithms for correlating the collective of peptide masses generated from a digestion of a pure protein with sequence databases. This technique is commonly referred to as peptide-mass fingerprinting (Henzel et al. 1993; James et al. 1993; Mann et al. 1993; Pappin et al. 1993; Yates et al. 1993). Peptide-mass searching is most simply described as a method to identify proteins already contained within a sequence database using an algorithm to match a set of peptide-masses generated using specific cleavage reagents (either enzymatic or chemical) from the protein of interest, with theoretical peptide masses calculated from each sequence entry in the database if the database sequences had been cleaved with the same specificity as the reagent in the experiment (see Fig. 10.6). A ranking (or score) is then calculated to provide a measure of the fit between the observed and expected peptide masses. A number of search programs can be found on the World Wide Web and most include examples to guide their use. An incomplete list of these sites can be found in Table 10.6.

Peptide-mass searches can be conducted in databases of full-length protein or gene sequences. They are rarely successful against translations of short nucleotide sequences (such as ESTs) as there is generally insufficient sequence information contained for the multiple peptide matches required for conclusive identification. The key to successful peptide-mass searching is employing the highest mass accuracy available (Chait et al. 1993; Mann et al. 1993; Fenyo et al. 1998). Ideally one should use internally calibrated monoisotopic masses for any peptide-mass search. For MALDI-MS this is generally only possible on instruments fitted with delayed-extraction/time-lag focusing ion sources. This is now a common application on commercial instruments available (e.g., Vestal et al.

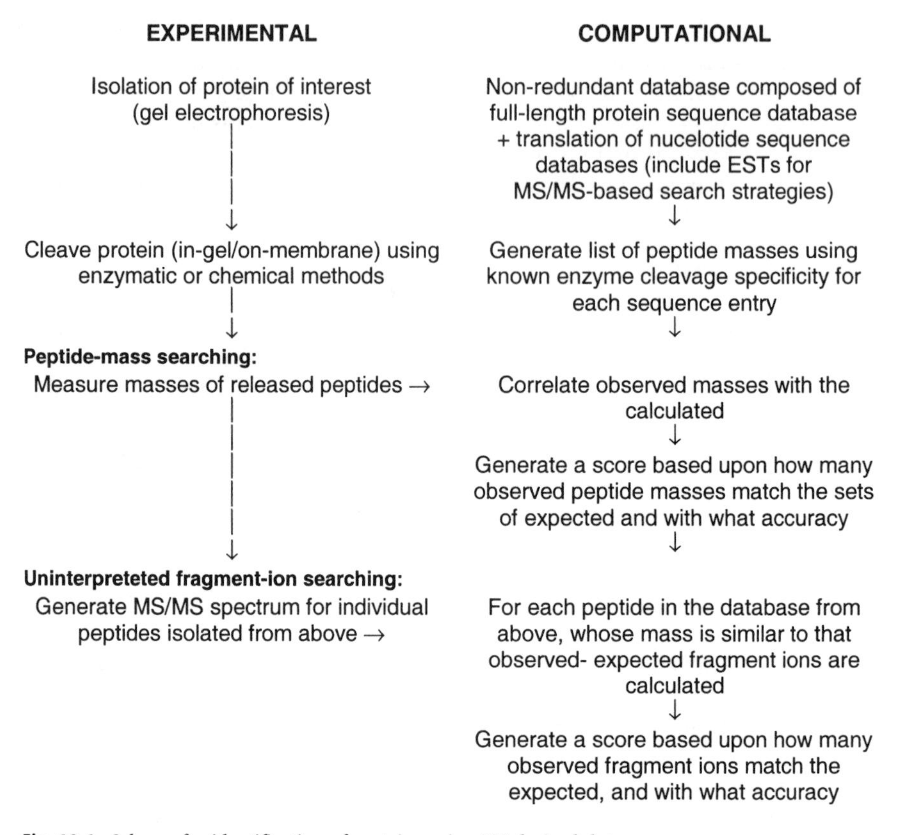

Fig. 10.6. Scheme for identification of proteins using MS derived data

1995). Instruments capable of 5 ppm accuracy can assign monoisotopic masses (the mass of the most abundant isotope for a certain peptide) and therefore employ very tight tolerances in peptide-mass searches which limits the number of spurious matches (Jensen et al. 1996a; Takach et al. 1997). If this option is not available, the researcher is advised to calibrate the spectrum internally to achieve the highest possible mass accuracy even when average masses are used.

Analysis of the results of peptide-mass searching experiments may not always be straightforward. In most cases there are input peptide masses that do not match with expected peptide masses of the highest ranking result. One should attempt to determine the origin of these masses, and the following includes some of the possible reasons for these 'orphan' masses (Patterson and Aebersold 1995):

1. The correct protein was identified by the search, but the masses are the result of post-translational or artifactual modification or post-translational processing. Plausible modifications should be considered tentative unless confirmed experimentally.
2. The correct protein was identified, but some peptides were derived from non-specific proteolysis or cleavage by a contaminating protease. This possibility

Table 10.6. Uniform Resource Locator's (URL's) for sites providing on-line protein identification using mass spectrometric-derived data

Features and comments	WWW Uniform Resource Locator (http://www...)	Resource
Peptide-mass search program	cbrg.inf.ethz.ch/ subsection3-1-3.html	CBRG, ETH-Zurich, Switzerland
Peptide-mass search program	expasy.ch/sprot/findmod.html	ExPASy Molecular Biology Server, Switzerland
Peptide-mass and fragment-ion search programs	mann.embl-heidelberg.de/ Services/PeptideSearch/ PeptideSearchIntro.html	EMBL Protein & Peptide Group, Heidelberg, Germany
Peptide-mass search program	mdc-berlin.de/~emu/ peptide-mass.html	Max Delbruck Centre, Berlin, Germany
Peptide-mass and fragment-ion search programs	prowl.rockefeller.edu/PROWL/prot-id-main.html	Rockefeller University, New York, USA
Peptide-mass and fragment-ion search programs	seqnet.dl.ac.uk/mowse.html	SEQNET, Daresbury, UK
Peptide-mass (MS-Fit) and fragment-ion (MS-Tag) search programs	prospector.ucsf.edu/	University of California, San Francisco, USA

can be tested by determining whether the protein candidate can produce the peptide masses without any assumptions as to the cleavage specificity

3. The correct protein was identified, but the 'pure' protein was contaminated with one or more additional proteins. Some programs can take this into account during the initial search, but in all cases if there are sufficient unmatched masses these can be used in an additional peptide-mass search.
4. A sequence homologue, or processing variant, from either the same or a different species was identified. Some search programs allow species-specific searches and that could eliminate this possibility, however, if confirmatory data is obtained, matches to proteins from other species can be useful, especially for scientists working with organisms whose genomes are relatively poorly characterized.
5. The match was a false positive! This is the most disturbing outcome and may be difficult to prove or disprove, particularly if the highest ranked protein did not have a high score; sometimes confidence can be gained from the difference between the first and second highest and subsequent scores; the protein may also not yet reside in the database and it may be truly novel. Several of the newer programs for peptide mass mapping take at least some of these possibilities into account and either allow for secondary searches with 'orphan' masses and/or use a scoring scale that confidently distinguishes between true and false positives.

In addition to the overriding importance of mass accuracy, the specificity of the cleavage is also critical. Trypsin, the most commonly used reagent for peptide-mass searching, specifically cleaves at sites C-terminal to Lys or Arg (if not followed by a Pro). However, like most proteolytic enzymes, trypsin, or contaminating proteases can cleave at other sites and it may not cleave its substrate to completion. While so-called missed cleavage sites are often included as a parameter for peptide-mass search programs, nonspecific cleavage is more difficult to accommodate in the search algorithms. In addition, trypsin will also autodigest. This is advantageous as the trypsin-derived peptides can be used for internal mass calibration. Trypsin – derived peptides can, however, if present in large excess over the peptides of interest, suppress ionization or obscure target peptides by overlapping in the spectrum. Endoproteinase LysC (Lys-X, some Asn-X) is another excellent enzyme for peptide-mass searching; while it can also miss cleavage sites and cleave nonspecifically, it has the advantage of far fewer autolysis products than trypsin.

A number of researchers have investigated methods to increase the confidence of peptide-mass assignments through incorporation of orthogonal data types not the second algorithm. Such data can be obtained from a second aliquot of the sample, in addition to the original sample employed for peptide-mass searching, or, in some instances, by manipulation of the probe of a MALDI-MS. The additional information helps to distinguish between database matches which cannot be differentiated by peptide-mass analysis alone. The orthogonal methods that have been employed are listed in Table 10.7 and include: site-specific chemical modification [e.g., methyl esterification, which adds +14 u for each acidic residue – Asp, Glu, or C-terminus – of the peptide (Pappin et al. 1995)]; or iodination, which adds +126 u for each tyrosine (Craig et al. 1995); determination of partial amino acid composition of the peptide [hydrogen/deuterium exchange in which each amino acid can exchange a defined number of its hydrogen' s for the heavier deuterium, and the total mass increase reflects the peptide composition –from 0 to 5 u per residue (James et al. 1994)] or through identification of immonium ions from MS/MS spectra; determination of the N-terminal residue of each peptide in the mixture by Edman degradation (Jensen et al. 1996b) [or the use of an aminopeptidase (Woods et al. 1995)]; additional enzymatic digestion, either a sub-digestion (Pappin et al. 1995) or a parallel digestion (James et al. 1994); or a

Table 10.7. Orthogonal approaches to increase the confidence of peptide-mass searching (For references see text)

Method	Residues modified	Mass increase (u)/residue
Methyl esterification	Asp, Glu, C-terminus	14
Iodination	Tyr	126
Hydrogen/deuterium exchange	All amino acids	From 0 to 5
Partial amino acid composition	Identification of some component residues	–
Enzymatic digestion	Sub-digestion in addition to original	–
one-step Edman degradation or aminopeptidase digestion	Identification of N-terminal residue(s)	–
Carboxypeptidase digestion	Identification of C-terminal residue(s)	–

carboxypeptidase digestion to determine the C-terminal residue(s) (often less informative) (Woods et al. 1995). A comprehensive study of the benefits of employing some of these orthogonal approaches can be found in a recent paper by Fenyo et al. (1998) in which the implications of incorporating orthogonal data types for the identification of proteins from *S. cerevisiae* has been examined.

Peptide-mass searching can be a powerful tool for extremely rapid and sensitive protein identification, particularly in species for which the genome is well characterized. To further increase the sample throughput researchers have begun automating sample processing for MALDI-MS and peptide-mass searching. Shevchenko et al. (1996a) demonstrated a semi-automated approach for a large scale yeast 2-DE proteome project which employed a two part protein identification strategy: firstly, a 32-port parallel enzyme digestion robot (Houthaeve et al. 1995) provided samples for automated loading onto a MALDI-MS plate for automated delayed extraction MALDI-MS using a small aliquot of the peptide digest and subsequent automated peptide-mass searching; secondly, if the first strategy did not conclusively identify the protein half of the remaining sample was subjected to nano-ESI-MS/MS to generate a sequence tag (Section 7.2) after it was reduced in volume and cleaned up using a microdesalting step (Shevchenko et al. 1996b). Peptide-sequence tags generated from a number of peptides in the mixture were then used to identify the protein. If this was unsuccessful, the remaining sample could be methyl esterified and subjected to another nano-ESI-MS/MS experiment providing the possibility of de novo sequence interpretation (Shevchenko et al. 1996a).

7.2 Searching Databases with Uninterpreted or PartiallyInterpreted MS/MS Spectra

Fragmentation of gas-phase peptide ions in IT or TQ MS often provides extensive sequence-specific information. Direct interpretation of these MS/MS spectra is not always simple, and in many cases the data from a single MS/MS spectrum may be insufficient to unambiguously determine the peptide sequence. Using an LC separation prior to the MS, peptides enter the instrument separated in time, and MS/MS spectra are generated from (many) different peptides. Since each peptide will generate a sequence-specific MS/MS spectrum, only a few spectra (minimally one) exhibiting good quality peptide fragmentation are needed for protein identification by correlation with a database entry. The use of programs that assist in or automatically perform database correlations has tremendously affected our ability to identify proteins rapidly and conclusively. At the same time, algorithms were being developed for peptide-mass searching; algorithms that employed somewhat similar logic but used MS/MS spectra for database searching were also developed (Eng et al. 1994; Mann and Wilm 1994; Clauser et al. 1995). The informational content of an MS/MS spectrum can provide rapid and unambiguous identification of a peptide from a sequence database because of the complementary redundant data present in an MS/MS spectrum which include:

(1) the mass of the intact peptide,
(2) sequence specific fragment-ions derived from the N-terminus (typically *b* series),
(3) sequence-specific fragment-ions from the C-terminus (typically *y* series), and
(4) sometimes internal fragment ions, including immonium ions (partial composition).

The MS/MS spectrum is a plot of the frequency of the peptide fragmentation, where the *m/z* value of the fragments represents specific amino acid sequence fragment occurrences. The more susceptible a bond to fragmentation the more intense that fragment ion will be in the spectrum. Therefore, some peptide sequences will yield complete N- and C-terminal ion series, and most will yield only some fragment ions. It is obvious that the confidence in protein identification increases with the increasing quality of the peptide MS/MS spectra used to search the database. An indicator of the quality of MS/MS spectra is the completeness of the *b* and *y* series ions, respectively. The quality of the spectra is dependent upon the process employed to generate the fragmentation. The least controlled fragmentation process is that of PSD in MALDI-MS. Fragmentation in TQ and IT MS instruments can be controlled precisely. Fortunately, knowledge of the mass of the peptide, of limited fragmentation information, and the cleavage specificity of the enzyme employed to generate the peptide has been shown to be sufficient to identify peptides by database searching. The first two publications describing algorithms for searching sequence databases with MS/MS spectra took different strategies. Mann and Wilm partially interpreted the MS/MS spectrum to generate a 'peptide-sequence tag' (Mann and Wilm 1994), whereas Yates and colleagues developed an automated approach for searching databases with uninterpreted MS/MS spectra (Eng et al. 1994; Yates et al. 1995b). In both cases the programs can be run in an error-tolerant mode whereby peptides carrying post–translational modification can often be identified. Subsequently, Clauser and colleagues published a program that matches masses from uninterpreted MS/MS spectra to sequence databases using a program called MS-Tag (Clauser et al. 1995). A demonstration of the use of these three approaches is described below, while a schematic representation of how identification is achieved with MS data is shown in Fig. 10.6.

The peptide-sequence tag (named PeptideSearch) approach of Mann and Wilm (1994) relies on partial manual interpretation of the MS/MS spectrum to generate the 'peptide-sequence tag'. The 'tag' is derived by interrogating the MS/MS spectrum to determine whether an ion series (either *b* or *y*) can be deduced. The mass difference between these ions defines specific amino acid residues that make up the partially interpreted sequence. This information essentially divides the peptide into three regions with potentially five pieces of information:
(1) the mass from the N-terminus to the interpreted sequence (the mass of the lowest ion in the ion series),
(2) the interpreted peptide sequence (anywhere from one to many residues), and
(3) the mass of the C-terminal region (the difference between the mass of the highest ion in the series and the intact mass). The cleavage specificity of the enzyme used provides two additional pieces of information if desired, i.e.,

(4) N-terminal and
(5) C-terminal specificity. However, these can be left out if the program is run in error-tolerant mode. The program can also be instructed to disregard the precursor ion mass information in cases in which the identification of unknown or modified amino acids residues is attempted. Collectively, these types of information are sufficiently specific to achieve unambiguous protein identification by searching sequence databases including EST databases.

A more automated approach to search sequence databases with MS/MS spectra was developed by Yates and colleagues (Eng et al. 1994; Yates et al. 1995b). The resulting program is referred to as SEQUEST. SEQUEST first generates a list of theoretical peptide masses for each entry in the database, using either enzyme cleavage specificity or the sum of contiguous amino acid residues; if no cleavage site specificity is indicated it determines which of these calculated masses match the experimentally determined peptide mass (within a stated tolerance), thereby generating a list of candidate peptides. In a second step, the program calculates the fragment ion masses expected for each of the candidate peptides, thus generating a predicted MS/MS spectrum. In a third step, the experimentally determined MS/MS spectrum is compared with the predicted spectra using correlation-analysis algorithms. Each comparison receives a score, and the highest scoring peptide(s) are then reported. Therefore MS/MS spectra are correlated with sequence databases without the need for any explicit determination of the peptide sequence. As we have described previously (Patterson and Aebersold 1995), the advantages of this approach include:
(1) possibility of identifying proteins in complex mixtures because each peptide that generates an MS/MS spectrum provides data for an independent database search. The power of this approach for the analysis of complex protein mixtures has since been determined (Yates et al. 1995a);
(2) if the approach is applied to a digest of a single protein the results are self-confirmatory, i.e., the same protein is identified independently several times
(3) when ions are selected in a data-dependent manner, the approach is easily automated and the MS/MS results can be automatically analyzed (Yates et al. 1995a, Ducret et al. 1998b); and
(4) peptides carrying specific post-translational modifications can be searched through a simple variation of the program, often resulting in identification of the specific site of modification and the peptide on which it is presented (Yates et al. 1995a).

In another variation on the theme of uninterpreted MS/MS fragment ion spectra searching, the program MS-Tag from Clauser and colleagues (1995) uses the fragment ion masses and the mass of the intact peptide similar to the manner described for SEQUEST, but without the correlation-analysis algorithms for matching. Peptides isobaric to the experimentally measured peptide are extracted from the sequence database (using enzyme cleavage specificity or not), and selected ion series are calculated for each peptide for matching with the MS/MS fragment ions from the experimentally determined peptide. The ions known to be generated by the MS employed are selected for the search. Only peptides that match a predetermined number of fragment ions are reported. The program can also be run in an error – tolerant mode and searches can include defined specific post-translational modifications.

Demonstration of the use of these three programs is shown with an example spectrum (Fig. 10.7). The MS/MS spectra were taken from an LC-MS/MS analysis of a digested, silver-stained yeast protein from a 2-D gel. Peptides were digested with trypsin as described in Table 10.2 and separated by capillary RP-HPLC on a 75-μm column. In a single analysis, individual eluting peptides were sequentially ionized, detected, and fragmented by CID. The information from a single MS/MS spectrum is shown in Fig. 10.7. The specific information that was manually extracted and input into each program is indicated. No information was extracted for the SEQUEST program as this is run in automatic mode. The SEQUEST parameters were set as follows: database searched was OWL (ftp://ncbi.nlm.nih.gov/repository/OWL/owl.fasta.Z), no enzyme specificity was indicated, peptide mass was average, no cysteine modification, and the possibility for methionine oxidized (+ 16 u) was indicated. SEQUEST registered six MS/MS spectra on doubly charged ions and one spectrum as a triply charged ion (as well as two keratin and two auto digestion trypsin peptides). All sequences identified were genuine tryptic peptides. For the use of PeptideSearch and MS-Tag programs, the operator must manually scan the numerous spectra generated during the entire LC-MS/MS run and select quality MS/MS spectra to search the databases. A rapid manual inspection of the MS/MS spectra identified four spectra which appeared appropriate for database searching. For each of the MS/MS spectra chosen, both programs identified the same protein as SEQUEST. In the example spectrum (Fig.10.7) the masses with an asterisk were entered in the MS-Tag program as well as the peptide mass 1039.24. Similar data were used for database searching the other three spectra. MS-Tag is a web-based program and results are returned through a web-browser, with the best match as the highest score. The program can also be run locally and can be obtained from the UCSF Mass Spectrometry Facility (http://donatello.ucsf.edu/). For each of the four spectra searched, MS-Tag returned the correct protein as the highest-ranking score; other peptides were also 'identified' for each of the four spectra, but with a lower score. Furthermore, since the same protein was identified for each individual search, protein identity was established at a high level of confidence. For the PeptideSearch program the peptide must be sequenced partially (usually at three residues) before the program can be run effectively. In the example shown (Fig. 10.7) the sequence is displayed in the spectrum (DTAE). In addition to the peptide mass (1039.24), trypsin was specified as the enzyme used to generate peptides and the masses of the N- and C-terminal 'tags' were added, as 509.5 and 926.9 u respectively. As with the other programs (except for SEQUEST) PeptideSearch can be run locally as well as over the Internet (http://www.mann.embl-heidelberg.de/Services/PeptideSearch/FR–PeptidePatternForm.html). For each of the four spectra, PeptideSearch returned the protein of interest ranked twice in first place and twice in second place. As with the other programs assignment of the correct protein was achieved. In summary, each program correctly identified the same protein. The programs differ in the amount of information which has to be manually determined and added to initiate the search. All of the above programs can be found on the World Wide Web (see Table 10.6).

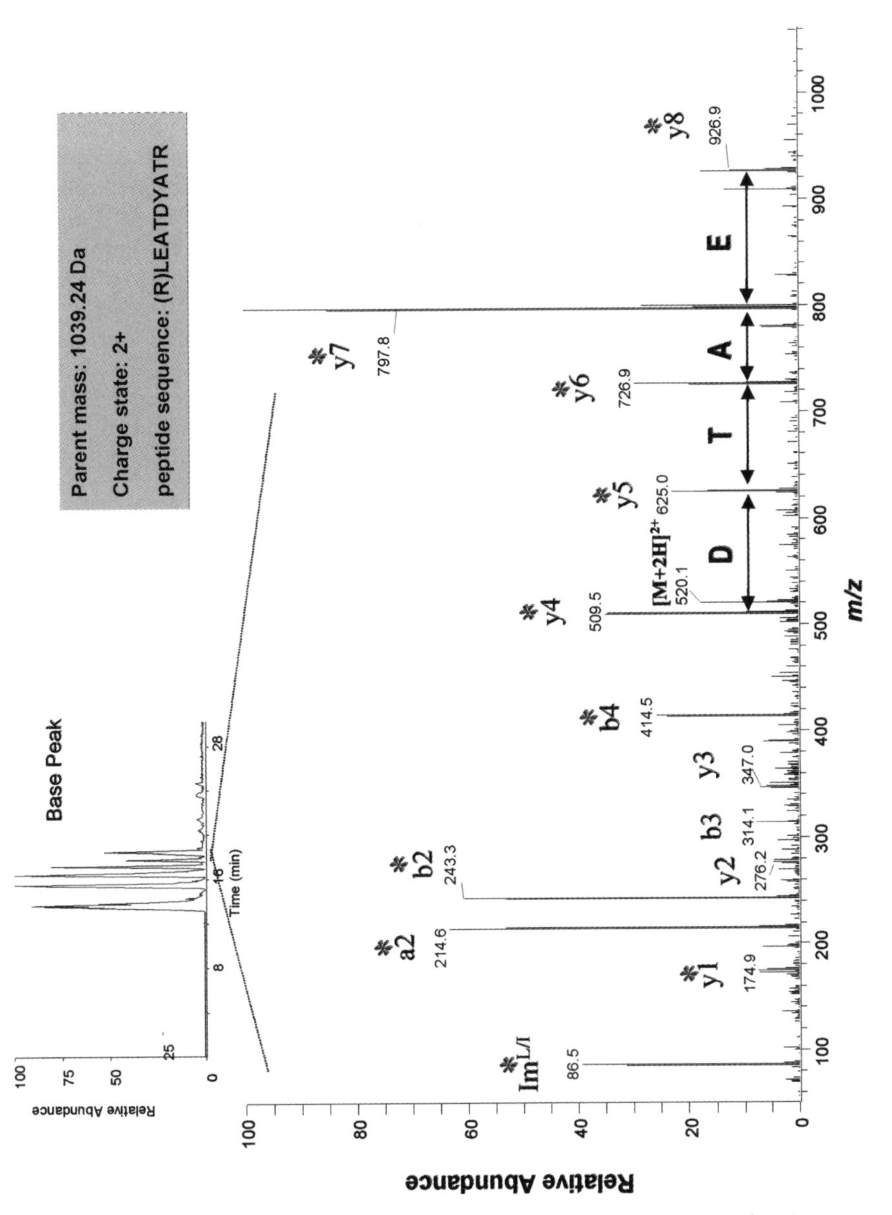

Fig. 10.7. CID spectrum of 'hypothetical protein', Putative Branched-Chain Amino Acid Aminotransferase (YJ9M-YEAST), generated by MS/MS on a TQMS. The protein was excised from a silver-stained gel, digested by trypsin, extracted and separated on a 75-μm RP-HPLC column according to protocols in the text. Information used for each database search program is indicated in the figure. No information was extracted for the SEQUEST program. MS-Tag used the doubly charged parent mass of 520.1 u as well as the ions marked with an *asterisk*. For PeptideSearch partial sequence information, (DTAE) the N- and C-terminal mass 'tags' (509.5 and 926.9) and the parent mass of 1039.24 u were used. The *inset* shows the base peak chromatogram of the entire run

7.3 Database Searching Strategies and de novo Sequencing

Search strategies are dependent upon the type of MS available to the researcher. If only peptide masses are able to be measured then peptide-mass searching is the predominant option, although one may still be able to employ methods that increase the information content of individual peptides (the so-called orthogonal approaches described previously). If selected peptides can be fragmented in the MS, by either PSD or CID, then uninterpreted or partially interpreted MS/MS searching can be performed. Of course, one would always like to be able to identify each peptide that has been generated in a digest mixture. This can be achieved through either partial sequence information or just the peptide mass (for tentative assignments). However, as stated above, correlative search strategies are only successful if the protein (or very close sequence homologue) to be identified exists in a sequence database. If this is not the case then the protein sequence will have to be determined de novo.

It is rare that the complete peptide sequence can be deduced de novo from an MS/MS fragment ion spectrum without further specific manipulations such as fragmentation of derivatized peptides. Moreover, with decreasing amounts of peptide fragmentation patterns become more obscure and difficult to interpret. Nevertheless, three approaches can be employed to simplify the task of de novo sequence interpretation from MS/MS spectra (see Table 10.8):

1. Two MS/MS spectra can be generated from the same peptide with and without methyl esterification (see Section 6.1). Methyl esterification will result in an additional 14 u being added to each carboxyl group in the peptide (including the C-terminus). This approach is simplest if there are no acidic residues in the peptide, less useful with increasing numbers of acidic residues, and least useful if there is one acidic residue at the N-terminus. If no acidic residues are present, then all of the C-terminal fragment ions will be mass shifted by +14 u compared with the underivatized MS/MS spectrum, allowing assignment of the *y* series ions.
2. The second approach also aims at identifying *b* and *y* ion series. Tryptic (or LysC) digestion is conducted in a buffer containing 50 % (v/v) $H_2{}^{18}O$/50 % (v/v) $H_2{}^{16}O$ (Schnolzer et al. 1996). Due to the fact that the proteolytic enzymes incorporate a water-derived oxygen into the C-terminal carboxyl group of the peptide, each peptide exists in two forms of equal concentration, one with the expected mass, the other with a mass 2 u higher due to the incorporation of

Table 10.8. Methods to assist in de novo sequencing by MS/MS (For references see text)

Method	Residues Method	Mass Increase (u)/residue
Methyl esterification	Asp, Glu, C-terminus	14
Enzymatic digestion in the presence of 50 % $H_2{}^{18}O$	C-terminal residue (all y series + 2 (or 4) u	50 % of peptide as + 2
Hydrogen/Deuterium exchange	All amino acids	From 0–5

^{18}O. Selection of both peptides in the mass spectrometer for fragmentation results in a MS/MS spectrum in which all of the C-terminal derived fragment ions exist as doublets separated by 2 u.

3. The use of hydrogen/deuterium exchange has been demonstrated to be another means of assisting in interpretation of PSD-MALDI-MS spectra (Spengler et al. 1993). However, it should be noted that de novo sequencing is greatly assisted if the entire MS/MS spectrum (including the immonium ions) is recorded with high-resolution and mass accuracy (Shevchenko et al. 1997).

The development of new methodologies and instruments with higher mass accuracy, higher sensitivity and sophisticated software will greatly amplify the rate at which protein identification can be achieved and enhance our ability to study proteins that are currently at untouchable low levels.

References

Adams MD, Kelley JM, Gocayne JD, Dubnick M, Polymeropoulos MH, Xiao H, Merril CR, Wu A, Olde B, Moreno RF et al (1991) Complementary DNA sequencing: expressed sequence tags and human genome project. Science 252: 1651–1656

Aebersold RH, Leavitt J, Saavedra RA, Hood LE, Kent SB (1987) Internal amino acid sequence analysis of proteins separated by one- or two-dimensional gel electrophoresis after in situ protease digestion on nitrocellulose. Proc Nat Acad Sci USA 84: 6970–6974

Beavis RC, Chait BT (1990) Rapid, sensitive analysis of protein mixtures by mass spectrometry. Proc Natl Acad Sci USA 87: 6873–6877

Bentz H, Chang RJ, Thompson AY, Glaser CB, Rosen DM (1990) Amino acid sequence of bovine osteoinductive factor. J Biol Chem 265, 5024–5029

Biemann K (1990) Appendix 5. Nomenclature for peptide fragment ions (positive ions). Methods Enzymol 193: 886–887

Boyle WJ, van der Geer P, Hunter T (1991) Phosphopeptide mapping and phosphoamino acid analysis by two-dimensional separation on thin-layer cellulose plates. Methods Enzymol 201: 110–149

Brown RS, Lennon JJ (1995) Mass resolution improvement by incorporation of pulsed ion extraction/ionization linear time-of-flight mass spectrometry. Anal Chem 67: 1998–2003

Chait BT, Wang R, Beavis RC, Kent SB (1993) Protein ladder sequencing. Science 262: 89–92

Clauser KR, Hall SC, Smith DM, Webb JW, Andrews LE, Tran HM, Epstein LB, Burlingame AL (1995) Rapid mass spectrometric peptide sequencing and mass matching for characterization of human melanoma proteins isolated by two-dimensional PAGE. Proc Natl Acad Sci USA 92: 5072–5076

Cooks RG, Glish GL, Kaiser RE, McLuckey SA (1991) Ion Trap Mass Spectrometry. Chem Eng News 69: 26–41

Cornish TJ, Cotter RJ (1994) A Curved Field Reflectron Time-Of-Flight Mass-Spectrometer For the Simultaneous Focusing Of Metastable Product Ions. Rapid Commun Mass Spectrom 8: 781–785

Courchesne PL, Luethy R, Patterson SD (1997) Comparison of in-gel and on-membrane digestion methods at low to sub-pmol level for subsequent peptide and fragment-ion mass analysis using matrix-assisted laser-desorption ionization mass spectrometry. Electrophoresis 18: 369–381

Courchesne PL, Patterson SD (1997) Manual microcolumn chromatography for sample clean-up before mass spectrometry. BioTechniques 22: 244–250

Craig AG, Fischer WH, Rivier JE, McIntosh JM, Gray WR (1995) MS based scanning methodologies applied to *Conus* venom. In: Crabb JW (ed) Techniques in Protein Chemistry VI. Academic Press, San Diego, pp 31–38

Davis MT, Stahl DC, Hefta SA, Lee TD (1995) A microscale electrospray interface for on-line, capillary liquid chromatography/tandem mass spectrometry of complex peptide mixtures. Anal Chem 67: 4549–4556

Ducret A, Bruun CF, Bures EJ, Marhaug G, Husby G, Aebersold R (1996) Characterization of human serum amyloid A protein isoforms separated by two-dimensional electrophoresis by liquid chromatography/electrospray ionization tandem mass spectrometry. Electrophoresis 17: 866–876

Ducret A, Bartone N, Haynes PA, Blanchard A, Aebersold R (1998a) A simplified gradient solvent delivery system for capillary liquid chromatography-electrospray ionization mass spectrometry. Anal Biochem 265:129–138

Ducret A, Oostveen v-I, Eng JK, Yates JR 3rd, Aebersold R (1998b) High throughput protein characterization by automated reverse-phase chromatography/electrospray tandem mass spectrometry. Protein Sci 7: 706–719

Eng JK, McCormack AL, Yates JR 3rd (1994) An approach to correlate tandem mass spectral data pf peptides with amino acid sequences in a protein database. J Am Soc Mass Spectrom 5: 976–989

Fenn JB, Mann M, Meng CK, Wong SF, Whitehouse CM (1989) Electrospray ionization for mass spectrometry of large biomolecules. Science 246: 64–71

Fenyo D, Qin J, Chait BT (1998) Protein identification using mass spectrometric information. Electrophoresis 19: 998–1005

Fernandez-Patron C, Calero M, Collazo PR, Garcia JR, Madrazo J, Musacchio A, Soriano F, Estrada R, Frank R, Castellanos S-LR, et al. (1995) Protein reverse staining: high-efficiency microanalysis of unmodified proteins detected on electrophoresis gels. Anal Biochem 224: 203–211

Garner MH, Garner WH, Gurd FR (1974) Recognition of primary sequence variations among sperm whale myoglobin components with successive proteolysis procedures. J Biol Chem 249: 1513–1518

Gevaert K, Demol H, Sklyarova T, Vandekerckhove J, Houthaeve T (1998) A peptide concentration and purification method for protein characterization in the subpicomole range using matrix assisted laser desorption/ionization postsource decay (MALDI-PSD) sequencing. Electrophoresis 19: 909–917

Gorg A, Obermaier C, Boguth G, Csordas A, Diaz JJ, Madjar JJ (1997) Very alkaline immobilized pH gradients for two-dimensional electrophoresis of ribosomal and nuclear proteins, Electrophoresis 18: 328–337

Gorg A, Postel W, Gunther S (1988) The current state of two-dimensional electrophoresis with immobilized pH gradients, Electrophoresis 9: 531–546

Griffin PR, Coffman JA, Hood LE, Yates JR 3rd (1991) Structural analysis of proteins by capillary HPLC electrospray tandem mass spectrometry. Int J Mass Spectrom Ion Processes 111: 131–149

Grunnet I, Knudsen J (1983) Medium-chain fatty acid synthesis by goat mammary-gland fatty acid synthetase. The effect of limited proteolysis. Biochem J 209: 215–222

Hager DA, Burgess RR (1980) Elution of proteins from sodium dodecyl sulfate-polyacrylamide gels, removal of sodium dodecyl sulfate, and renaturation of enzymatic activity: results with sigma subunit of Escherichia coli RNA polymerase, wheat germ DNA topoisomerase, and other enzymes. Anal Biochem 109: 76–86

Hellman U, Wernsted C, Gonez J, Heldin CH (1995) Improvement of an in-gel digestion procedure for the micropreparation of internal protein-fragments for amino acid sequencing. Anal Biochem 224: 451–455

Henzel WJ, Billeci TM, Stults JT, Wong SC, Grimley C, Watanabe C (1993) Identifying proteins from two-dimensional gels by molecular mass searching of peptide fragments in protein sequence databases. Proc Nat Acad Sci USA 90: 5011–5015

Hess D, Covey TC, Winz R, Brownsey RW, Aebersold R (1993) Analytical and micropreparative peptide mapping by high performance liquid chromatography/electrospray mass spectrometry of proteins purified by gel electrophoresis. Protein Sci 2: 1342–1351

Houthaeve T, Gausepohl H, Mann M, Ashman K (1995) Automation of micro-preparation and enzymatic cleavage of gel electrophoretically separated proteins. FEBS Lett 376: 91–94

Hunkapiller MW, Lujan E, Ostrander F, Hood LE (1983) Isolation of microgram quantities of proteins from polyacrylamide gels for amino acid sequence analysis. Methods Enzymol 91: 227–236

Hunt DF, Alexander JE, McCormack AL, Martino PA, Michel H, Shabanowitz J, Sherman N, Moseley MA, Jorgenson JW, Tomer KB (1991) Mass spectrometric methods for protein and peptide sequence analysis. In: Villafranca JJ (ed) Techniques in Protein Chemistry II. Academic Press, San Diego, pp 441–465

Hunt DF, Henderson RA, Shabanowitz J, Sakaguchi K, Michel H, Sevilir N, Cox AL, Appella E, Engelhard VH (1992) Characterization of peptides bound to the class I MHC molecule HLA-A2.1 by mass spectrometry. Science 255: 1261–1263

James P, Quadroni M, Carafoli E, Gonnet G (1993) Protein identification by mass profile fingerprinting. Biochem Biophys Res Commun 195: 58–64

James P, Quadroni M, Carafoli E, Gonnet G (1994) Protein identification in DNA databases by peptide mass fingerprinting. Protein Sci 3: 1347–1350

Jeno P, Mini T, Moes S, Hintermann E, Horst M (1995) Internal sequences from proteins digested in polyacrylamide gels. Anal Biochem 224: 75–82

Jensen ON, Podtelejnikov A, Mann M (1996a) Delayed extraction improves specificity in database searches by matrix-assisted laser desorption/ionization peptide maps. Rapid Commun Mass Spectrom 10: 1371–1378

Jensen ON, Vorm O, Mann M (1996b) Sequence patterns produced by incomplete enzymatic digestion or one-step Edman degradation of peptide mixtures as probes for protein database searches. Electrophoresis 17: 938–944

Jensen ON, Wilm M, Shevchenko A, Mann M (1998) Peptide sequencing of 2-DE gel-isolated proteins by nanoelectrospray tandem mass spectrometry. In: Link AJ (ed) Methods in molecular biology 2-D proteome analysis protocols. Humana Press, Totowa, New Jersey

Jonscher KR, Yates JR 3rd (1997) The quadrupole ion trap mass spectrometer–a small solution to a big challenge. Anal Biochem 244: 1–15

Karas M, Hillenkamp F (1988) Laser desorption ionization of proteins with molecular masses exceeding 10 000 daltons. Anal Chem 60: 2299–2301

Karlsson KE, Novotny M (1988) Separation efficiency of slurry-packed liquid chromatography microcolumns with very small inner diameters. Anal Chem 60: 1662–1665

Kaufmann R, Kirsch D, Spengler B (1994) Sequencing of peptides in a time-of-flight mass spectrometer: evaluation of postsource decay following matrix-assisted laser desorption ionization (MALDI). Int J Mass Spectrom Ion Processes 131: 355–385

Kawasaki H, Emori Y, Suzuki K (1990) Production and separation of peptides from proteins stained with Coomassie brilliant blue R-250 after separation by sodium dodecyl sulfate-polyacrylamide gel electrophoresis. Anal Biochem 191: 332–336

Kawasaki H, Suzuki K (1990) Separation of peptides dissolved in a sodium dodecyl sulfate solution by reversed-phase liquid chromatography: Removal of sodium dodecyl sulfate from peptides using an ion-exchange precolumn. Anal Biochem 186: 264–268

Kennedy RT, Jorgenson JW (1989) Preparation and evaluation of packed capillary liquid chromatography columns with inner diameter from 20 to 50 um. Anal Chem 61: 1128–1135

Kirchner M, Fernandez J, Shakey QA, Gharahdaghi F, Mische S (1996) Enzymatic digestion of PVDF-bound proteins: A survey of sixteen non-ionic detergents. In: Marshal D (ed) Techniques in Protein Chemistry VII. Academic Press, San Diego, pp 287–298

Klarskov K, Roecklin D, Bouchon B, Sabati'e J, Van, D-A, Bischoff R (1994) Analysis of recombinant Schistosoma mansoni antigen rSmp28 by on-line liquid chromatography-mass spectrometry combined with sodium dodecyl sulfate polyacrylamide gel electrophoresis. Anal Biochem 216: 127–134

Klose J (1975) Protein mapping by combined isoelectric focusing and electrophoresis of mouse tissues. A novel approach to testing for induced point mutations in mammals. Humangenetik 26: 231–243

Konigsberg W, Goldstein J, Hill RJ (1963) The structure of human haemoglobin VII. The digestion of the beta chain of human haemoglobin with pepsin. J Biol Chem 238: 2028–2033

Laemmli UK (1970) Cleavage of structural proteins during the assembly of the head of bacteriophage T4. Nature 227: 680–685

Lecchi P, Caprioli RM (1995) Matrix-assisted laser desorption mass spectrometry for peptide mapping. In: Hancock WS (ed) New Methods in Peptide Mapping For the Characterization of Proteins. CRC Press, Boca Raton, Florida, pp 219–240

Lill U, Schreil A, Henschen A, Eggerer H (1984) Hysteretic behaviour of citrate synthase. Site-directed limited proteolysis. Eur J Biochem 143: 205–212

Mann M, Wilm M (1994) Error-tolerant identification of peptides in sequence databases by peptide sequence tags. Anal Chem 66: 4390–4399

Mann M, Hojrup P, Roepstorff P (1993) Use of mass spectrometric molecular weight information to identify proteins in sequence databases. Biol Mass Spectrom 22: 338–345

McCormack AL, Schieltz DM, Goode B, Yang S, Barnes G, Drubin D, Yates JR 3rd (1997) Direct analysis and identification of proteins in mixtures by LC/MS/MS and database searching at the low-femtomole level. Anal Chem 69: 767–776

McLuckey SA, Van Berkel GJ, Goeringer DE, Glish GL (1994) Ion trap mass spectrometry. Using high-pressure ionization. Anal Chem 66: 737A-743A

Merril CR, Switzer RC, Van, K-ML (1979) Trace polypeptides in cellular extracts and human body fluids detected by two-dimensional electrophoresis and a highly sensitive silver stain. Proc Nat Acad Sci USA 76: 4335–4339

Moritz RL, Eddes J, Ji H, Reid GE, Simpson RJ (1995) Rapid separation of proteins and peptides using conventional silica-based supports: Identification of 2-D gel proteins following in-gel proteolysis. In: Crabb JW (ed) Techniques in Protein Chemistry VI. Academic Press, San Diego, pp 311–319

Moritz RL, Eddes JS, Reid GE, Simpson RJ (1996) S-pyridylethylation of intact polyacrylamide gels and in situ digestion of electrophoretically separated proteins: a rapid mass spectrometric method for identifying cysteine-containing peptides. Electrophoresis 17: 907–917

Nelson RW, Dogruel D, Williams P (1994) Mass determination of human immunoglobulin IgM using matrix-assisted laser desorption/ionization time-of-flight mass spectrometry. Rapid Commun Mass Spectrom 8: 627–631

O'Farrell PH (1975) High resolution two-dimensional electrophoresis of proteins. J Biol Chem 250: 4007–4021

Oostveen v-I, Ducret A, Aebersold R (1997) Colloidal silver staining of electroblotted proteins for high sensitivity peptide mapping by liquid chromatography-electrospray ionization tandem mass spectrometry. Anal Biochem 247: 310–318

Ortiz ML, Calero M, Fernandez P-C, Patron CF, Castellanos L, Mendez E (1992) Imidazole-SDS-Zn reverse staining of proteins in gels containing or not SDS and microsequence of individual unmodified electroblotted proteins. FEBS Lett 296: 300–304

Pappin DJC, Hojrup P, Bleasby AJ (1993) Rapid identification of proteins by peptide-mass fingerprinting. Curr Biol 3: 327–332

Pappin DJC, Rahman D, Hansen HF, Bartlet-Jones M, Jeffery W, Bleasby AJ (1995) Chemistry, mass spectrometry and peptide-mass databases: Evolution of methods for the rapid identification and mapping of cellular proteins. In: Burlingame AL, Carr SA (eds) Mass Spectrometry in the Biological Sciences. Humana Press, Totowa, New Jersey, pp 135–150

Patterson SD (1994) From electrophoretically separated protein to identification: strategies for sequence and mass analysis. Anal Biochem 221: 1–15

Patterson SD, Aebersold R (1995) Mass spectrometric approaches for the identification of gel-separated proteins. Electrophoresis 16: 1791–1814

Perides G, Kuhn S, Scherbarth A, Traub P (1987) Probing of the structural stability of vimentin and desmin-type intermediate filaments with Ca2+-activated proteinase, thrombin and lysine-specific endoproteinase Lys-C. Eur J Cell Biol 43: 450–458

Poncz L, Dearborn DG (1983) The resistance to tryptic hydrolysis of peptide bonds adjacent to N epsilon,N-dimethyllysyl residues. J Biol Chem 258: 1844–1850

Rabilloud T (1992) A comparison between low background silver diammine and silver nitrate protein stains. Electrophoresis 13: 429–39

Rabilloud T, Kieffer S, Procaccio V, Louwagie M, Courchesne PL, Patterson SD, Martinez P, Garin J, Lunardi J (1998) Two-dimensional electrophoresis of human placental mitochondria and protein identification by mass spectrometry: Toward a human mitochondrial proteome. Electrophoresis 19: 1006–1014

Rosenfeld J, Capdevielle J, Guillemot JC, Ferrara P (1992) In-gel digestion of proteins for internal sequence analysis after one- or two-dimensional gel electrophoresis. Anal Biochem 203: 173–179

Sanchez J-C, Ravier F, Pasquali C, Frutiger S, Paquet N, Bjellqvist B, Hochstrasser DF, Hughes GJ (1992) Improving the detection of proteins after transfer to polyvinylidene difluoride membranes. Electrophoresis 13: 715–717

Schnolzer M, Jedrzejewski P, Lehmann WD (1996) Protease-catalyzed incorporation of O-18 into peptide-fragments and its application for protein sequencing by electrospray and matrix-assisted laser desorption/ionization mass-spectrometry. Electrophoresis 17: 945–953

Schwartz JC, Jardine I (1996) Quadrupole ion trap mass spectrometry. Methods in Enzymol 270: 552–586

Shevchenko A, Jensen ON, Podtelejnikov AV, Sagliocco F, Wilm M, Vorm O, Mortensen P, Shevchenko A, Boucherie H, Mann M (1996a) Linking genome and proteome by mass spectrometry: large-scale identification of yeast proteins from two dimensional gels. Proc Natl Acad Sci USA 93: 14440–14445

Shevchenko A, Wilm M, Vorm O, Mann M (1996b) Mass spectrometric sequencing of proteins silver-stained polyacrylamide gels. Anal Chem 68: 850–858

Shevchenko A, Chernushevich I, Ens W, Standin KG, Thomson B, Wilm M, Mann M (1997) Rapid 'de novo' peptide sequencing by a combination of nanoelectrospray, isotopic labeling and a quadrupole/time-of-flight mass spectrometer. Rapid Commun Mass Spectrom 11: 1015–1024

Spackman DH, Stein WH, Moore S (1960) The disulfide bonds of ribonuclease. J Biol Chem 235: 648–659

Spengler B, Lutzenkirchen F, Kaufmann R (1993) On-target deuteration for peptide sequencing by laser mass spectrometry. Org Mass Spectrom 28: 1482–1490

Steinberg TH, Haugland RP, Singer VL (1996a) Applications of SYPRO orange and SYPRO red protein gel stains. Anal Biochem 239: 238–245

Steinberg TH, Jones LJ, Haugland RP, Singer VL (1996b) SYPRO orange and SYPRO red protein gel stains: one-step fluorescent staining of denaturing gels for detection of nanogram levels of protein. Anal Biochem 239: 223–237

Takach EJ, Hines WM, Patterson DH, Juhasz P, Falick AM, Vestal ML, Martin SA (1997) Accurate mass measurements using MALDI-TOF with delayed extraction. J Protein Chem 16: 363–369

Tomasselli AG, Frank R, Schiltz E (1986) The complete primary structure of GTP:AMP phosphotransferase from beef heart mitochondria. FEBS Lett 202: 303–308

Vestal ML, Juhasz P, Martin SA (1995) Delayed extraction matrix-assisted laser desorption time-of-flight mass spectrometry. Rapid Commun Mass Spectrom 9: 1044–1050

Vorm O, Roepstorff P, Mann M (1994) Improved resolution and very high-sensitivity in MALDI TOF of matrix surfaces made by fast evaporation. Anal Chem 66: 3281–3287

Ward LD, Simpson RJ (1991) Micropreparative protein isolation from polyacrylamide gels following detection by high-resolution dynamic imaging: application to microsequencing. Peptide Res 4: 187–193

Wiley WC, McLaren IH (1953) Time-of-flight mass spectrometer with improved resolution. Rev Sci Instrum 26: 1150–1157

Wilkins MR, Gasteiger E, Sanchez J-C, Bairoch A, Hochstrasser DF (1998) Two-dimensional gel electrophoresis for proteome projects: the effects of protein hydrophobicity and copy number. Electrophoresis, 19: 1501–1505

Wilm M, Mann M (1996) Analytical properties of the nanoelectrospray ion source. Anal Chem 68: 1–8

Wilm M, Shevchenko A, Houthaeve T, Breit S, Schweigerer L, Fotsis T, Mann M (1996) Femtomole sequencing of proteins from polyacrylamide gels by nano-electrospray mass spectrometry. Nature 379: 466–469

Woods AS, Huang AYC, Cotter RJ, Pasternack GR, Pardoll DM, Jaffee EM (1995) Simplified high-sensitivity sequencing of a major histocompatibility complex class I-associated immunoreactive peptide using matrix-assisted laser-desorption ionization mass-spectrometry. Anal Biochem 226: 15–25

Xiang F, Beavis RC (1994) A method to increase contaminant tolerance in protein matrix-assisted laser desorption/ionization by the fabrication of thin protein-doped polycrystalline films. Rapid Commun Mass Spectrom 8: 199–204

Yates JR 3rd, Speicher S, Griffin PR, Hunkapiller T (1993) Peptide mass maps: A highly informative approach to protein identification. Anal Biochem 214: 397–408

Yates JR 3rd, Eng JK, McCormack AL (1995a) Mining genomes: correlating tandem mass spectra of modified and unmodified peptides to sequences in nucleotide databases. Anal Chem 67: 3202–3210

Yates JR 3rd, Eng JK, McCormack AL, Schieltz D (1995b) Method to correlate tandem mass spectra of modified peptides to amino acid sequences in the protein database. Anal Chem 67: 1426–1436

Yates JR 3rd, Carmack E, Hays L, Link AJ, Eng JK (1998) Automated protein identification using microcolumn liquid chromatography-tandem mass spectrometry. In: Link AJ (ed) Methods in Molecular Biology, 2-D proteome analysis protocols. Humana Press Totowa, New Jersey

Zaluzec EJ, Gage DA, Allison J, Watson JT (1994) Direct matrix-assisted laser-desorption ionization mass-spectrometric analysis of proteins immobilized on nylon-based membranes. J Am Soc Mass Spectrom 5: 230–237

Zewert T, Harrington M (1992a) Polyethyleneglycol methacrylate 200 as an electrophoresis matrix in hydroorganic solvents. Electrophoresis 13: 824–831

Zewert T, Harrington M (1992b) Polyhydroxy and polyethyleneglycol (meth)acrylate polymers: Physical properties and general studies for their use as electrophoresis markers. Electrophoresis 13: 817–824

Zewert TE, Harrington MG (1993) Protein electrophoresis. Curr Opin Biotech 4: 3–8

Zhang H, Andren PE, Caprioli RM (1995) Micro-preparation procedure for high-sensitivity matrix-assisted laser desorption ionization mass spectrometry. J Mass Spectrom 30: 1768–1771

Zhang W, Czernik AJ, Yungwirth T, Aebersold R, Chait BT (1994) Matrix-assisted laser desorption mass spectrometric peptide mapping of proteins separated by two-dimensional gel electrophoresis: determination of phosphorylation in synapsin I. Protein Sci 3: 677–686

CHAPTER 11

Mass Spectrometry of Intact Proteins from Two-Dimensional PAGE

CH. ECKERSKORN[1] and K. STRUPAT[2]

1 Introduction

Many important cellular processes depend on co- and post-translational modifications of proteins. These alterations, which can affect the size, charge, conformation and the stability of proteins, not only increase the number of different protein species per cell, but also add to the complexity of the cellular protein pattern, hence, making proteome analysis rather difficult. Due to its separation power and its good reproducibility, two-dimensional polyacrylamide gel electrophoresis has evolved as a crucial technique to study quantitatively and qualitatively protein expression and, thus, has become essential in proteomics. A wide range of techniques have been developed to identify electrophoretically separated proteins with partial sequence information and/or peptide masses obtained after proteolytic digestions directly in the gel or on a blotting membrane in databases (see Chaps. 9 and 10, this Vol.). Combined with the techniques described in this chapter for determining the mass of the mature protein, a powerful strategy is provided to investigate how co- and post-translational modifications influence protein structure and function and whether the expression of particular isoforms is under developmental or disease control.

Molecular mass determination of proteins from SDS-gels is derived from electrophoretic mobilities which are dependent upon
(1) variabilities in binding of SDS due to clusters of acidic or basic amino acids,
(2) modification of the protein with carbohydrate groups; and
(3) incomplete unfolding even in the presence of the denaturant.

In particular, partial retention of secondary structures in the presence of SDS has been shown to dramatically alter the electrophoretic mobility (Dianoux et al. 1992). Therefore, many proteins migrate anomalously under conditions of SDS-PAGE and apparent mass values may be in error by more than 30 % (e.g. Wilkins et al. 1996).

State-of-the-art mass spectrometry is the most accurate technique for molecular mass determination. The breakthrough in mass spectrometry of proteins

[1] TopLab, Company for Applied Biotechnology, Innovation Center of Biotechnology, 82152 Martinsried, Germany.
[2] Institute for Medical Physics and Biophysics, University of Münster, 48149 Münster, Germany.

came in 1988, when two novel methods were introduced, enabling the transfer of large intact proteins into the gaseous phase of a mass spectrometer. In electrospray ionization mass spectrometry (ESI-MS) protein ionization is achieved during the formation of microscopic droplets of a protein-containing solution which is pumped through a small orifice at low flow rates within an electric field at atmospheric pressure (Fenn et al. 1989). Because of this ionization procedure, ESI can be readily coupled to liquid chromatography or capillary electrophoresis. In matrix-assisted laser desorption/ionization mass spectrometry (MALDI-MS), sample ions are formed from a solid state (Karas and Hillenkamp 1988). The analyte (proteins) is deposited on a target by co-crystallization with a matrix (laser photon absorbing organic compounds) and, subsequently, transferred to the high-vacuum ionization chamber. Ionization is induced by short pulses of laser light which is focused on the sample target. The mass (m) to charge (z) ratio (m/z) of ions formed by either ionization methods is then measured in the attached mass analyzer. Both methods can achieve mass accuracies of up to 0.01 % when analyzing proteins with a mass of up to 30–40 kDa and show slightly lower accuracies above this value (Chait and Kent 1992).

The most important prerequisite for mass spectrometry is the sample preparation step. Ideal solvent systems for the analyte molecules are composed of a dilute organic acid in water (e.g. 0.1 % formic acid, acetic acid) and an organic solvent, e.g. 20–60 % methanol or acetonitrile. Any other nonvolatile additives like buffer salts and detergents influence the ionization process in a way that results in a decreased mass accuracy and a lower sensitivity, including complete suppression of the ion signal. Proteins are difficult analytes for mass spectrometry due to their large size and the enormous variability in their amino acid composition that influences their solubility and structure. After 2-D PAGE separation, all proteins are concentrated in individual spots as SDS-protein complexes within the polyacrylamide gel matrix. In principle, there are three possible ways to interface gel electrophoresis with mass spectrometry (see Fig. 11.1).

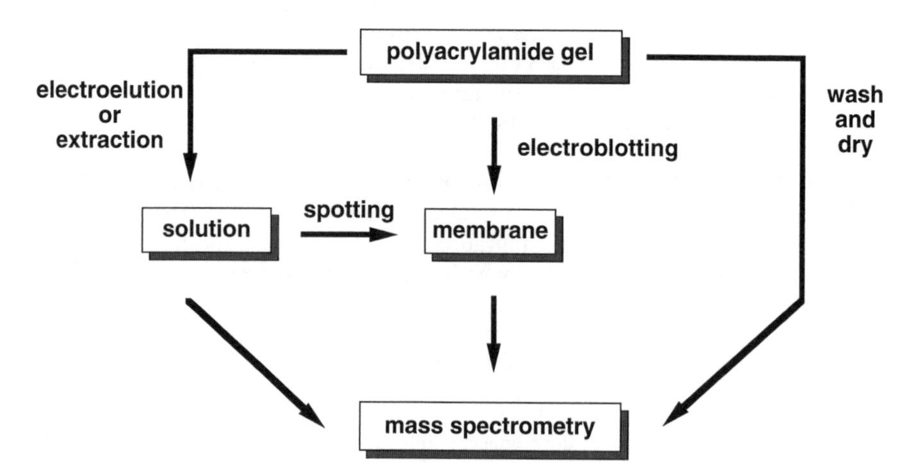

Fig. 11.1. Possible methods to prepare electrophoretically separated proteins for mass spectrometry

2 Elution of Proteins from Gels for Mass Spectrometry

The oldest and most popular methods to recover gel-separated proteins into a buffer solution involve elution from the polyacrylamide gel matrix by diffusion (Hager and Burgess 1980), electroelution (Hunkapillar et al. 1983; Simpson et al. 1987), acid extraction or organic solvent extraction (Feick and Shiozawa 1990). From the multiplicity of reports using different procedures for eluting proteins from gel slices it becomes obvious that no single method has proven to be entirely satisfactory. The elution technique has principal drawbacks, such as low and inconsistent sample recoveries (especially with hydrophobic and/or larger proteins), and contamination with large amounts of sodium dodecyl sulfate (SDS), buffer salts and gel-derived compounds. This contamination interferes with mass spectrometry and additional purification steps are mandatory. However, many proteins are insoluble in the absence of a detergent, particularly after they have been denatured during gel electrophoresis. Therefore, removal of SDS by cold acetone precipitation prior to ESI-MS as described by Le Maire et al. (1993) may not be generally applicable. To circumvent the need for protein precipitation Haebel et al. (1995) prepared a protein for MALDI-MS by using SDS-free buffer during electroelution. As a rule of thumb, if a protein is sufficiently soluble and available in high picomole amounts, the protein can be purified for molecular mass determination by standard application procedures for MALDI-MS or ESI-MS (Dunphy et al. 1993; Yanase et al. 1994 ; Mortz et al. 1996).

3 Mass Spectrometry of Proteins Electroblotted onto Polymer Membranes

Currently, the most rapid and efficient method for isolating electrophoretically separated proteins and especially for the small quantities present in 2-D polyacrylamide gels is electroblotting onto suitable polymer membranes. Due to high blotting efficiencies, the complete protein pattern is retained on the membrane surface without any noticeable diffusion (Jungblut et al. 1990). Evaluation of commercially available blotting membranes showed that the efficiency of electroblotting is dependent upon structural parameters of the membrane, including specific surface area, pore size distribution and pore volumes. Almost quantitative retention ($> 90\,\%$) of proteins was obtained for membranes with a high specific surface area and narrow pores (Eckerskorn and Lottspeich 1993). An advantage of the electroblotting procedure is that the proteins immobilized on membranes are completely free of any salts because of their migration to the corresponding electrodes during the electroblotting process. Other components could be washed off after the transfer. Therefore, during the past 10 years electroblotting was often the method of choice for high-sensitive N-terminal sequence analysis and sensitive amino acid composition analysis.

In standard preparations for MALDI-MS, a dilute protein solution is mixed with a concentrated matrix solution. The assumed functions of the matrix are to absorb the energy from the laser pulse, to embed and isolate the proteins and to assist in their ionization. Best results are obtained if the proteins are homogeneously distributed and fully embedded in an environment of matrix molecules (Hillenkamp et al. 1991). Considering the tight interaction of the immobilized proteins with the hydrophobic surface of the blotting membrane, a special technique for a suitable matrix incubation is a prerequisite for interfacing electrophoresis and mass spectrometry.

3.1 Scanning IR-MALDI Mass Spectrometry of Electroblotted Proteins

The feasibility of MALDI-MS of electroblotted proteins was first described in 1992 by Eckerskorn et al. and was further evaluated and developed as a standard procedure for 2-D separated proteins (Strupat et al. 1994 a and b; Eckerskorn et al. 1997). It turned out that the preparation of the blotted proteins was surprisingly easy when dicarbonic acids (e.g. succinic acid) were used in combination with a 2.94 µm infrared (IR) laser. Immediately after electroblotting the wet membrane was incubated in a matrix solution typically for 15 min and then left to dry under the ambient laboratory temperature and humidity. Membranes which had been allowed to dry before incubation with the matrix solutions yielded vastly inferior results. For MALDI-MS small pieces of the matrix-incubated membrane containing the proteins of interest were excised with a razor blade and mounted onto the sample support by a conductive double adhesive tape. They were then transferred to the vacuum chamber of the mass spectrometer.

The first experiments (Strupat et al. 1994a) showed that results obtained with IR-MALDI (Er-YAG laser and succinic acid as matrix) were considerably superior to that of all UV laser/matrix combinations tested in terms of single shot intensities, signal-to-noise ratio and shot-to-shot reproducibility. This is essentially a wavelength effect, as has been confirmed by the investigation of the same membrane sample incubated in 2,5-DHB which performs equally well in UV- and IR-MALDI. The superiority of IR- versus UV-MALDI was attributed to the fact that electroblotted proteins are distributed throughout, mostly on the top area of the 50- to 150-µm-thick fibrillar membrane structure. Due to the much larger penetration and ablation depth of IR-light, more analyte molecules were available for analysis in IR-MALDI. It was, furthermore, found that depth profiling by multiple irradiation of the same spot is possible, revealing differences between different membranes and locations across a protein band.

The accessible mass range was investigated by electrophoresis of a high molecular weight standard on SDS-polyacrylamide gel. Representative MALDI mass spectra (sum of ten single spectra) from trypsin inhibitor and galactosidase and, for comparison, a single spectra for BSA desorbed from the PVDF membrane after electroblotting are shown in Fig. 11.2. Generally, the obtained spectra are not significantly different from corresponding MALDI mass spectra obtained by standard preparation procedure. The presence of the galactosidase dimer at 230 kDa gives rise to the hope that even larger proteins can be analyzed after SDS-PAGE and electroblotting.

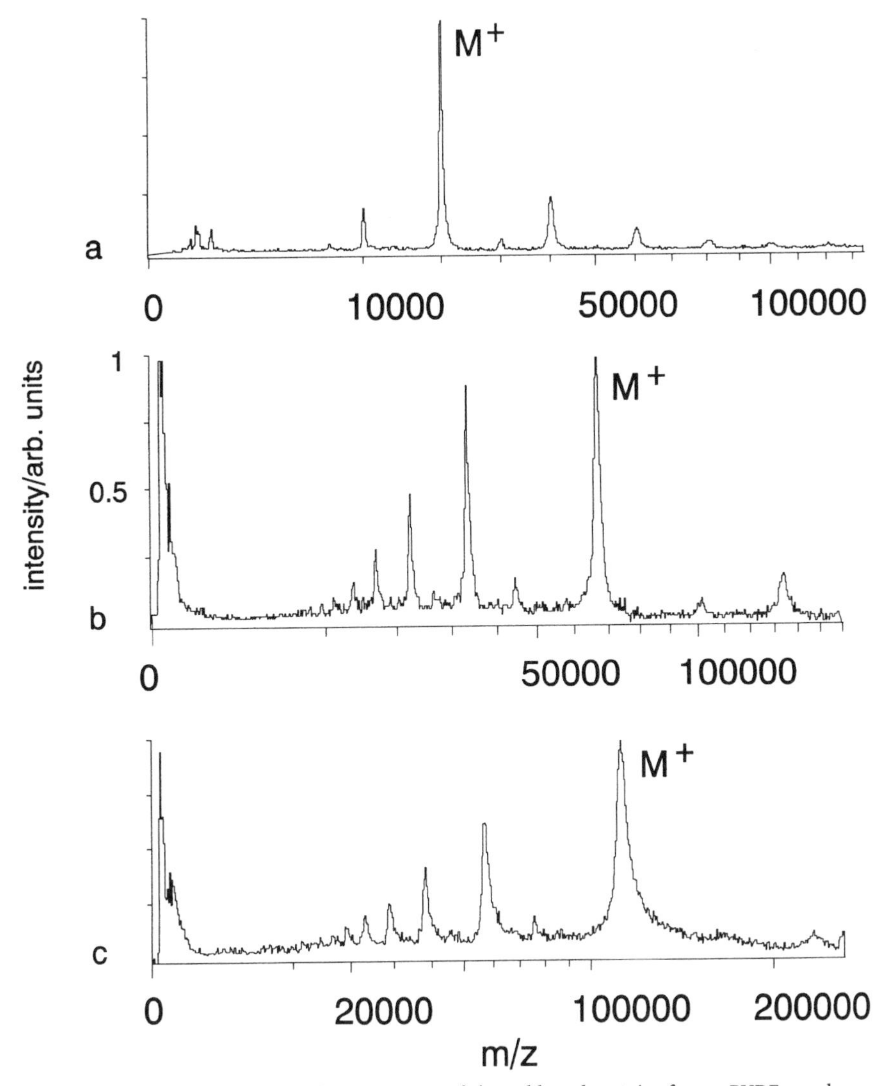

Fig. 11.2. Representative IR-MALDI mass spectra of electroblotted proteins from a PVDF membrane (Immobilon PSQ) **a** trypsin inhibitor after incubation with adipic acid; sum of ten single shot spectra **b** one single shot spectra of BSA after incubation with succinic acid **c** galactosidase after incubation with succinic acid; sum of 10 single shot spectra

The analysis of a variety of different electroblotted proteins showed that the quality of spectra was independent of protein size, hydrophobicity, or posttranslational modification. For example, IR-MALDI-MS of the individual proteins of a electrophoretically separated and electroblotted Sendai Virus D52 yielded excellent spectra for the water-soluble polymerase P, as well as for the highly glycosylated transmembrane protein HN or the fusion protein F (Strupat et al. 1994b).

A prerequisite for mass spectrometric analysis of electroblots is the preservation of the lateral electrophoretic resolution throughout the embedding process in the matrix. When the laser beam is scanned along the axis of protein migration, the width obtained of the scanned protein bands agreed well with the width obtained for the corresponding stained protein bands of a reference blot (Strupat et al. 1994a). This indicates that no substantial diffusion of protein is induced by the MALDI preparation procedure. This novel approach, called scanning IR-MALDI MS, was successfully applied to 1-D separations (SDS-PAGE), e.g. to identify the differential expression of proteins during the circadian cycle (Mittag et al. 1997). For 2-D PAGE, mass contour plots from selected areas of the corresponding electroblots were obtained as pseudo-images of the gel separation by scanning the membrane surface step by step with IR-MALDI MS (Eckerskorn et al. 1997). Transthyretin (TTR), a spot in the mass range of 13 kDa, was chosen from a 2-D separation of human blood plasma to compare the local protein distribution as determined by scanning IR-MALDI-MS to that measured by standard laser densitometry of an identical silver-stained spot (Fig. 11.3). As can be seen in Fig. 11.3, the local distributions are identical within the accuracy of the measurement. A nearly circular symmetric distribution of equal diameter is found by both methods, with the maximum intensity in the center of the spot. The mass of the TTR protein averaged over sum spectra taken from the 69 squares that resulted in protein signals was 13 756 ± 19 daltons. This compares well to the value of 13 760.4 daltons which was calculated from the deduced amino acid sequence of the gene's nucleotide sequence and reflects a homogeneous protein without posttranslational modifications. Similar agreement between local protein distribution as determined by the densitometry and the scanning mass spectrometry was found for all other spots investigated.

The sensitivity and mass resolution of scanning IR-MALDI-MS are demonstrated in Fig. 11.4 in another area of human blood plasma (Eckerskorn et al. 1997). Spot A, a protein from the α2 chain of haptoglobin, has a partially overlapping satellite spot X that is barely visible in the corresponding silver-stained gel. The partial overlap is evident in spectrum 2 (Fig. 11.4) as well as the full separation of the two proteins outside the overlap area in spectra 2 and 3. The corresponding mass contour plots resolve clearly two spots within sharp borderlines. The low signal intensity of protein X is due to its low protein concentration which is close to the detection limit of silver staining, but the molecular mass peaks of protein X were reproducible across the whole spot area and are clearly discerned in the spectra. It should also be noted that the determined mass of protein X (19 750 daltons) exceeds the mass of haptoglobin spot A (16 355 daltons) by 3 kDa in spite of its higher electrophoretic mobility.

Spot D, also a relative low contrast spot, has an elongated shape along the mass axis in the silver-stained gel (Fig. 11.4). The IR-MALDI scan revealed a continuous increase in protein mass along the axis of decreasing electrophoretic mobility. Further characterization of the spot showed homogeneous N-terminal sequences, indicating a series of human serum albumin fragments with staggered C termini. The mass resolution of the TOF mass analyzers is limited to a value of ≈ 100 in this mass range. The relatively broad peaks in spectra 4a-c (Fig. 11.4), therefore, represent nonresolved signals of two or more components present in

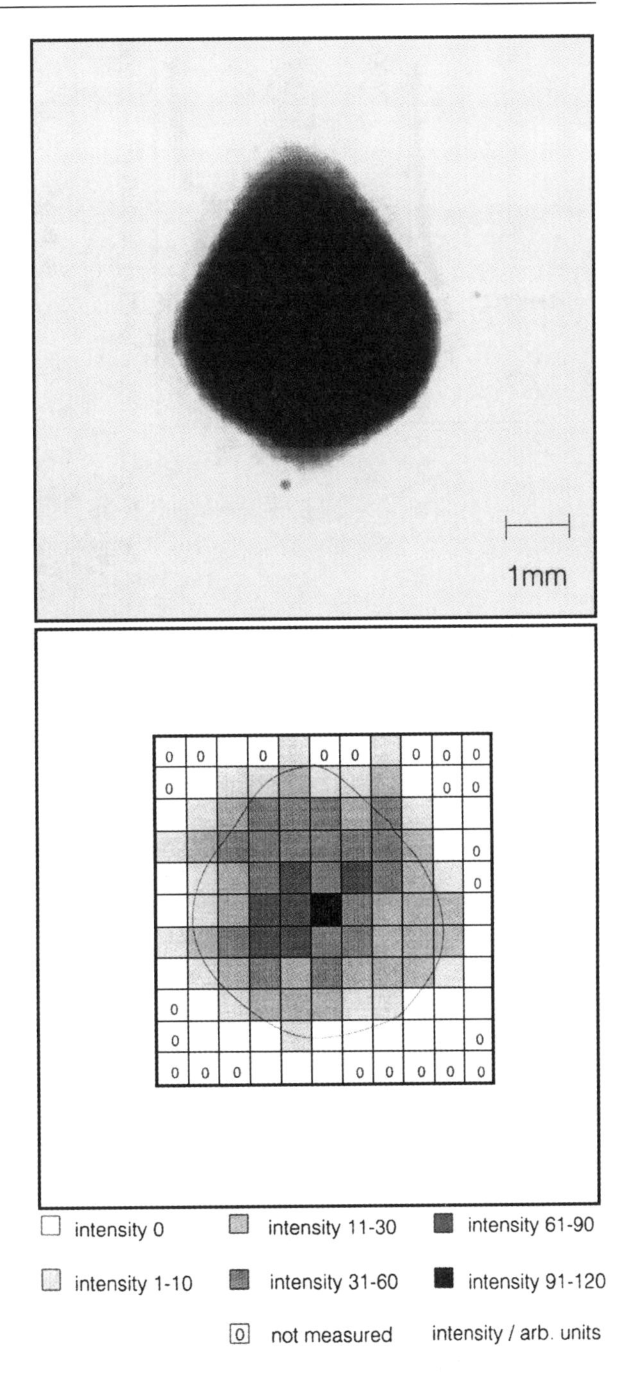

Fig. 11.3. Laser densitograph (*top*) and 2-D gray scale histogram (*bottom*) of the local distribution of relative signal intensities of the TTR mass signals (*peak areas*) as determined by scanning IR-MALDI-MS. For the mass spectrometry experiment, the investigated blot area was subdivided into an array of 121 (= 11×11) $500 \times 500\ \mu m^2$ squares and signal intensitiy was averaged over ten spectra for each square. The *contour line* of the laser densitograph was transferred to the mass contour plot for ease of comparison. (Reprinted from Eckerskorn et al. 1997 with permission from Analytical Chemistry)

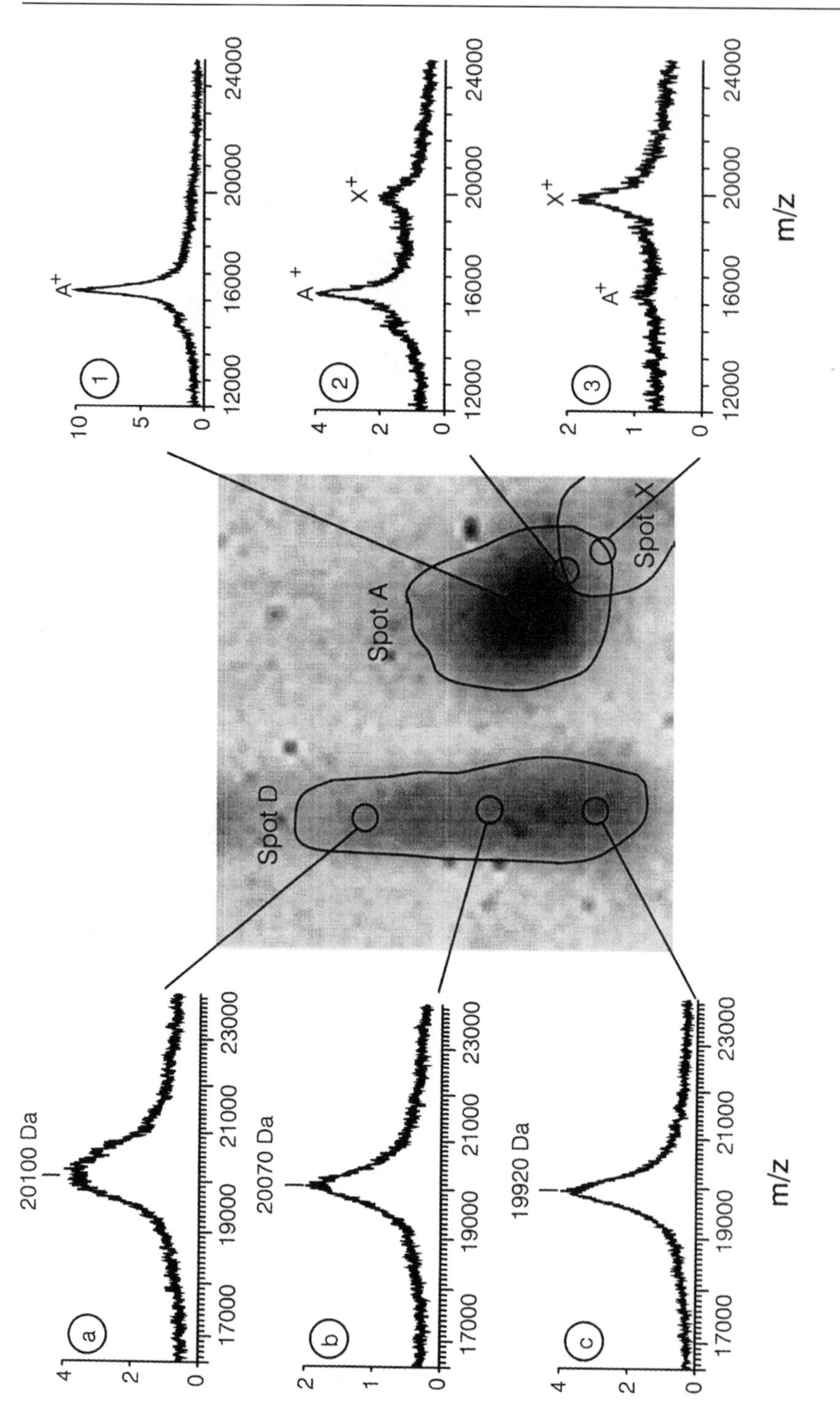
a
20100 Da
b
20070 Da
c
19920 Da
17000
19000
21000
23000
m/z
Spot D
Spot A
Spot X
1
2
3
A+
X+
12000
16000
20000
24000
m/z

the sample area which were not completely resolved by SDS-PAGE separation. This clearly demonstrates the limitation of this technique as it now stands. An increased mass resolution is desirable and, most probably, achievable with modern developments in TOF mass spectrometers such as delayed extraction.

3.2 UV-MALDI Mass Spectrometry of Electroblotted Proteins

Most of the commercially available MALDI mass spectrometers are equipped with UV lasers, which makes it desirable to introduce a preparation technique for electroblotted proteins which is applicable to UV-MALDI-MS. In the past, proteins that were directly applied from solutions (spotted) onto membranes based on PVDF (Karas et al. 1990; Vestling and Fenselau 1994), nylon (Zaluzec et al. 1994) and polyethylene (Blackledge and Alexander 1995) could be desorbed easily with UV-MALDI after matrix incubation of the membrane. While results of spotted proteins were encouraging, the spectra quality obtained from proteins electroblotted onto membranes and desorbed by UV-MALDI was poor in terms of reproducibility, peak width, and signal-to-noise ratios of protein mass signals (Strupat et al. 1994a; Vestling and Fenselau 1994; Blais et al. 1996). A reason for the lower spectra quality in UV-MALDI-MS might be related to the different membrane-protein interaction in the spotted and electroblotted preparations. During the blotting process, formation of tight hydrophobic interactions between protein and membrane surface occurs, preventing an efficient resolubilization of the immobilized proteins under the conditions used for preparing samples for UV-MALDI-MS. Better results have been recently reported by Patterson (1995) for cytochrome C which was electroblotted onto a hydrophilic PVDF membrane modified with quarternary amines (Immobilon CD). This is most likely due to the charged groups on the membrane surface which allow a more efficient matrix incubation. A systematic investigation by Schreiner et al. (1996) with standard proteins electroblotted onto uncharged PVDF membranes revealed that immobilized proteins could be partially detached from the membrane surface by preincubation of the immobilized proteins in aqueous dilute acids and organic modifiers prior to incubation in the matrix. This treatment permits the embedding of the resolubilized proteins into the crystalline lattice of the matrix molecules within the pore volume of the membrane in a way comparable to a standard preparation. Consequently, the local resolution of the electrophoretic separation is not preserved.

◄

Fig. 11.4. Densitometer plots and contours of spots *A*, *X*, and *D* with mass spectra from selected locations. All mass spectra are the sum of ten single shot spectra from the same location. For explanation see text. (Reprinted from Eckerskorn et al. 1997 with permission from Analytical Chemistry)

4 UV-MALDI Mass Spectrometry of Proteins Directly from Polyacrylamide Gels

The most straightforward method would be the direct mass measurement of proteins within the polyacrylamide gel matrix. This, however, is circumscribed by principal problems: after the 1-D or 2-D electrophoretic separation the proteins are in SDS-containing buffer. The gel with the included protein has to be washed, incubated with MALDI matrix solution and, finally, dried down. In contrast to electroblotted proteins, where most of the proteins are adsorbed on top of the membrane surface, proteins in the gels are more or less evenly distributed throughout the polyacrylamide gel matrix and have only limited accessibility to the desorption/ionization process. As discussed by Strupat et al. (1994a), no signals could be obtained from proteins embedded in 1-mm-thick polyacrylamide gels either by UV- or IR-MALDI MS. Recently, Ogorzalek-Loo et al. (1996, 1997) presented data for the direct desorption and ionization of proteins from thin-layer IEF gels by UV-MALDI-MS. After soaking the gels in sinapic acid, they were analyzed with a UV laser scanning down the IEF strip. The preparation yielded definite "sweet spots" for good spectra, making it important to explore bands horizontally, as well as vertically. These results indicate that a portion of the proteins could be extracted to gel surface and, thus, be accessible for MALDI-MS. Best results were obtained for native IEF and proteins < 30 kDa. The combination of native IEF and scanning UV-MALDI-MS was referred to by the authors as "virtual 2-D gel electrophoresis".

5 Conclusion

Most of the existing impact of proteomics involves the mapping of complex protein mixtures by 2-D PAGE. In contrast to common protein dyes for protein detection, the molecular mass as an intrinsic molecular property identifies a protein spot as an individual protein and provides distinction of different proteins by their molecular masses as a further dimension. The results of scanning IR-MALDI-MS demonstrate that in principle an electroblotted 2-D protein pattern could be converted into a mass contour plot. This new combination of 2D-PAGE and MALDI-MS yields a 3-D presentation, where the x and y coordinates correspond to the lateral separation of the 2-D PAGE and the z value to the intensities of the molecular masses of individual proteins. This may afford in the near future an alternative to standard detection methods, with a much improved mass accuracy allowing inter alia a better study of post translational modifications.

References

Blackledge JA, Alexander AJ (1995) Polyethylene membrane as a sample support for direct matrix-assisted laser desorpton/ionization mass spectrometric analysis of high mass proteins. Anal Chem 67: 843–848

Blais JC, Nagnan-Le-Meillour P, Bolbach G, Tabet JC (1996) MALDI-TOFMS identification of odorant binding proteins (OBPs) electroblotted onto poly(vinylidene difluoride) membranes. Rapid Commun Mass Spectrom 10: 1–4

Chait BT, Kent SBH (1992) Weighing naked proteins: practical, high-accuracy mass measurement of peptides and proteins. Science 257: 1885–1894

Dianoux AC, Stasia MJ, Garin J, Gagnon J, Vignais PV (1992) The 23-kilodalton protein, a substrate of protein kinase C, in bovine neutrophil cytosol is a member of the S100 family. Biochemistry 31: 5898–5905

Dunphy JC, Busch KL, Buchanan MV (1993) Rapid extraction and structural characterization of biomolecules in agarose gels by laser desorption Fourier transform mass spectrometry. Anal Chem 65: 1329–1335

Eckerskorn C, Lottspeich F (1993) Structural characterization of blotting membranes and the influence of membrane parameters for electroblotting and subsequent amino acid sequence analysis of proteins. Electrophoresis 14: 831–838

Eckerskorn C, Strupat K, Karas M, Hillenkamp F, Lottspeich F (1992) Mass spectrometric analysis of blotted proteins after gel electrophoretic separation by matrix-assisted laser desorption/ionization. Electrophoresis 13: 664–665

Eckerskorn C, Strupat K, Schleuder D, Sanchez JC, Hochstrasser D, Lottspeich F, Hillenkamp F (1997) Analysis of proteins by direct-scanning infrared-MALDI mass spectrometry after 2D-PAGE separation and electroblotting. Anal Chem 69: 2888–2892

Feick RG, Shiozawa JA (1990) A high-yield method for the isolation of hydrophobic proteins and peptides from polyacrylamide gels for protein sequencing. Anal Biochem 187: 205–211

Fenn JB, Mann M, Meng CK, Wong SF, Whitehouse CM (1989) Electrospray ionization for mass spectrometry of large biomolecules. Science 246: 64–71

Haebel S, Jensen C, Andersen SO, Roepstorff P (1995) Isoforms of a cuticular protein from larvae of the meal beetle, *Tenebrio molitor*, studied by mass spectrometry in combination with Edman degradation and two-dimensional polyacrylamide gel electrophoresis. Protein Sci 4: 394–404

Hager DA, Burgess RR (1980) Elution of proteins from sodium dodecyl sulfate-polyacrylamide gels, removal of sodium dodecyl sulfate, and renaturation of enzymatic activity: results with sigma subunit of Escherichia coli RNA polymerase, wheat germ DNA topoisomerase, and other enzymes. Anal Biochem 109: 76–86

Hillenkamp F, Chait BT, Beavis RC, Karas M (1991) Matrix-assisted laser desorption/ionization mass spectrometry of biopolymers. Anal Chem 63: 1193A-1203A

Hunkapillar MW, Lujan E, Ostrander F, Hood LE (1983) Isolation of microgram quantities of proteins from polyacrylamide gels for amino acid sequence analysis. Methods Enzymol 91: 227–236

Jungblut P, Eckerskorn C, Lottspeich F, Klose J (1990) Blotting efficiency investigated by using two-dimensional electrophoresis, hydrophobic membranes and proteins from different sources. Electrophoresis 11: 581–588

Karas M, Hillenkamp F (1988) Laser desorption ionization of proteins with molecular masses exceeding 10 000 daltons. Anal Chem 60: 2299–2301

Karas M, Bahr U, Ingendoh AI, Nordhof E, Stahl DC, Strupat K, Hillenkamp F (1990) Principles and applications of UV- laser desorption/ ionization mass spectrometry. Anal Chim Acta 241: 175–185

Le Maire M, Deschamps S, Moller J, Le Caer JP, Rossier J (1993) Electrospray ionization mass spectrometry on hydrophobic peptides electroeluted from sodium dodecyl sulfate-polyacrylamide gel electrophoresis: application to the topology of the sarcoplasmic reticulum Ca^{2+} ATPase. Anal Biochem 214: 50–57

Mittag M, Eckerskorn C, Strupat K, Hastings JW (1997) Differential translational initiation of *lbp* mRNA is caused by a 5' upstream open reading frame. FEBS Lett 411: 245–250

Mortz E, Sareneva T, Haebel S, Julkunen I, Roepstorff P (1996) Mass spectrometric characterization of glycosylated interferon-gamma variants separated by gel electrophoresis. Electrophoresis 17: 907–917

Ogorzalek-Loo RR, Stevenson TI, Mitchell C, Loo JA, Andrews P (1996) Mass spectrometry of proteins directly from polyacrylamide gels. Anal Chem 68: 1910–1917

Ogorzalek-Loo RR, Mitchell C, Stevenson TI, Martin SA, Wade MH, Juhasz P, Patterson DH, Peltier JM, Loo JA, Andrews P (1997) Sensitivity and mass accuracy for proteins analyzed directly from polyacrylamide gels: Implications for proteome mapping. Electrophoresis 18: 382–390

Patterson SD (1995) Matrix-assisted laser desorption/ionization mass spectrometric approaches for the identification of gel-separated proteins in the 5–50 pmol range. Electrophoresis 16: 1104–1114

Schreiner M, Strupat K, Lottspeich F, Eckerskorn C (1996) Ultraviolet matrix assisted laser desorption/ionization-mass spectrometry of electroblotted proteins. Electrophoresis 17: 954–961

Simpson RJ, Moritz RL, Nice EE, Grego B (1987) A high-performance liquid chromatography procedure for recovering subnanomole amounts of protein from SDS-gel electroeluates for gas-phase sequence analysis. Eur J Biochem 165: 21–29

Strupat K, Karas M, Hillenkamp F, Eckerskorn C, Lottspeich, F (1994a) Matrix-assisted laser desorption/ionization mass spectrometry of proteins electroblotted after polyacrylamide gel electrophoresis. Anal Chem 66: 464–470

Strupat K, Eckerskorn C, Karas M, Hillenkamp F (1994b) Aspects in mass spectrometric analysis of electroblotted proteins after gel electrophoretic separation by matrix-assisted laser desorption/ionization. Proc 42nd ASMS Conf on Mass Spectrometry and Allied Topics, Chicago, 946pp

Vestling M, Fenselau C (1994) Poly(vinylidene difluoride) membranes as the interface between laser desorption mass spectrometry, gel electrophoresis, and in situ proteolysis. Anal Chem 66: 471–477

Wilkins MR, Pasquali C, Smith I, Williams KL, Hochstrasser DF (1996) From proteins to proteomes: large scale protein identification by two-dimensional electrophoresis and amino acid analysis. Bio/Technology 14:61–65

Yanase H, Cahill S, Martin de Llano JJ, Manning LR, Schneider K, Chait BT, Vandegriff KD, Winslow RM, Manning JM (1994) Properties of a recombinant human hemoglobin with aspartic acid 99(beta), an important intersubunit contact site, substituted by lysine. Protein Sci 3: 1213–1223

Zaluzec EJ, Gage DA, Allison J, Watson JT (1994) Direct matrix-assisted laser desorption/ionization mass specrometric analysis of proteins immobilized on nylon-based membranes. J Am Soc Mass 5: 230–237

Subject Index

Druck: Strauss Offsetdruck, Mörlenbach
Verarbeitung: Schäffer, Grünstadt